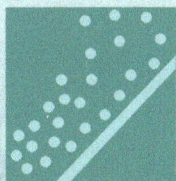

# Complexity, Criticality and Computation (C³)

Edited by
Mikhail Prokopenko

Printed Edition of the Special Issue Published in *Entropy*

www.mdpi.com/journal/entropy

MDPI

# Complexity, Criticality and Computation (C³)

## Special Issue Editor
Mikhail Prokopenko

MDPI • Basel • Beijing • Wuhan • Barcelona • Belgrade

**MDPI**

*Special Issue Editor*
Mikhail Prokopenko
The University of Sydney
Australia

*Editorial Office*
MDPI AG
St. Alban-Anlage 66
Basel, Switzerland

This edition is a reprint of the Special Issue published online in the open access journal *Entropy* (ISSN 1099-4300) from 2016–2017 (available at: http://www.mdpi.com/journal/entropy/special_issues/criticality_computation).

For citation purposes, cite each article independently as indicated on the article page online and as indicated below:

Author 1; Author 2. Article title. *Journal Name* **Year**, *Article number*, page range.

**First Edition 2017**

ISBN 978-3-03842-514-4 (Pbk)
ISBN 978-3-03842-515-1 (PDF)

# Table of Contents

# About the Special Issue Editor

**Mikhail Prokopenko**, Professor, is the Director of Complex Systems Research Group (Faculty of Engineering and IT) and the Director of the Centre for Complex Systems at the University of Sydney. He also leads the postgraduate Program in Complex Systems. Mikhail has a strong international reputation in complex self-organising systems and computational intelligence, with over 160 publications, patents, and edited books. He received a PhD in Computer Science (Australia), MA in Economics (USA), and MSc in Applied Mathematics (USSR). Over the last decade, Prof. Prokopenko has co-organised the series of International Workshops on Guided Self-Organization, and was a keynote speaker at several international events. In 2016 Mikhail led team Gliders2016 to World Championship at the 20th International RoboCup competition(SimLeague2D).

He is the Chief Editor for Computational Intelligence section of Frontiers Robotics and AI journal, a senior member of IEEE and a fellow of The Royal Society of New South Wales.

**MDPI**

*Editorial*

# Complexity, Criticality and Computation

**Mikhail Prokopenko**

Centre for Complex Systems, Faculty of Engineering and IT, The University of Sydney, Sydney, NSW 2006, Australia; mikhail.prokopenko@sydney.edu.au

Received: 3 August 2017; Accepted: 3 August 2017; Published: 4 August 2017

**Keywords:** complex systems; critical dynamics; distributed computation; information; phase transitions; information thermodynamics; guided self-organization

What makes a system "complex"? Is it merely the number of components it integrates, a nonlinear nature of the dependencies and feedbacks among its parts, or an unpredictable behavior it exhibits over time? The term "complexity" was initially applied generically to express the lack of predictability, reflecting on the self-organization of a synergistic macroscopic behavior out of interactions between the constituent microscopic parts, and the emergence of global patterns. Without a doubt, by now the concept has acquired a fairly definitive meaning, describing a distinct field of research and education and a new approach to science and engineering. There are abundant examples showing that the enterprise of Complex Systems, having achieved a substantial level of maturity, reaches back into our everyday lives, revealing patterns of complexity that should be considered without employing a reductionist logic [1].

Similarly, the idea of criticality was originally motivated by studies of various crises and disruptive events, as well as sensitivities to initial conditions, but over time has developed into a precise field: critical dynamics. Research into critical dynamics is typically focused on the behavior of dynamical spatiotemporal systems during phase transitions where scale invariance prevails and symmetries break. Crucially, such behavior can be understood in terms of the control and order parameters. For instance, a second-order phase transition in a ferromagnetic system, separating two qualitatively different behaviors, can be reached by controlling the temperature parameter: the "disordered" and isotropic (symmetric) high-temperature phase is characterized by the absence of net magnetization, while the "ordered" and anisotropic (less symmetric) low-temperature phase can be described by an order parameter, the net magnetization vector defining a preferred direction in space. Critical phenomena have become associated with the physics of critical points, such as fractal behavior, the divergence of the correlation length, power-law divergences (e.g., the divergence of the magnetic susceptibility in the ferromagnetic phase transition), universality of relevant critical exponents, and so on. Now, these precise theoretical notions begin to reconnect with their motivating applied studies of crisis modeling, forecasting, and response. There is a growing awareness that complexity is strongly related to criticality, and many examples of self-organizing complex systems can be found in applications managing complexity specifically at critical regimes.

A similar loop originating in practical studies, maturing to an exact science with precise but narrow definitions, and then reaching back to applied scenarios, can be seen in the realm of distributed computation. These days, complex systems can be viewed as distributed information-processing systems, in the domains ranging from systems biology and artificial life to computational neuroscience, digital circuitry, and transport networks [2]. Consciousness emerging from neuronal activity and interactions, cell behavior resultant from gene regulatory networks, and flocking and swarming behaviors are all examples of global system behavior emerging as a result of the local interactions of the individuals (neurons, genes, animals). Can these interactions be seen as a generic computational process? This question shapes the third component of our Special Issue, linking computation to complexity and criticality.

The issue begins with three papers which deal with the foundational aspects of information processing in complex systems [3–5]. The study of Allen et al. [3] describes two quantitative indices that summarize the structure of a complex system: (i) its complexity profile, based on the multivariate mutual information at a given scale or higher, and (ii) the marginal utility of information, characterizing the extent to which a system can be described using limited amounts of information. Information is understood to have a scale equal to the multiplicity (or redundancy) at which it arises, and so the analysis shows how these indices capture the multi-scale structure of complex systems. The work of Chicharro and Panzeri [4] also deals with the redundant aspects of information: it extends the framework of mutual information decomposition, based on the construction of information gain lattices, separating the information into the unique, redundant, and synergy components. In doing so, the work proposes a new construction of information gain and loss lattices. The framework developed by Biehl et al. [5] presents a novel formal analysis of the specific and complete local integration of entities within distributed dynamical systems (e.g., Bayesian networks), and puts it in the context of measures of complexity and information integration, as well as multi-information. The analysis presented in this paper goes to the core of complexity phenomenon, seen through the lens of synergistic integrative organization, viewing entities as patterns occurring within a spatiotemporal trajectory.

A cross-disciplinary connection between information-theoretic and game-theoretic aspects of complexity and computation is explored in the study of Harré [6], which focuses on the mutual information between previous game states and an agent's next action. This reveals a novel connection between the computational principles of logic gates, the structure of games, and the agents' decision strategies.

Nonlinear dynamics and inherent feedbacks typical in complex systems are considered in the next investigation by Zhang et al. [7], which addresses the problem of achieving and maintaining consensus in second-order multi-agent systems. This problem is pertinent to several scenarios, such as distributed control in networks of mass-spring systems, synchronization of coupled harmonic oscillators, and stability analysis of power systems. The study produces an adaptive consensus protocol for the problem's variant with an exogenous disturbance generated by an unknown exogenous system.

The next four papers [8–11] are placed in the complexity–criticality–computation overlap which is central to our issue. The study of Erten et al. [8] continues the information-theoretic theme by applying the information dynamics framework to studies of critical thresholds during epidemics. The approach uses the transfer entropy as a measure of distributed communications during a network-wide contagion seen as computation, as well as the active information storage as a measure of the corresponding distributed memory. The results for finite-size systems identify a critical interval, rather than an exact critical threshold. The methods described by Roli et al. [9] also detect criticality; that is, they distinguish between different phases separated by a critical regime. The approach is centered on the relevance index—an information-theoretic ratio relating the multi-information (or integration) measure to the mutual information between a subsystem and the rest of the system. The reported results demonstrate that the relevance index is consistently maximized at the critical regime. A phase transition-like behavior is investigated in the paper by Kramer et al. [10] as well. Their work identifies qualitative changes such as macroscopic spatiotemporal pattern formation in dynamics of Cellular Automata, by varying the inertia—an inner resistance to changes within cells—as the control parameter. In an ecological context, the inertia is related to an impairment and competition between species. The study by Mayer [11] illustrates the effects of critical connectivity in echo state networks and identifies under which conditions the recurrent connectivity is achieved. The results are contrasted with alternative approaches considering the dynamics near the "edge of chaos". The overall approach opens a way to organize reservoirs of neuronal connections as recurrent filters with a memory compression feature.

The three final studies are also biologically motivated. Continuing with the topic of neuronal connectivity, the paper by Kunert-Graf [12] attempts to identify the source of complexity in the biological neuronal network of *C. elegans*, the only organism for which its "connectome" is known. Using a suitably defined measure (once again based on information theory), the study argues that the somatic nervous system of *C. elegans* is much more complex than a random graph with the same degree distribution. The complexity and efficiency of solutions evolved by nature is a source of inspiration for another study, in which Kwiecień and Pasieka [13] use a computational swarm optimization algorithm to solve a travel planning problem. The presented approach is found to outperform the particle swarm optimization algorithm. The analysis presented by Farnsworth [14] brings the subject of distributed information processing to "the far end of the complexity gradient", centering the discussion on the question of free-will in artificial agents. Not surprisingly, this thought-provoking examination highlights the role of information in shaping the interactions and dynamics among patterns, as well as the distribution of matter and energy in space and time. The work concludes with the conjecture that free-will—which currently remains a property of living things—may still be attained in synthetic robots.

The contributions to this special issue show that the overlap between complexity, criticality, and computation provides fertile ground with both theoretical and practical dimensions. Considering complex systems as dynamical systems performing distributed computation suggests a unifying perspective, which reveals key thermodynamic and information-processing components, as well as their behavior near critical regimes. These components (e.g., collective memory, long-range communications, and synergistic modifications), together with the consequent physical fluxes [15], can be quantified and optimized. In spirit of Guided Self-Organization [2], the resultant dynamics can then be guided towards desired regions of the corresponding state-spaces, combining the power and efficiency of the self-organization so abundant in nature with the accuracy and reliability of traditional design approaches.

**Acknowledgments:** I express our thanks to the authors of the above contributions, and to the journal Entropy and MDPI for their support during this work.

**Conflicts of Interest:** The author declares no conflict of interest.

## References

1. Prokopenko, M. (Ed.) *Advances in Applied Self-Organizing Systems*; Springer: London, UK, 2008.
2. Prokopenko, M. (Ed.) *Guided Self-Organisation: Inception*; Springer: London, UK, 2014.
3. Allen, B.; Stacey, B.C.; Bar-Yam, Y. Multiscale Information Theory and the Marginal Utility of Information. *Entropy* **2017**, *19*, 273. [CrossRef]
4. Chicharro, D.; Panzeri, S. Synergy and Redundancy in Dual Decompositions of Mutual Information Gain and Information Loss. *Entropy* **2017**, *19*, 71. [CrossRef]
5. Biehl, M.; Ikegami, T.; Polani, D. Specific and Complete Local Integration of Patterns in Bayesian Networks. *Entropy* **2017**, *19*, 230. [CrossRef]
6. Harré, M. Utility, Revealed Preferences Theory, and Strategic Ambiguity in Iterated Games. *Entropy* **2017**, *19*, 201. [CrossRef]
7. Zhang, X.; Zhu, Q.; Liu, X. Consensus of Second Order Multi-Agent Systems with Exogenous Disturbance Generated by Unknown Exosystems. *Entropy* **2016**, *18*, 423. [CrossRef]
8. Erten, E.Y.; Lizier, J.T.; Piraveenan, M.; Prokopenko, M. Criticality and Information Dynamics in Epidemiological Models. *Entropy* **2017**, *19*, 194. [CrossRef]
9. Roli, A.; Villani, M.; Caprari, R.; Serra, R. Identifying Critical States through the Relevance Index. *Entropy* **2017**, *19*, 73. [CrossRef]
10. Kramer, K.; Koehler, M.; Fiore, C.E.; da Luz, M.G.E. Emergence of Distinct Spatial Patterns in Cellular Automata with Inertia: A Phase Transition-Like Behavior. *Entropy* **2017**, *19*, 102. [CrossRef]
11. Mayer, N.M. Echo State Condition at the Critical Point. *Entropy* **2017**, *19*, 3. [CrossRef]
12. Kunert-Graf, J.M.; Sakhanenko, N.A.; Galas, D.J. Complexity and Vulnerability Analysis of the *C. elegans* Gap Junction Connectome. *Entropy* **2017**, *19*, 104. [CrossRef]
13. Kwiecień, J.; Pasieka, M. Cockroach Swarm Optimization Algorithm for Travel Planning. *Entropy* **2017**, *19*, 213. [CrossRef]

14. Farnsworth, K.D. Can a Robot Have Free Will? *Entropy* **2017**, *19*, 237. [CrossRef]
15. Prokopenko, M.; Einav, I. Information thermodynamics of near-equilibrium computation. *Phys. Rev. E* **2015**, *91*, 062143. [CrossRef] [PubMed]

*Article*

# Multiscale Information Theory and the Marginal Utility of Information

**Benjamin Allen** [1,2,*]**, Blake C. Stacey** [3,4] **and Yaneer Bar-Yam** [4]

[1] Department of Mathematics, Emmanuel College, Boston, MA 02115, USA
[2] Program for Evolutionary Dynamics, Harvard University, Cambridge, MA 02138, USA
[3] Department of Physics, University of Massachusetts-Boston, Boston, MA 02125, USA; bstacey@sunclipse.org
[4] New England Complex Systems Institute, Cambridge, MA 02139, USA; yaneer@necsi.edu
[*] Correspondence: benjcallen@gmail.com

Received: 28 February 2017; Accepted: 9 June 2017; Published: 13 June 2017

**Abstract:** Complex systems display behavior at a range of scales. Large-scale behaviors can emerge from the correlated or dependent behavior of individual small-scale components. To capture this observation in a rigorous and general way, we introduce a formalism for multiscale information theory. Dependent behavior among system components results in overlapping or shared information. A system's structure is revealed in the sharing of information across the system's dependencies, each of which has an associated scale. Counting information according to its scale yields the quantity of scale-weighted information, which is conserved when a system is reorganized. In the interest of flexibility we allow information to be quantified using any function that satisfies two basic axioms. Shannon information and vector space dimension are examples. We discuss two quantitative indices that summarize system structure: an existing index, the complexity profile, and a new index, the marginal utility of information. Using simple examples, we show how these indices capture the multiscale structure of complex systems in a quantitative way.

**Keywords:** complexity; complex systems; entropy; information; scale

---

## 1. Introduction

The field of complex systems seeks to identify, understand and predict common patterns of behavior across the physical, biological and social sciences [1–7]. It succeeds by tracing these behavior patterns to the structures of the systems in question. We use the term "structure" to mean the totality of quantifiable relationships, or *dependencies*, among the components comprising a system. Systems from different domains and contexts can share key structural properties, causing them to behave in similar ways. For example, the central limit theorem tells us that the sum over many independent random variables yields an aggregate value whose probability distribution is well-approximated as a Gaussian. This helps us understand systems composed of statistically independent components, whether those components are molecules, microbes or human beings. Likewise, different chemical elements and compounds display essentially the same behavior near their respective critical points. The critical exponents which encapsulate the thermodynamic properties of a substance are the same for all substances in the same universality class, and membership in a universality class depends upon structural features such as dimensionality and symmetry properties, rather than on details of chemical composition [8].

Outside of known universality classes, identifying the key structural features that dictate the behavior of a system or class of systems often relies upon an *ad hoc* leap of intuition. This becomes particularly challenging for complex systems, where the set of system components is not only large, but also interwoven and resistant to decomposition. Information theory [9,10] holds promise as a general tool for quantifying the dependencies that comprise a system's structure [11]. We can consider the amount of information that would be obtained from observations of any component or set of components. Dependencies mean that one observation can be fully or partially inferred from another, thereby reducing the amount of joint information present in a set of components, compared to the amount that would be present without those dependencies. Information theory allows one to quantify not only fixed or rigid relationships among components, but also "soft" relationships that are not fully determinate, e.g., statistical or probabilistic relationships.

However, traditional information theory is primarily concerned with amounts of independent bits of information. Consequently, each bit of non-redundant information is regarded as equally significant, and redundant information is typically considered irrelevant, except insofar as it provides error correction [10,12]. These features of information theory are natural in applications to communication, but present a limitation when characterizing the structure of a physical, biological, or social system. In a system of celestial bodies, the same amount of information might describe the position of a moon, a planet, or a star. Likewise, the same amount of information might describe the velocity of a solitary grasshopper, or the mean velocity of a locust swarm. A purely information-theoretic treatment has no mechanism to represent the fact that these observables, despite containing the same amount of information, differ greatly in their significance.

Overcoming this limitation requires a multiscale approach to information theory [13–18]—one that identifies not only the *amount* of information in a given observable but also its *scale*, defined as the number or volume of components to which it applies. Information describing a star's position applies at a much larger scale than information describing a moon's position. In shifting from traditional information theory to a multiscale approach, redundant information becomes not irrelevant but crucial: Redundancy among smaller-scale behaviors gives rise to larger-scale behaviors. In a locust swarm, measurements of individual velocity are highly redundant, in that the velocity of all individuals can be inferred with reasonable accuracy by measuring the velocity of just one individual. The multiscale approach, rather than collapsing this redundant information into a raw number of independent bits, identifies this information as large-scale and significant precisely because it is redundant across many individuals.

The multiscale approach to information theory also sheds light on a classic difficulty in the field of complex systems: to clarify what a "complex system" actually is. Naïvely, one might think to define complex systems as those that display the highest complexity, as quantified using Shannon information or other standard measures. However, the systems deemed the most "complex" by these measures are those in which the components behave independently of each other, such as ideal gases. Such systems lack the multiscale regularities and interdependencies that characterize the systems typically studied by complex systems researchers. Some theorists have argued that true complexity is best viewed as occupying a position between order and randomness [19–21]. Music is, so the argument goes, intermediate between still air and white noise. But although complex systems contain both order and randomness, they do not appear to be mere blends of the two. A more satisfying answer is that complex systems display behavior across a wide range of scales. For example, stock markets can exhibit small-scale behavior, as when an individual investor sells a small number of shares for reasons unrelated to overall market activity. They can also exhibit large-scale behavior, e.g., a large institutional investor sells many shares [22], or many individual investors sell shares simultaneously in a market panic [23].

Formalizing these ideas requires a synthesis of statistical physics and information theory. Statistical physics [24–27]—in particular the renormalization group of phase transitions [28,29]—provides a notion of scale in which individual components acting in concert can be considered equivalent to larger-scale units. Information theory provides the tool of *multivariate mutual information*: the information shared among an arbitrary number of variables (also called *interaction information* or *co-information*) [13–17,30–39]. These threads were combined in the *complexity profile* [13–18], a quantitative index of structure that characterizes the amount of information applying at a given scale or higher. In the context of the complexity profile, the multivariate mutual information of a set of $n$ variables is considered to have scale $n$. In this way, information is understood to have scale equal to the multiplicity (or redundancy) at which it arises—an idea which is also implicit in other works on multivariate mutual information [40,41].

Here we present a mathematical formalism for multiscale information theory, for use in quantifying the structure of complex systems. Our starting point is the idea discussed in the previous paragraph, that information has scale equal to the multiplicity at which it arises. We formalize this idea mathematically and generalize it in two directions: First, we allow each system component to have an arbitrary intrinsic scale, reflecting its inherent size, volume, or multiplicity. For example, the mammalian muscular system includes both large and small muscles, corresponding to different scales of environmental challenge (e.g., pursuing prey and escaping from predators versus chewing food) [42]. Scales are additive in our formalism, in the sense that a set of components acting in perfect coordination is formally equivalent to a single component with scale equal to the sum of the scales of the individual components. This equivalence can greatly simplify the representation of a system. Consider, for example, an avalanche consisting of differently-sized rocks. To represent this avalanche within the framework of traditional (single-scale) information theory, one must either neglect the differences in size (thereby diminishing the utility of the representation) or else model each rock by a collection of myriad statistical variables, each corresponding to a equally-sized portion. Our formalism, by incorporating scale as a fundamental quantity, allows each rock to be represented in a direct and physically meaningful way.

Second, in the interest of generality, we use a new axiomatized definition of information, which encompasses traditional measures such as Shannon information as well as other quantifications of freedom or indeterminacy. In this way, our formalism is applicable to system representations for which traditional information measures cannot be used.

Using these concepts of information and scale, we identify how a system's joint information is distributed across each of its *irreducible dependencies*—relationships among some components conditional on all others. Each irreducible dependency has an associated scale, equal to the sum of the scales of the components included in this dependency. This formalizes the idea that any information pertaining to a system applies at a particular scale or combination of scales. Multiplying quantities of information by the scales at which they apply yields the *scale-weighted information*, a quantity that is conserved when a system is reorganized or restructured.

We use this multiscale formalism to develop quantitative indices that summarize important aspects of a system's structure. We generalize the complexity profile to allow for arbitrary intrinsic scales and a variety of information measures. We also introduce a new index, the *marginal utility of information* (MUI), which characterizes the extent to which a system can be described using limited amounts of information. The complexity profile and the MUI both capture a tradeoff of complexity versus scale that is present in all systems.

Our basic definitions of information, scale, and systems are presented in Sections 2–4, respectively. Sections 5 formalizes the multiscale approach to information theory by defining the information and scale of each of a system's dependencies. Sections 6 and 7 discuss our two indices of structure. Section 8 establishes a mathematical relationship between these two indices for a special class of systems. Section 9 applies our indices of structure to the noisy voter model [43]. We conclude in Sections 10 and 11 by discussing connections between our formalism and other work in information theory and complex systems science.

## 2. Information

We begin by introducing a generalized, axiomatic notion of information. Conceptually, information specifies a particular entity out of a set of possibilities and thus enables us to describe or characterize that entity. Information measures such as Shannon information quantify the amount of resources needed in this specification. Rather than adopting a specific information measure, we consider that the amount of information may be quantified in different ways, each appropriate to different contexts.

Let $A$ be the set of components in a system. An *information function*, $H$, assigns a nonnegative real number to each subset $U \subset A$, representing the amount of information needed to describe the components in $U$. (Throughout, the subset notation $U \subset A$ includes the possibility that $U = A$.) We require that an information function satisfy two axioms:

- *Monotonicity:* The information in a subset $U$ that is contained in a subset $V$ cannot have more information than $V$, that is, $U \subset V \Rightarrow H(U) \leq H(V)$.
- *Strong subadditivity:* Given two subsets, the information contained in both cannot exceed the information in each of them separately minus the information in their intersection:

$$H(U \cup V) \leq H(U) + H(V) - H(U \cap V). \tag{1}$$

Strong subadditvity expresses how information combines when parts of a system ($U$ and $V$) are regarded as a whole ($U \cup V$). Information regarding $U$ may overlap with information regarding $V$ for two reasons. First, $U$ and $V$ may share components; this is corrected for by subtracting $H(U \cap V)$. Second, constraints in the behavior of non-shared components may reduce the information needed to describe the whole. Thus, information describing the whole may be reduced due to overlaps or redundancies in the information applying to different parts, but it cannot be increased.

In contrast to other axiomatizations of information, which uniquely specify the Shannon information [9,44–46] or a particular family of measures [47–50], the two axioms above are compatible with a variety of different measures that quantify information or complexity:

- *Microcanonical or Hartley entropy:* For a system with a finite number of joint states, $H_0(U) = \log m$ is an information function, where $m$ is the number of joint states available to the subset $U$ of components. Here, information content measures the number of yes-or-no questions which must be answered to identify one joint state out of $m$ possibilities.
- *Shannon entropy:* For a system characterized by a probability distribution over all possible joint states, $H(U) = -\sum_{i=1}^{m} p_i \log p_i$ is an information function, where $p_1, \ldots, p_m$ are the probabilities of the joint states available to the components in $U$ [9]. Here, information content measures the number of yes-or-no questions which must be answered to identify one joint state out of all the joint states available to $U$, where more probable states can be identified more concisely.
- *Tsallis entropy:* The Tsallis entropy [51,52] is a generalization of the Shannon entropy with applications to nonextensive statistical mechanics. For the same setting as in Shannon entropy, Tsallis entropy is defined as $H_q(U) = -\sum_{i=1}^{m} p_i^q (p_i^{1-q} - 1)/(1 - q)$ for some parameter $q \geq 0$. Shannon entropy is recovered in the limit $q \to 1$. Tsallis entropy is an information function for $q \geq 1$ (but not for $q < 1$); this follows from Proposition 2.1 and Theorem 3.4 of [53].
- *Logarithm of period:* For a deterministic dynamic system with periodic behavior, an information function $L(U)$ can be defined as the logarithm of the period of a set $U$ of components (i.e., the time it takes for the joint state of these components to return to an initial joint state) [54]. This information function measures the number of questions which one should expect to answer in order to locate the position of those components in their cycle.
- *Vector space dimension:* Consider a system of $n$ components, each of whose state is described by a real number. Then the joint states of any subset $U$ of $m \leq n$ components can be described by points in some linear subspace of $\mathbb{R}^m$. The minimal dimension $d(U)$ of such a subspace is an information function, equal to the number of coordinates one must specify in order to identify the joint state of $U$.
- *Matroid rank:* A matroid consists of a set of elements called the *ground set*, together with a *rank function* that takes values on subsets of the ground set. Rank functions are defined to include the monotonicity and strong subadditivity properties [55], and generalize the notion of vector subspace dimension. Consequently, the rank function of a matroid is an information function, with the ground set identified as the set of system components.

In principle, measurements of algorithmic complexity may also be regarded as information functions. For example, when a subset $U$ can be encoded as a binary string, the algorithmic complexity $H(U)$ can be quantified as the length of the shortest self-delimiting program producing this string, with respect to some universal Turing machine [56]. Information content then measures the number of machine-language instructions which must be given to reconstruct $U$. Algorithmic complexity—at least under certain formulations—obeys versions of the monotonicity and strong subadditivity axioms [56,57]. However, while conceptually clean, this definition is difficult to apply quantitatively. First, the algorithmic complexity is only defined up to a constant which depends on the choice of universal Turing machine. Second, as a consequence of the halting problem, algorithmic complexity can only be bounded, not computed exactly.

## 3. Scale

A defining feature of complex systems is that they exhibit nontrivial behavior on multiple scales [1,13,14]. While the term "scale" has different meanings in different scientific contexts, we use the term scale here in the sense of the number of entities or units acting in concert, with each involved entity potentially weighted according to a measure of importance.

For many systems, it is reasonable to regard all components as having *a priori* equal scale. In this case we may choose the units of scale so that each component has scale equal to 1. This convention was used in previous work on the complexity profile [13–17]. However, it is in many cases necessary to represent the components of a system as having different intrinsic scales, reflecting their built-in size, multiplicity or redundancy. For example, in a system of many physical bodies, it may be natural to identify the scale of each body as a function of its mass, reflecting the fact that each body comprises many molecules moving in concert. In a system of investment banks [58–60], it may be desirable to assign weight to each bank according to its volume of assets. In these cases, we denote the *a priori* scale of a system component $a$ by a positive real number $\sigma(a)$, defined in terms of some meaningful scale unit.

Scales are additive, in the sense that a set of completely interdependent components can be replaced by a single component whose scale is equal to the sum of the scales of the individual components. We describe this property formally in Section 5.4 and in Appendix B.

## 4. Systems

We formally define a *system* $\mathcal{A}$ to comprise three elements:

- A finite set $A$ of components,
- An information function $H_{\mathcal{A}}$, giving the information in each subset $U \subset A$,
- A scale function $\sigma_{\mathcal{A}}$, giving the intrinsic scale of each compoent $a \in A$.

The choice of information and scale functions will reflect how the system is modeled mathematically, and the kind of statements we can make about its structure. We omit the subscripts from $H$ and $\sigma$ when only one system is under consideration.

In this work, we treat the three elements of a system as unchanging, even though the system itself may be dynamic (existing in a sequence of states through time). A dynamic system can be represented as a set of time histories, or—using the approach of ergodic theory—by defining a probability distribution over states with probabilities corresponding to frequencies of occupancy over extended periods of time. The methods outlined here could also be used to explore the dynamics or histories of a system's structure, using information and scale functions whose values vary as relationships change within a system over time. However, our current work focuses only on the static or time-averaged properties of a system.

In requiring that the set $A$ of components be finite, we exclude, for example, systems represented as continuum fields, in which each point in a continuous space might be regarded as a component. While the concepts of multiscale information theory may still be useful in thinking about such systems, the mathematical representation of these concepts presents challenges that are beyond the scope of this work.

We shall use four simple systems as running examples. Each consists of three binary random variables, each having intrinsic scale one.

- *Example A: Three independent components.* Each component is equally likely to be in state 0 or state 1, and the system as a whole is equally likely to be in any of its eight possible states.
- *Example B: Three completely interdependent components.* Each component is equally likely to be in state 0 or state 1, but all three components are always in the same state.
- *Example C: Independent blocks of dependent components.* Each component is equally likely to take the value 0 or 1; however, the first two components always take the same value, while the third can take either value independently of the coupled pair.
- *Example D: The* $2 + 1$ *parity bit system.* The components can exist in the states 110, 101, 011, or 000 with equal probability. In each state, each component is equal to the parity (0 if even; 1 if odd) of the sum of the other two. Any two of the component are statistically independent of each other, but the three as a whole are constrained to have an even sum.

We define a *subsystem* of $\mathcal{A} = (A, H_A, \sigma_A)$ as a triple $\mathcal{B} = (B, H_B, \sigma_B)$, where $B$ is a subset of $A$, $H_B$ is the restriction of $H_A$ to subsets of $B$, and $\sigma_B$ is the restriction of $\sigma_A$ to elements of $B$.

## 5. Multiscale Information Theory

Here we formalize the multiscale approach to information theory. We begin by introducing notation for dependencies. We then identify how information is shared across irreducible dependencies, generalizing the notion of multivariate mutual information [13–17,30–39] to an arbitrary information function. We then define the scale of a dependency and introduce the quantity of scale-weighted information. Finally, we formalize the key concepts of independence and complete interdependence.

### 5.1. Dependencies

A *dependency* among a collection of components $a_1, \ldots, a_m$ is the relationship (if any) among these components such that the behavior of some of the components is in part obtainable from the behavior of others. We denote this dependency by the expression $a_1; \ldots; a_m$. This expression represents a relationship, rather than a number or quantity. We use a semicolon to keep our notation consistent with information theory (in particular, with multivariate mutual information; see Section 5.2).

We can identify a more general concept of *conditional dependencies*. Consider two disjoint sets of components $a_1, \ldots, a_m$ and $b_1, \ldots, b_k$. The conditional dependency $a_1; \ldots; a_m | b_1, \ldots, b_k$ represents the relationship (if any) between $a_1, \ldots, a_m$ such that the behavior of some of these components can yield improved inferences about the behavior of others, relative to what could be inferred from the behavior of $b_1, \ldots, b_k$. We call this the dependency of $a_1, \ldots, a_m$ *given* $b_1, \ldots, b_k$, and we say $a_1, \ldots, a_m$ are *included* in this dependency, while $b_1, \ldots, b_k$ are *excluded*.

We call a dependency *irreducible* if every system component is either included or excluded. We denote the set of all irreducible dependencies of a system $\mathcal{A}$ by $\mathfrak{D}_{\mathcal{A}}$. A system's dependencies can be organized in a Venn diagram, which we call a *dependency diagram* (Figure 1).

The relationship between the components and dependencies of $\mathcal{A}$ can be captured by a mapping, which we denote by $\delta$, from $A$ to subsets of $\mathfrak{D}_{\mathcal{A}}$. A component $a \in A$ maps to the set of irreducible dependencies that include $a$ (or in visual terms, the region of the dependency diagram that corresponds to component $a$). For example, in a system of three components $a, b, c$, we have

$$\delta(a) = \{a; b; c, \quad a; b | c, \quad a; c | b, \quad a | b, c\}. \tag{2}$$

We extend the domain of $\delta$ to subsets of components, by mapping each subset $U \subset A$ onto to the set of all irreducible dependencies that include at least one element of $U$; for example,

$$\delta(\{a,b\}) = \{a;b;c, \quad a;b|c, \quad a;c|b, \quad b;c|a, \quad a|b,c, \quad b|a,c\}. \tag{3}$$

Visually, $\delta(\{a,b\})$ is the union of the circles representing $a$ and $b$ in the dependency diagram. Finally, we extend the domain of $\delta$ to dependencies, by mapping the dependency $a_1;\dots;a_m|b_1,\dots,b_k$ onto the set of all irreducible dependencies that include $a_1,\dots,a_m$ and exclude $b_1,\dots,b_k$; for example,

$$\delta(a|c) = \{a;b|c, \quad a|b,c\}. \tag{4}$$

Visually, $\delta(a|c)$ consists of the regions corresponding to $a$ but not to $c$.

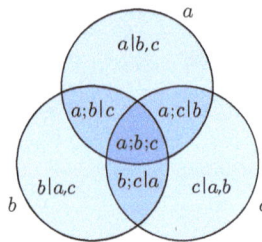

**Figure 1.** The dependency diagram of a system with three components, $a$, $b$ and $c$, represented by the interiors of the three circles. The seven irreducible dependencies shown above correspond to the seven interior regions of the Venn diagram encompassed by the boundaries of the three circles. Irreducible dependencies are shaded according to their scale, assuming that each component has scale one. Reducible dependencies such as $a|b$ are not shown.

### 5.2. Information Quantity in Dependencies

Here we define the shared information, $I_A(x)$, of a dependency $x$ in a system $A$. $I$ generalizes the multivariate mutual information [13–17,30–39] to an arbitrary information function $H$. We note that $H$ and $I$ characterize the same quantity—information—but are applied to different kinds of arguments: $H$ is applied to subsets of components of $A$, while $I$ is applied to dependencies.

The values of $I$ on irreducible dependencies $x$ of $A$ are uniquely determined by the system of equations

$$\sum_{x \in \delta(U)} I(x) = H(U) \qquad \text{for all subsets } U \subset A \tag{5}$$

As $U$ runs over all subsets of $A$, the resulting system of equations determines the values $I(x)$, $x \in \mathfrak{D}_A$, in terms of the values $H(U)$, $U \subset A$. The solution is an instance of the inclusion-exclusion principle [61], and can also be obtained by Gaussian elimination. An explicit formula obtained in the context of Shannon information [32] applies as well to any information function. Figure 2 shows the information in each irreducible dependecy for our four running examples.

We extend $I$ to dependencies that are not irreducible by defining the shared information $I(x)$ to be equal to the sum of the values of $I(y)$ for all irreducible dependencies $y$ encompassed by a dependency $x$:

$$I(x) = \sum_{y \in \delta(x)} I(y). \tag{6}$$

Our notation corresponds to that of Shannon information theory. For example, in a system of two components $a$ and $b$, solving (5) yields

$$I(a;b) = H(a) + H(b) - H(a,b). \tag{7}$$

Above, $H(a, b)$ is shorthand for $H(\{a, b\})$; we use similar shorthand throughout. Equation (7) coincides with the classical definition of mutual information [9,10], with $H$ representing joint Shannon information. Similarly, $I(a_1|b_1, \ldots, b_k)$ is the conditional entropy of $a_1$ given $b_1, \ldots, b_k$, and $I(a_1; a_2|b_1, \ldots, b_k)$ is the conditional mutual information of $a_1$ and $a_2$ given $b_1, \ldots, b_k$. For a dependency including more than two components, $I(x)$ is the multivariate mutual information (also called interaction information or co-information) of the dependency $x$ [30–33,36–39].

For any information function $H$, we observe that the information of one component conditioned on others, the conditional information $I(a_1|b_1, \ldots, b_k)$ is nonnegative due to the monotonicity axiom. Likewise, the mutual information of two components conditioned on others, $I(a_1; a_2|b_1, \ldots, b_k)$, is nonnegative due to the strong subadditivity axiom. However, the information shared among three or more components can be negative. This is illustrated in running Example **D**, for which the tertiary shared information $I(a; b; c)$ is negative (Figure 2D). Such negative values appear to capture an important property of dependencies, but their interpretation is the subject of continuing discussion [34,36,37,62,63].

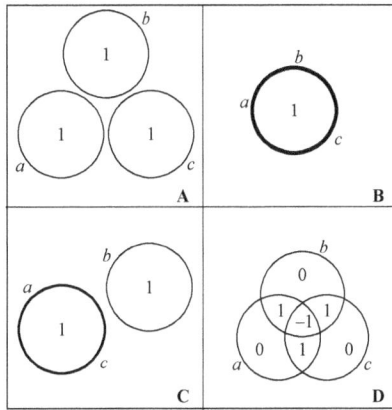

**Figure 2.** Dependency diagrams for our running example systems: (**A**) three independent bits; (**B**) three completely interdependent bits; (**C**) independent blocks of dependent bits; and (**D**) the $2 + 1$ parity bit system. Regions of information zero in (**A–C**) are not shown.

### 5.3. Scale-Weighted Information

Multiscale information theory is based on the principle that any information about a system should be understood as applying at a specific scale. Information shared among a set of components—arising from dependent behavior among these components—has scale equal to the sum of the scales of these components. This principle was first discussed in the context of the complexity profile [13–17], and is also implicit in other works on multivariate information theory [40,41], as we discuss in Section 10.2.

To formalize this principle, we define the scale $s(x)$ of an irreducible dependency $x \in \mathfrak{D}_A$ to be equal to the total scale of all components included in $x$:

$$s(x) = \sum_{\substack{a \in A \\ x \text{ includes } a}} \sigma(a). \tag{8}$$

The information in an irreducible dependency $x \in \mathfrak{D}_A$ is understood to apply at scale $s(x)$. Large-scale information pertains to many components and/or to components of large intrinsic scale; whereas small-scale information pertains to few components, and/or components of small intrinsic scale. In running Example **C** (Figure 2C), the bit of information that applies to components $a$ and $c$ has scale 2, while the bit applying to component $b$ has scale 1.

The overall significance of a dependency in a system depends on both its information and its scale. It therefore natural to weight quantities of information by their scale. We define the *scale-weighted information* $S(x)$ of an irreducible dependency $x$ to be the scale of $x$ times its information quantity

$$S(x) = s(x)I(x). \tag{9}$$

Extending this definition, we define the scale-weighted information of any subset $U \subset \mathfrak{D}_A$ of the dependence space to be the sum of the scale-weighted information of each irreducible-dependency in this subset:

$$S(U) = \sum_{x \in U} S(x) = \sum_{x \in U} s(x)I(x). \tag{10}$$

The scale-weighted information of the entire dependency space $\mathfrak{D}_A$—that is, the scale-weighted information of the system $\mathcal{A}$—is equal to the sum of the scale-weighted information of each component:

$$S(\mathfrak{D}_A) = \sum_{a \in A} \sigma(a)H(a). \tag{11}$$

As we show in Appendix A, this property arises directly from the fact that scale-weighted information counts redundant information according to its multiplicity or total scale.

According to Equation (11), the total scale-weighted information $S(\mathfrak{D}_A)$ does not change if the system is reorganized or restructured, as long as the information $H(a)$ and scale $\sigma(a)$ of each individual component $a \in A$ is maintained. The value $S(\mathfrak{D}_A)$ can therefore be considered a conserved quantity. The existence of this conserved quantity implies a tradeoff of information versus scale, which can be illustrated using the example of a stock market. If investors act largely independently of each other, information overlaps are minimal. The total amount of information is large, but most of this information is small-scale—applying only to a single investor at a time. On the other hand, in a market panic, there is much overlapping or redundant information in their actions—the behavior of one can be largely inferred from the behavior of others [23]. Because of this redundancy, the amount of information needed to describe their collective behavior is low. This redundancy also makes this collective behavior large-scale and highly significant.

### 5.4. Independence and Complete Interdependence

Components $a_1, \ldots, a_k \in A$ are *independent* if their joint information is equal to the sum of the information in each separately:

$$H(a_1, \ldots, a_k) = H(a_1) + \ldots + H(a_k). \tag{12}$$

In running Example **C**, components $a$, $b$, and $c$ are independent. This definition generalizes standard notions of independence in information theory, linear algebra, and matroid theory.

We extend the notion of independence to subsystems: subsystems $\mathcal{B}_i = (B_i, H_{\mathcal{B}_i}, \sigma_{\mathcal{B}_i})$ of $\mathcal{A}$, for $i = 1, \ldots, k$, are defined to be independent of one another if

$$H_A(B_1 \cup \ldots \cup B_k) = H_{\mathcal{B}_1}(B_1) + \ldots + H_{\mathcal{B}_k}(B_k). \tag{13}$$

We recall from Section 4 that $H_{\mathcal{B}_i}$ is the restriction of $H_A$ to subsets of $B_i$. In running Example **C**, the subsystem comprised of components $a$ and $c$ is independent of the subsystem comprised of component $b$.

Independence has the following *hereditary property* [64]: if subsystems $\mathcal{B}_1, \ldots, \mathcal{B}_k$ are independent, then all components and subsystems of $\mathcal{B}_i$ are independent of all components and subsystems of $\mathcal{B}_j$, for all $j \neq i$. We prove the hereditary property of independence from our axioms in Appendix C.

At the opposite extreme, we define a set of components $U \subset A$ to be *completely interdependent* if $H(a) = H(U)$ for any component $a \in U$. In words, any information applying to any component in $U$ applies to all components in $U$.

A set $U \subset A$ of completely interdependent components can be replaced by a single component of scale $\sum_{a \in U} \sigma(a)$ to obtain an equivalent, reduced representation of the system. Thus, in running Example **C**, the set $\{a, c\}$ is completely interdependent, and can be replaced by a single component of scale two. We show in Appendix B that replacements of this kind preserve all relevant quantities of information and scale.

## 6. Complexity Profile

We now turn to quantitative indices that summarize a system's structure. One such index is the complexity profile [13–17], which concretizes the observation that a complex system exhibits structure at multiple scales. We define the complexity profile of a system $\mathcal{A}$ to be a real-valued function $C_{\mathcal{A}}(y)$ of a positive real number $y$, equal to the total amount of information at scale $y$ or higher in $\mathcal{A}$:

$$C_{\mathcal{A}}(y) = I(\{x \in \mathfrak{D}_{\mathcal{A}} : s(x) \geq y\}). \tag{14}$$

Equation (14) generalizes previous definitions of the complexity profile [13–17], which use Shannon information as the information function and consider all components to have scale one.

The complexity profile reveals the levels of interdependence in a system. For systems where components are highly independent, $C(0)$ is large and $C(y)$ decreases sharply in $y$, since only small amounts of information apply at large scales in such a system. Conversely, in rigid or strongly interdependent systems, $C(0)$ is small and the decrease in $C(y)$ is shallower, reflecting the prevalence of large-scale information, as shown in in Figure 3. We plot the complexity profiles of our four running examples in Figure 4.

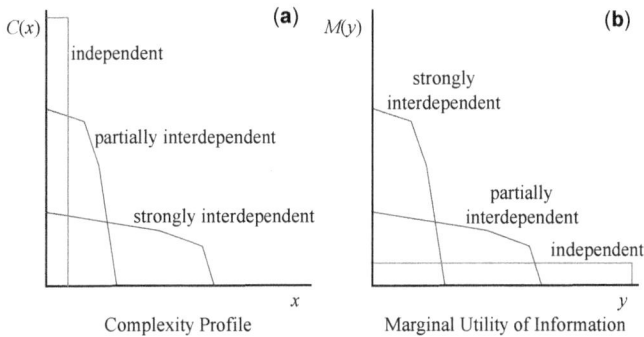

**Figure 3.** Schematic illustration of the (**a**) complexity profile (CP) and (**b**) marginal utility of information (MUI) for systems with varying degrees of interdependence among components. If the components are independent, all information applies at scale 1, so the complexity profile has $C(1)$ equal to the number of components and $C(x) = 0$ for $x > 1$. As the system becomes more interdependent, information applies at successively larger scales, resulting in a shallower decrease of $C(x)$. For the MUI, if components are independent, the optimal description scheme describes only a single component at a time, with marginal utility 1. As the system becomes more interdependent, information overlaps allow for more efficient descriptions that achieve greater marginal utility. For both the CP and MUI, the total area under the curve is equal to the total scale-weighted information $S(\mathfrak{D})$, which is preserved under reorganizations of the system. The CP and MUI are not reflections of each other in general, but they are for an important class of systems (see Section 8).

Previous works have developed and applied an explicit formula for the complexity profile [13,14,16,17,35] for cases where all components have equal intrinsic scales, $\sigma(a) = 1$ for all $a \in A$. To construct this formula, we first define the quantity $Q(j)$ as the sum of the joint information of all collections of $j$ components:

$$Q(j) = \sum_{i_1,\ldots,i_j} H(a_{i_1},\ldots,a_{i_j}). \tag{15}$$

The complexity profile can then be expressed as

$$C(k) = \sum_{j=N-k}^{N-1} (-1)^{j+k-N} \binom{j}{j+k-N} Q(j+1), \tag{16}$$

where $N = |A|$ is the number of components in $A$ [13,14]. The coefficients in this formula can be inferred from the inclusion-exclusion principle [61]; see [13] for a derivation.

## Complexity Profile

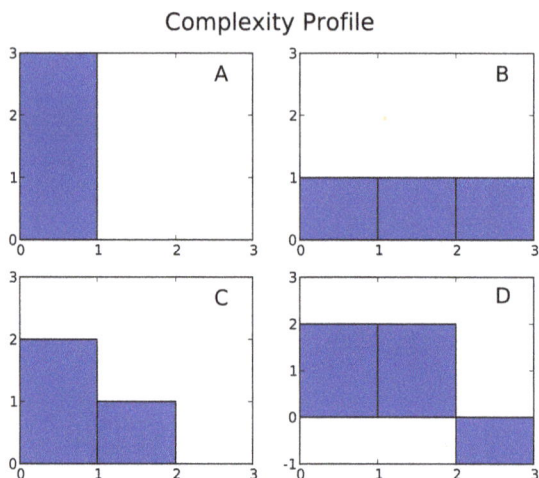

**Figure 4.** (**A–D**) Complexity profile $C(k)$ for Examples **A** through **D**. Note that the total (signed) area bounded by each curve equals $S(\mathcal{D}_A) = \sum_{a \in A} H(a) = 3$. For Example **D** (the parity bit), the information at scale 3 is negative.

The complexity profile has the following properties:

1. *Conservation law:* The area under $C(y)$ is equal to the total scale-weighted information of the system, and is therefore independent of the way the components depend on each other [13]:

$$\int_0^\infty C(y) \, dy = S(\mathcal{D}_A). \tag{17}$$

This result follows from the conservation law for scale-weighted information, Equation (11), as shown in Appendix A.

2. *Total system information:* At the lowest scale $y = 0$, $C(y)$ corresponds to the overall joint information: $C(0) = H(A)$. For physical systems with the Shannon information function, this is the total entropy of the system, in units of information rather than the usual thermodynamic units.

3. *Additivity:* If a system $A$ is the union of two independent subsystems $B$ and $C$, the complexity profile of the full system is the sum of the profiles for the two subsystems, $C_A(y) = C_B(y) + C_C(y)$. We prove this property Appendix D.

Due to the combinatorial number of dependencies for an arbitrary system, calculation of the complexity profile may be computationally prohibitive; however, computationally tractable approximations to the complexity profile have been developed [15].

The complexity profile has connections to a number of other information-theoretic characterizations of structure and dependencies among sets of random variables [38,40,41,65–67], as we discuss in Section 10.2. What distinguishes the complexity profile from these other approaches is the explicit inclusion of scale as an axis complementary to information.

## 7. Marginal Utility of Information

Here we introduce an new index characterizing multiscale structure: the *marginal utility of information* (MUI), denoted $M(y)$. The MUI quantifies how well a system can be characterized using a limited amount of information.

To obtain this index, we first ask how much scale-weighted information (as defined in Section 5.3) can be represented using $y$ or fewer units of information. We call this quantity the *maximal utility of information*, denoted $U(y)$. For small values of $y$, an optimal characterization will convey only large-scale features of the system. As $y$ increases, smaller-scale features will be progressively included. For a given system $\mathcal{A}$, the maximal amount of scale-weighted information that can be represented, $U(y)$, is constrained not only by the information limit $y$, but also by the pattern of information overlaps in $\mathcal{A}$—that is, the structure of $\mathcal{A}$. More strongly interdependent systems allow for larger amounts of scale-weighted information to be described using the same amount of information $y$.

We define the marginal utility of information as the derivative of maximal utility: $M(y) = U'(y)$. $M(y)$ quantifies how much scale-weighted information each additional unit of information can impart. The value of $M(y)$, being the derivative of scale-weighted information with respect to information, has units of scale. $M(y)$ declines steeply for rigid or strongly interdependent systems, and shallowly for weakly interdependent systems.

We now develop the formal definition of $U(y)$. We call any entity $d$ that imparts information about system $\mathcal{A}$ a *descriptor* of $\mathcal{A}$. The utility of a descriptor will be defined as a quantity of the form

$$u = \sum_{a \in A} \sigma(a) I(d;a). \tag{18}$$

For this to be a meaningful expression, we consider each descriptor $d$ to be an element of an augmented system $\mathcal{A}^{\dagger} = (A^{\dagger}, H_{\mathcal{A}^{\dagger}})$, whose components include $d$ as well as the original components of $\mathcal{A}$, which is a subsystem of $\mathcal{A}^{\dagger}$. The amount of information that $d$ conveys about any subset $V \subset A$ of components is given by

$$
\begin{aligned}
I(d;V) &= I_{\mathcal{A}^{\dagger}}(d;V) \\
&= H_{\mathcal{A}^{\dagger}}(d) + H_{\mathcal{A}^{\dagger}}(V) - H_{\mathcal{A}^{\dagger}}(\{d\} \cup V).
\end{aligned}
\tag{19}
$$

For example, the amount that $d$ conveys about a component $a \in A$ can be written $I(d;a) = H(d) + H(a) - H(d,a)$. $I(d;A)$ is the total information $d$ imparts about the system. Because the original system $\mathcal{A}$ is a subsystem of $\mathcal{A}^{\dagger}$, the augmented information function $H_{\mathcal{A}^{\dagger}}$ coincides with $H_{\mathcal{A}}$ on subsets of $A$.

The quantities $I(d;V)$ are constrained by the structure of $\mathcal{A}$ and the axioms of information functions. Applying these axioms, we arrive at the following constraints on $I(d;V)$:

(i)    $0 \leq I(d;V) \leq H(V)$ for all subsets $V \subset A$.
(ii)   For any pair of nested subsets $W \subset V \subset A$, $0 \leq I(d;V) - I(d;W) \leq H(V) - H(W)$.
(iii)  For any pair of subsets $V, W \subset A$,

$$I(d;V) + I(d;W) - I(d;V \cup W) - I(d;V \cap W) \leq H(V) + H(W) - H(V \cup W) - H(V \cap W).$$

To obtain the maximum utility of information, we interpret the values $I(d;V)$ as variables subject to the above constraints. We define $U(y)$ as the maximum value of the utility expression, Equation (18), as $I(d; V)$ vary subject to constraints (i)–(iii) and that the total information $d$ imparts about $\mathcal{A}$ is less than or equal to $y$: $I(d; A) \leq y$.

$U(y)$ characterizes the maximal amount of scale-weighted information that could in principle be conveyed about $\mathcal{A}$ using $y$ or less units of information, taking into account the information-sharing in $\mathcal{A}$ and the fundamental constraints on how information can be shared. $U(y)$ is well-defined since it is the maximal value of a linear function on a bounded set. Moreover, elementary results in linear programming theory [68] imply that $U(y)$ is piecewise linear, increasing and concave in $y$.

The above results imply that the marginal utility of information, $M(y) = U'(y)$, is piecewise constant, positive and nonincreasing. The MUI thus avoids the issue of counterintuitive negative values. The value of $M(y)$ can be understood as the additional scale units that can be described by an additional bit of information, given that the first $y$ bits have been optimally utilized. Code for computing the MUI has been developed and is available online [69].

The marginal utility of information has the following additional properties:

1. *Conservation law:* The total area under the curve $M(y)$ equals the total scale-weighted information of the system:
$$\int_0^\infty M(y)\, dy = S(\mathfrak{D}_A). \tag{20}$$

   This property follows from the observation that, since $M(y)$ is the derivative of $U(y)$, the area under this curve is equal to the maximal utility of any descriptor, which is equal to $S(\mathfrak{D}_A)$ since utility is defined in terms of scale-weighted information.

2. *Total system information:* The marginal utility vanishes for information values larger than the total system information, $M(y) = 0$ for $y > H(A)$, since, for higher values, the system has already been fully described.

3. *Additivity:* If $\mathcal{A}$ separates into independent subsystems $\mathcal{B}$ and $\mathcal{C}$, then
$$U_{\mathcal{A}}(y) = \max_{\substack{y_1+y_2=y \\ y_1,y_2 \geq 0}} \left( U_{\mathcal{B}}(y_1) + U_{\mathcal{C}}(y_2) \right). \tag{21}$$

The proof follows from recognizing that, since information can apply either to $\mathcal{B}$ or to $\mathcal{C}$ but not both, an optimal description allots some amount $y_1$ of information to subsystem $\mathcal{B}$, and the rest, $y_2 = y - y_1$, to subsystem $\mathcal{C}$. The optimum is achieved when the total maximal utility over these two subsystems is maximized. Taking the derivative of both sides and invoking the concavity of $U$ yields a corresponding formula for the marginal utility $M$:
$$M_{\mathcal{A}}(y) = \min_{\substack{y_1+y_2=y \\ y_1,y_2 \geq 0}} \max \left\{ M_{\mathcal{B}}(y_1), M_{\mathcal{C}}(y_2) \right\}. \tag{22}$$

Equations (21) and (22) are proven in Appendix E. This additivity property can also be expressed as the reflection (generalized inverse) of $M$. For any piecewise-constant, nonincreasing function $f$, we define the reflection $\tilde{f}$ as
$$\tilde{f}(x) = \max\{y : f(y) \leq x\}. \tag{23}$$

A generalized inverse [15] is needed since, for piecewise constant functions, there exist $x$-values for which there is no $y$ such that $f(y) = x$. For such values, $\tilde{f}(x)$ is the largest $y$ such that $f(y)$ does not exceed $x$. This operation is a reflection about the line $f(y) = y$, and applying it twice recovers the original function. If $\mathcal{A}$ comprises independent subsystems $\mathcal{B}$ and $\mathcal{C}$, the additivity property, Equation (22), can be written in terms of the reflection as
$$\tilde{M}_{\mathcal{A}}(x) = \tilde{M}_{\mathcal{B}}(x) + \tilde{M}_{\mathcal{C}}(x). \tag{24}$$

Equation (24) is proven in Appendix E.

Plots of the MUI for our four running examples are shown in Figure 5. The most interesting case is the parity bit system, Example **D**, for which the marginal utility is

$$M(y) = \begin{cases} \frac{3}{2} & 0 \leq y \leq 2 \\ 0 & y > 2. \end{cases} \tag{25}$$

The optimal description scheme leading to this marginal utility is shown in Figure 6. The marginal utility of information $M(y)$ captures the intermediate level of interdependency among components in Example **D**, in contrast to the maximal independence and maximal interdependence in Examples **A** and **B**, respectively (Figure 5). For an $N$-component generalization of Example **D**, in which each component acts as a parity bit for all others, we show in Appendix F that the MUI is given by

$$M(y) = \begin{cases} \frac{N}{N-1} & 0 \leq y \leq N-1 \\ 0 & y > N-1. \end{cases} \tag{26}$$

The MUI is similar in spirit to, and can be approximated by, principal components analysis, Information Bottleneck methods [70–73], and other methods that characterize the best possible description of a system using limited resources [66,74–79]. We discuss these connections in Section 10.3.

## Marginal Utility of Information

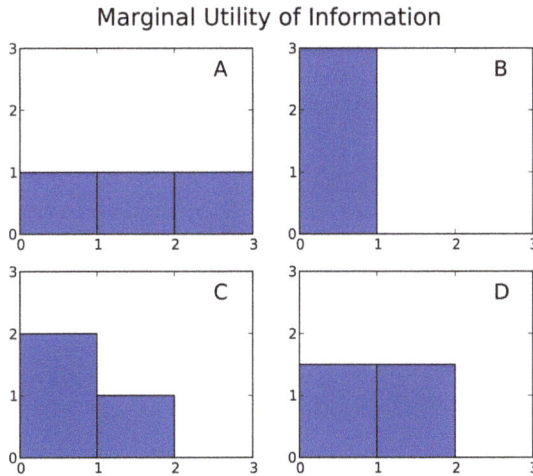

**Figure 5.** (**A–D**) Marginal Utility of Information for Examples **A** through **D**. The total area under each curve is $\int_0^\infty M(y)\,dy = S(\mathfrak{D}) = 3$. For Example **A**, all components are independent, and there is no more efficient description scheme than to describe one component at a time, with marginal utility 1. In Example **B**, the system state can be communicated with a single bit, with marginal utility 3. For Example **C**, the most efficient description scheme describes the fully correlated pair first (marginal utility 2), followed by the third component (marginal utility 1). The MUI for Example **C** can also be deduced from the additivity property, Equation (22). Examples **A–C** are all independent block systems; it follows from the results of Section 8 that their MUI functions are reflections (generalized inverses) of the corresponding complexity profiles shown in Figure 4. For Example **D**, the optimal description scheme is illustrated in Figure 6, leading to a marginal utility of $M(y) = 3/2$ for $0 \leq y \leq 2$ and $M(y) = 0$ for $y > 2$.

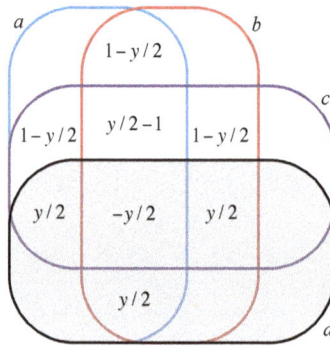

**Figure 6.** Information overlaps in the parity bit system, Example **D** of Figure 2, augmented with a descriptor $d$ having information content $y \leq 2$ and maximal utility. Symmetry considerations imply that such a descriptor must convey an equal amount of information about each of the three components $a$, $b$ and $c$. Constraints (i)–(iv) then yield that the amount described about each component must equal $y/2$ for $0 \leq y \leq 2$, and 1 for $y > 2$. Thus the maximal utility is $U(y) = 3y/2$ for $0 \leq y \leq 2$, and 3 for $y > 2$, leading to the marginal utility given in Equation (25) and shown in Figure **5D**.

## 8. Reflection Principle for Systems of Independent Blocks

For systems with a particularly simple structure, the complexity profile and the MUI turn out to be reflections (generalized inverses) of each other. The simplest case is a system consisting of a single component $a$. In this case, according to Equation (14), the complexity profile $C(x)$ has value $H(a)$ for $0 \leq x \leq \sigma(a)$ and zero for $x > \sigma(a)$:

$$C(x) = H(a)\Theta(\sigma(a) - x). \tag{27}$$

Above, $\Theta(y)$ is a step function with value 1 for $y \geq 0$ and 0 otherwise. To compute the marginal utility of information for this system, we observe that a descriptor with maximal utility has $I(d; a) = \min\{y, H(a)\}$ for each value of the informational constraint $y$, and it follows that

$$M(y) = \sigma(a)\Theta(H(a) - y). \tag{28}$$

We observe that $C(x)$ and $M(x)$ are reflections of each other: $C(x) = \tilde{M}(x)$, where $\tilde{M}(x)$ is defined in Equation (23).

We next consider a system whose components are all independent of each other. Additivity over independent subsystems (Property 3 in Sections 6 and 7), together with Equations (27) and (28), implies

$$C(x) = \tilde{M}(x) = \sum_{a \in A} H(a)\Theta(\sigma(a) - x). \tag{29}$$

Thus the reflection principle holds for systems of independent components.

More generally, one can consider a system of "independent blocks"—that is, a system that can be partitioned into independent subsystems, where the components of each subsystem are completely interdependent (see Section 5.4 for definitions.) Running example **C** is such a system, in that it can be partitioned into independent subsystems with component sets $\{a, c\}$ and $\{b\}$, and each of these sets is completely interdependent. We show in Appendix B that any set of completely interdependent components can be replaced by a single component, with scale equal to the sum of the scales of the replaced components, without altering the complexity profile or MUI. Thus, for systems of independent blocks, each block can be collapsed into a single component, whereupon Equation (29) applies and the reflection principle holds.

We have thus established that for any system of independent blocks, the complexity profile and the MUI are reflections of each other $C(x) = \bar{M}(x)$. However, this relationship does not hold for every system. $C(x)$ and $M(y)$ are not reflections of each other in the case of Example **D**, and, more generally, for a class of systems that exhibit negative information, as shown in Equation (26).

## 9. Application to Noisy Voter Model

As an application of our framework, we compute the marginal utility of information for the noisy voter model [43] on a complete graph. This model is a stochastic process with $N$ "voters". Each voter, $i = 1, \ldots, N$, can exist in one of two states, which we label $\eta_i \in \{-1, 1\}$. Each voter updates its state at Poisson rate 1. With probability $u$ it chooses $\pm 1$ with equal probability; otherwise, with probability $1 - u$, it copies a random other individual. Here $u \in [0, 1]$ is noise (or mutation) parameter that mediates the level of interdependence among voters. It follows that voter $i$ flips its state (from $+1$ to $-1$ or vice versa) at Poisson rate

$$\lambda_i = \frac{u}{2} + \frac{1 - u}{N - 1} \sum_{j=1}^{N} \frac{|\eta_j - \eta_i|}{2}. \tag{30}$$

For $u \ll 1$, voters are typically in consensus; for $u = 1$, all voters behave independently. The noisy voter model is mathematically equivalent to the Moran model [80] of neutral drift with mutation, and to a model of financial behavior on networks [23,81].

This model has a stationary distribution in which the number $m$ of voters in the $+1$ state has a beta-binomial distribution (Figure 7) [18,82]:

$$P(m) \propto \frac{\Gamma(N + M - m)\Gamma(M + m)}{\Gamma(N + 1 - m)\Gamma(1 + m)}, \qquad M = \frac{(N - 1)u}{2(1 - u)}. \tag{31}$$

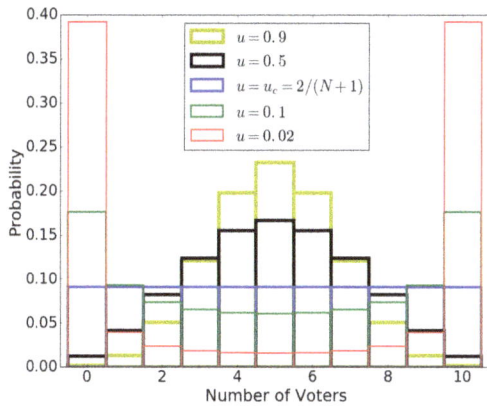

**Figure 7.** Stationary probability distribution for the noisy voter model [43] on a complete graph of size 10. Plot shows the probability of finding a given number $i$ of voters in the $+1$ state, for different values of the noise (mutation) parameter $u$, according to Equation (31). For small values of $u$, voters are typically in consensus ($m = 0$ or $m = N$); as $u$ increases their behavior becomes more independent.

For small $u$, $P$ is concentrated on the "consensus" states $0$ and $N$, converging to the uniform distribution on these states as $u \to 0$. For large $u$, $P$ exhibits a central tendency around $m = N/2$, and converges to the binomial distribution with $p = 1/2$ as $u \to 1$. These two modes are separated by the critical value of $u_c = 2/(N+1)$, at which $M = 1$ and $P$ becomes the uniform distribution on $\{0,\ldots,N\}$. This is the scenario in which mutation exactly balances the uniformizing effect of faithful copying. As $u \to 0$, $P$ converges to the uniform distribution on the states 0 and $N$.

The noisy voter model on a complete graph possesses *exchange symmetry*, meaning its behavior is preserved under permutation of its components. As a consequence, if subsets $U$ and $V$ have the same cardinality, $|U| = |V|$, then they have the same information, $H(U) = H(V)$. The information function is therefore fully characterized by the quantities $H_1,\ldots,H_N$, where $H_n$ is the information in each subset with $n \leq N$ components.

To calculate the MUI for systems with exchange symmetry, it suffices to consider descriptors that also possess exchange symmetry, so that $I(d;U) = I(d;V)$ whenever $|U| = |V|$. Denoting by $I_n$ the information that a descriptor imparts about a subset of size $n$, constraints (i)–(iii) of Section 7 reduce to

(i)  $\quad 0 \leq I_n \leq H_n$ for all $n \in \{1,\ldots,N\}$,

(ii)  $\quad 0 \leq I_n - I_{n-1} \leq H_n - H_{n-1}$ for all $n \in \{1,\ldots,N\}$,

(iii)  $\quad I_n + I_m - I_{n+m-\ell} - I_\ell \leq H_n + H_m - H_{n+m-\ell} - H_\ell$ for all $n, m, \ell \in \{1,\ldots,N\}$.

The maximum utility of information $U(y)$ is the maximum value of $NI_1$, subject to (i)–(iii) above and $I_N \leq y$. Since the number of constraints is polynomial in $N$, the maximum utility—and therefore the MUI—are readily computable.

The complexity profile and MUI for this model (Figure 8) both capture how the interdependence of voters is mediated by the noise parameter $u$. For small $u$, conformity among voters leads to large-scale information (positive $C(x)$ for large $x$) and enables efficient partial descriptions of the system (large $M(y)$ for small $y$). For large $u$, weak dependence among voters means that most information applies at small scale ($C(x)$ decreases rapidly) and optimal descriptions cannot do much better than to describe each component singly ($M(y) = 1$ for most values of $y$). Unlike the case of independent block systems (Section 8), the complexity profile and MUI are not reflections of each other for this model, but the reflection of each is qualitatively similar to the other.

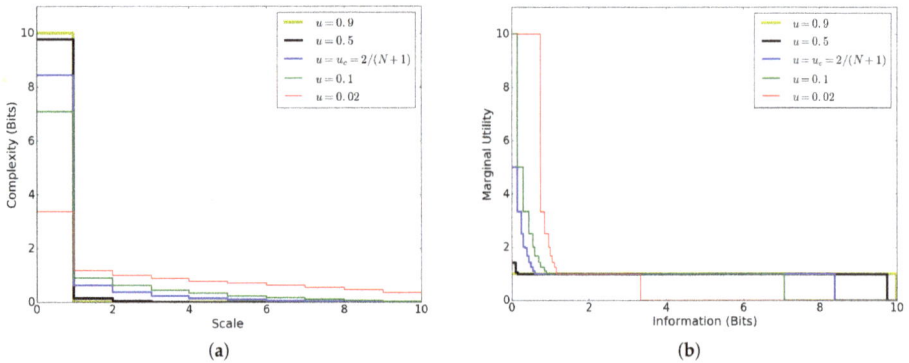

**Figure 8.** (**a**) Complexity profile and (**b**) marginal utility of information (MUI) for the noisy voter model [43] on a complete graph of size 10. The MUI is approximated by computing the (exact) maximal utility of information $U(y)$ at a discrete set of points, and then approximating $M(y) = U'(y) \approx \Delta M/\Delta y$. For small $u$, since voters are largely coordinated, much of their collective behavior can be described using small amounts of large-scale information. This leads to positive values of $C(x)$ for large $x$, and large values of $M(y)$ for small $y$. For large $u$, voters are largely independent, and therefore most information applies to one voter at a time. In this case it follows that $C(x)$ decreases rapidly to zero, and $M(y) = 1$ for most values of $y$. For both indices, the area under each curve is 10, which is the sum of the Shannon information of each voter (i.e., the total scale-weighted information), as guaranteed by Equation (20). For all values of $u$, the MUI appears to take a subset of the values $10/n$ for $n = 1,\ldots,10$.

## 10. Discussion

### 10.1. Potential Applications

Here we have presented a multiscale extension of information theory, along with two quantitative indices, as tools the analysis of structure in complex systems. Our generalized, axiomatic definition of information enables this framework to be applied to a variety of physical, biological, social, and economic systems. In particular, we envision applications to spin systems [1,17,83,84], gene regulatory systems [85–91], neural systems [92–94], biological swarming [95–97], spatial evolutionary dynamics [82,98,99], and financial markets [22,59,81,100–105].

In each of these systems, multiscale information theory enables the analysis of such questions as: Do components (genes, neurons, investors, etc.) behave largely independently, or are their behaviors strongly interdependent? How significant are intermediate scales of organization such as the genetic pathway, the cerebral lobe, or the financial sector? Can other "hidden" scales of organization be identified? Do the scales of behavior vary across different instances of a particular kind of system (e.g., gene regulation in stem versus differentiated cells; neural systems across taxa)? And how do the scales of organization in these systems compare to the scales of challenges they face?

The realization of these applications faces a computational hurdle: a full characterization of a system's structure requires the computation of $2^N$ informational quantities. Thus for large real-world systems, efficient approximations must be developed. Fortunately, there already exist efficient approximations to the complexity profile [15], and approximations to the MUI can be obtained using restricted description schemes as we discuss in Section 10.3 below.

### 10.2. Multivariate and Multiscale Information

The idea of using entropy or information to quantify a system's structure has deep roots. One of the earliest and most influential attempts was Schrödinger's concept of *negative entropy* or *negentropy* [106,107], which he introduced to quantify the extent to which a living system deviates from maximum entropy. Negentropy can be expressed in our formalism as $J = \sum_{a \in A} H(a) - H(A)$. This same quantity is known in other contexts as *multi-information* [38,65,66], *integration* [40] or *intrinsic information* [67], and is used to characterize the extent of statistical dependence among a set of random variables. Supposing that each component of our system has scale one, we have the following equivalent characterizations of this quantity:

$$J = S(\mathfrak{D}_A) - I(\mathfrak{D}_A) = \sum_{k=1}^{\infty} C(k) - C(1) = \sum_{k=2}^{\infty} C(k). \tag{32}$$

Equation (32) makes clear that while $J$ quantifies the deviation of a system from full independence, it does not identify the scale at which this deviation arises: all information at scales 2 and higher is subsumed into a single quantity.

Other proposed information-theoretic measures of structure also aggregate over scales in various ways. For example, the excess entropy, as defined by Ay et al. [41], is equal to $C(2)$, the amount of information that applies at scales 2 and higher. Another example is the complexity measure of Tononi et al. [40], which is defined for an $N$-component system as

$$T = \sum_{k=1}^{N} \left( \frac{1}{\binom{N}{k}} \sum_{\substack{U \subset A \\ |U|=k}} H(U) - \frac{k}{N} H(A) \right). \tag{33}$$

Using Equation (5), we can re-express $T$ as a weighted sum of the information in each irreducible dependency, where each dependency $x$ is weighted by a particular combinatorial function of its scale $s(x)$:

$$T = \sum_{x \in \mathcal{D}_A} I(x) \left( \sum_{k=1}^{N-s(x)} \frac{\binom{N-s(x)}{k}}{\binom{N}{k}} - \frac{N-1}{2} \right). \tag{34}$$

In this equation it is assumed that each it is assumed that each component has scale 1, so that $s(x)$ equals the number of components included in dependency $x$.

These previous measures can be understood as attempts to capture the idea that complex systems are not merely the sums of their parts—that is, they exhibit multiple scales of organization. We argue that this idea is best captured by making the notion of scale explicit, as a complementary axis to information. Doing so provides a formal basis for ideas that are implicit in earlier approaches.

The amount of information $I$ that we assign to each of a system's dependencies is known in the context of Shannon information as the multivariate mutual information or interaction information [30–33,37–39]. The use of multivariate mutual information is sometimes criticized [36,62], in part because it yields negative values that may be difficult to interpret. Such negative values arise for the complexity profile, but are avoided by the MUI. The value of $M(y)$ is always nonnegative and has a consistent interpretation as the additional scale units describable by an additional bit of information. The notion of descriptors also avoids negative values, since the information that a descriptor $d$ provides about a subset $U$ of components is always nonnegative: $I(d; U) \geq 0$.

Finally, some recent work raises the important question of whether measures based on shared information suffice to characterize the structure of a system. James and Crutchfield [63] exhibit pairs of systems of random variables that have qualitatively different probabilistic relationships, but the same joint Shannon information $H(U)$ for each subset $U$ of variables. As a consequence, the shared information $I$, complexity profile $C$, and marginal utility of information $M$, are the same for the two systems in such a pair. These examples demonstrate that probabilistic relationships among variables need not be determined by their information-theoretic measures, raising the question as to whether structure can be defined in terms of those measures. This conclusion, however, is dependent on the definition of the variables that are used to identify system states. To illustrate, if we take a particular system and combine variables into fewer ones with more states, the fewer information-theoretic measures that are obtained in the usual way become progressively less distinguishing of system probabilities. In the extreme case of a single variable having all of the possible states of the system, there is only one information measure. In the reverse direction, we have found [108] that a given system of random variables can be augmented with additional variables, the values of which are completely determined by the original system, in such a way that the probabilistic structure of the original system is uniquely determined by information overlaps in the augmented system. Thus, in this sense, moving to a more detailed representation can reveal relationships that are obscured in higher-level representations, and information theory may be sufficient to define structure in a general way.

### 10.3. Relation of the MUI to Other Measures

Our new index of structure, the MUI, is philosophically similar to data-reduction or dimensionality reduction techniques like principal component analysis, multidimensional scaling and detrended fluctuation analysis [74,76]; to the Information Bottleneck methods of Shannon information theory [70–73]; to Kolmogorov structure functions and algorithmic statistics in Turing-machine-based complexity theory [77–79]; to Gell-Mann and Lloyd's "effective complexity" [75]; and to Schneiderman et al.'s "connected information" [66]. All of these methods are mathematical techniques for characterizing the most important behaviors of the system under study. Each is an implementation of the idea of finding the best possible partial description of a system, where the resources available for this description (bits, coordinates, etc.) are constrained.

The essential difference from these previous measures is that the MUI is not tied to any particular method for generating partial descriptions. Rather, the MUI is defined in terms of optimally effective descriptors: for each possible amount of information invested in describing the system, the MUI considers the descriptor that provides the best possible theoretical return (in terms of scale-weighted information) on that investment. These returns are limited only by the structure of the system being described and the the fundamental constraints on information as encapsulated by our axioms.

In some applied contexts, it may be difficult or impossible to realize these theoretical maxima, due to constraints beyond those imposed by the axioms of information functions. It is often useful in these contexts to consider a particular "description scheme", in which descriptors are restricted to be of a particular form. Many of the data reduction and dimensionality reduction techniques described above can be understood as finding an optimal description of limited information using a specified description scheme. In these cases, the maximal utility found using the specified description scheme is in general less than the theoretical optimum. Calculating the marginal utility under a particular description scheme yields an approximation to the MUI.

### 10.4. Multiscale Requisite Variety

The discipline of cybernetics, an ancestor to modern control theory, used Shannon's information theory to quantify the difficulty of performing tasks, a topic of relevance both to organismal survival in biology and to system regulation in engineering. Ashby [109] considered scenarios in which a regulator device must protect some important entity from the outside environment and its disruptive influences. Successful regulation implies that if one knows only the state of the protected component, one cannot deduce the environmental influences; i.e., the job of the regulator is to minimize mutual information between the protected component and the environment. This is an information-theoretic statement of the idea of homeostasis. Ashby's "Law of Requisite Variety" states that the regulator's effectiveness is limited by its own information content, or *variety* in cybernetic terminology. An insufficiently flexible regulator will not be able to cope with the environmental variability.

Multiscale information theory enables us to overcome a key limitation of the requisite variety concept. In the framework of traditional cybernetics [109], each action of the environment requires a specific, unique reaction on the part of the regulator. This framework neglects the important difference between large-scale and fine-scale impacts. Systems may be able to absorb fine-scale impacts without any specific response, whereas responses to large-scale impacts are potentially critical to survival. For example, a human being can afford to be indifferent to the impact of a single molecule, whereas a falling rock (which may be regarded as the collective motion of many molecules) cannot be neglected. Ashby's Law does not make this distinction; indeed, there is no framework for this distinction in traditional information theory, since the molecule and the rock can be specified using the same amount of information.

This limitation can be overcome by a multiscale generalization of Ashby's Law [14], in which the responses of the system must occur at a scale appropriate to the environmental challenge. To protect against infection, for example, organisms have physical barriers (e.g., skin), generic physiological responses (e.g., clotting, inflammation) and highly specific adaptive immune responses, involving interactions among many cell types, evolved to identify pathogens at the molecular level. The evolution of immune systems is the evolution of separate large- and small-scale countermeasures to threats, enabled by biological mechanisms for information transmission and preservation [110]. By allowing for arbitrary intrinsic scales of components, and a range of different information functions, our work provides an expanded mathematical foundation for the multiscale generalization of Ashby's Law.

### 10.5. Mechanistic versus Informational Dependencies

Information-theoretic measures of a system's structure are essentially descriptive in nature. The tools we have proposed are aimed at identifying the scales of behavior of a system, but not necessarily the causes of this behavior. Importantly, causal influences at one scale can produce correlations at another. For example, the interactions in an Ising spin system are pairwise in character: the energy of a state depends only on the relative spins of neighboring pairs. These pairwise couplings can, however, give rise to long-range patterns [27]. Similarly, in models of coupled oscillators, dyadic physical interactions can lead to global synchronization [111]. Thus local interactions can create large-scale collective behavior.

## 11. Conclusions

Information theory has made, and will continue to make, formidable contributions to all areas of science. We argue that, in applying information theory to the study of complex systems, it is crucial to identify the scales at which information applies, rather than collapsing redundant or overlapping information into a raw number of independent bits. The multiscale approach to information theory falls squarely within the tradition of statistical physics—itself born of a marriage between probability theory and classical mechanics. By providing a general axiomatic framework for multiscale information theory, along with quantitative indices, we hope to deepen, clarify, and expand the mathematical foundations of complex systems theory.

**Acknowledgments:** We are grateful to Ryan G. James for developing code to compute the MUI.

**Author Contributions:** Benjamin Allen, Blake C. Stacey, and Yaneer Bar-Yam conceived the project; Benjamin Allen and Blake C. Stacey performed mathematical analysis; Benjamin Allen, Blake C. Stacey, and Yaneer Bar-Yam wrote the paper. All authors have read and approved the final version of the manuscript.

**Conflicts of Interest:** The authors declare no conflict of interest.

## Appendix A. Total Scale-Weighted Information

Here we prove two results regarding the total scale-weighted information of a system, $S(\mathfrak{D}_A)$. First we prove Equation (11), which shows that $S(\mathfrak{D}_A)$ depends only on the information and scale of each individual component:

**Theorem A1.** *For any system $A$,*

$$S(\mathfrak{D}_A) = \sum_{a \in A} \sigma(a) H(a). \tag{A1}$$

**Proof.** The proof amounts to a rearrangement of summations. We begin with the definition of scale-weighted information,

$$S(\mathfrak{D}_A) = \sum_{x \in \mathfrak{D}_A} s(x) I(x). \tag{A2}$$

Substituting the definition of $s(x)$, Equation (8), and rearranging yields

$$
\begin{aligned}
S(\mathfrak{D}_A) &= \sum_{x \in \mathfrak{D}_A} \left( \sum_{\substack{a \in A \\ x \text{ includes } a}} \sigma(a) \right) I(x) \\
&= \sum_{a \in A} \sigma(a) \sum_{\substack{x \in \mathfrak{D}_A \\ x \text{ includes } a}} I(x) \\
&= \sum_{a \in A} \sigma(a) I(\delta_a) \\
&= \sum_{a \in A} \sigma(a) H(a). \quad \square
\end{aligned}
$$

Next we prove Equation (17) showing that the area under the complexity profile is equal to $S(\mathfrak{D}_A)$:

**Theorem A2.** *For any system $A$,*

$$\int_0^\infty C(y)\, dy = S(\mathfrak{D}_A). \tag{A3}$$

**Proof.** We begin by substituting the definition of $C(y)$:

$$\int_0^\infty C(y)\,dy = \int_0^\infty I(\{x \in \mathcal{D}_A : \sigma(x) \geq y\})\,dy$$

$$= \int_0^\infty \left( \sum_{\substack{x \in \mathcal{D}_A \\ y \leq \sigma(x)}} I(x) \right) dy.$$

We then interchange the sum and integral on the right-hand side and apply Theorem A1:

$$\int_0^\infty C(y)\,dy = \sum_{x \in \mathcal{D}_A} \left( I(x) \int_0^{\sigma(x)} dy \right)$$

$$= \sum_{x \in \mathcal{D}_A} \sigma(x) I(x)$$

$$= S(\mathcal{D}_A). \quad \square$$

## Appendix B. Complete Interdependence and Reduced Representations

We mentioned in Section 5.4 that if a set of components is completely interdependent, they can be replaced by a single component, with scale equal to the sum of the scales of the replaced components. Here we define this replacement formally, and show that it preserves all quantities of shared information and scale-weighted information.

Let $\mathcal{A} = (A, H_A, \sigma_A)$ be a system. We begin by recalling that a set of components $U \subset A$ is *completely interdependent* if $H(a) = H(U)$ for each $a \in U$. It follows from the monotonicity axiom that $H(V) = H(U)$ for any nonempty subset $V \subset U$. The following lemma shows that, in evaluating the information function $H$, the entire set $U$ can be replaced by any subset thereof:

**Lemma A1.** *Let $U \subset A$ be a set of completely interdependent components. For any nonempty $V \subset U$ and any $W \subset A$,*

$$H(V \cup W) = H(U \cup W).$$

**Proof.** Applying the strong subadditivity axiom to the sets $U$ and $V \cup (W \setminus U)$, and invoking the fact that $H(V) = H(U)$, we obtain

$$H(U \cup W) \leq H(U) + H(V \cup (W \setminus U)) - H(V) = H(V \cup (W \setminus U)). \tag{A4}$$

But the monotonicity axiom implies the inequalities $H(V \cup (W \setminus U)) \leq H(V \cup W) \leq H(U \cup W)$. Combining with (A4) yields $H(V \cup (W \setminus U)) = H(V \cup W) = H(U \cup W)$. $\quad \square$

Now, for a system $\mathcal{A} = (A, H_A, \sigma_A)$ with a set $U$ of completely interdependent components, let us define a reduced system $\mathcal{A}^* = (A^*, H_{A^*}, \sigma_{A^*})$ in which the set $U$ has been replaced by a single component $u$. The reduced set of components is $A^* = (A \setminus U) \cup \{u\}$. The information $H_{A^*}(V)$, for $V \subset A^*$, is defined by

$$H_{A^*}(V) = \begin{cases} H_A(U \cup (V \setminus \{u\})) & \text{if } u \in V \\ H_A(V) & \text{if } u \notin V. \end{cases} \tag{A5}$$

Component $u$ of $\mathcal{A}^*$ has scale equal to the sum of the scales of components in $U$, while all other components maintain their scale:

$$\sigma_{A^*}(a) = \begin{cases} \sum_{b \in U} \sigma_A(b) & \text{if } a = u \\ \sigma_A(a) & \text{if } a \neq u. \end{cases} \tag{A6}$$

The following theorem shows that shared information and scale-weighted information are preserved in moving from $\mathcal{A}$ to $\mathcal{A}^*$:

**Theorem A3.** *Let $U = \{u_1, \ldots, u_k\} \subset A$ be a set of completely interdependent components of $\mathcal{A} = (A, H_A, \sigma_A)$, with $A \setminus U = \{a_1, \ldots, a_m\}$. Let $\mathcal{A}^* = (A^*, H_{A^*}, \sigma_{A^*})$ be the reduced system described above. Then the shared information $I_A$ and $I_{A^*}$ of the original and reduced systems, respectively, are related by*

$$I_A(u_1; \ldots; u_k; a_1; \ldots; a_\ell | a_{\ell+1}, \ldots, a_m) = I_{A^*}(u; a_1; \ldots, a_\ell | a_{\ell+1}, \ldots, a_m)$$

$$I_A(a_1; \ldots; a_\ell | u_1, \ldots, u_k, a_{\ell+1}, \ldots, a_m) = I_{A^*}(a_1; \ldots; a_\ell | u, a_{\ell+1}, \ldots, a_m) \qquad (A7)$$

$$I_A(u_1; \ldots; u_p; a_1; \ldots; a_\ell | u_{p+1}, \ldots, u_k, a_{\ell+1}, \ldots, a_m) = 0 \qquad \text{for } 1 \le p \le k - 1.$$

*The above equations also hold with the shared information $I_A$ and $I_{A^*}$ replaced by the scale-weighted information $S_A$ and $S_{A^*}$, respectively.*

In other words, if the irreducible dependency $x$ of $\mathcal{A}$ includes either all elements of $U$ or no elements of $U$, then, upon collapsing the elements of $U$ to the single component $u$ to obtain the dependency $x^*$ of $\mathcal{A}^*$, one has $I_{A^*}(x^*) = I_A(x)$ and $S_{A^*}(x^*) = S_A(x)$. If $x$ includes some elements of $U$ and excludes others, then $I_A(x) = S_A(x) = 0$. Thus all nonzero quantities of shared information and scale-weighted information are preserved upon collapsing the set $U$ to the single component $u$.

**Proof.** Define the function $J$ on the irreducible dependencies of $\mathcal{A}$ by

$$J(u_1; \ldots; u_k; a_1; \ldots; a_\ell | a_{\ell+1}, \ldots, a_m) = I_A(u_1; \ldots; u_k; a_1; \ldots; a_\ell | a_{\ell+1}, \ldots, a_m)$$
$$- I_{A^*}(u; a_1; \ldots, a_\ell | a_{\ell+1}, \ldots, a_m)$$
$$J(a_1; \ldots; a_\ell | u_1, \ldots, u_k, a_{\ell+1}, \ldots, a_m) = I_A(a_1; \ldots; a_\ell | u_1, \ldots, u_k, a_{\ell+1}, \ldots, a_m)$$
$$- I_{A^*}(a_1; \ldots; a_\ell | u, a_{\ell+1}, \ldots, a_m)$$
$$J(u_1; \ldots; u_p; a_1; \ldots; a_\ell | u_{p+1}, \ldots, u_k, a_{\ell+1}, \ldots, a_m) = I_A(u_1; \ldots; u_p; a_1; \ldots; a_\ell | u_{p+1}, \ldots, u_k, a_{\ell+1}, \ldots, a_m).$$

In light of Equation (5), the values of $J$ are the unique solution to the system of equations

$$\sum_{x \in \delta(V)} J(x) = \begin{cases} H_A(V) - H_{A^*}(\{u\} \cup V) & \text{if } V \cap U \ne \varnothing \\ H_A(V) - H_{A^*}(V) & \text{if } V \cap U = \varnothing, \end{cases} \qquad (A8)$$

as $V$ runs over subsets of $A$. But Lemma A1 and Equation (A5) imply that the right-hand side of Equation (A8) is zero for each $V \subset A$. Therefore, $J(x) = 0$ for each $x \in \mathcal{D}_A$, and Equation (A7) follows. The claim regarding scale-weighted information then follows from Equations (8), (9) and (A6). $\square$

Theorem A3 shows that all nonzero quantities of shared information and scale-weighted information are preserved when collapsing a set of completely dependent components into a single component. It follows that the complexity profile and MUI are also preserved under this collapsing operation.

### Appendix C. Properties of Independence

Here we prove fundamental properties of independent subsystems, which will be used in Appendices D and E to demonstrate the additivity properties of the complexity profile and MUI. Our first target is the *hereditary property of independence* (Theorem A4), which asserts that subsystems of independent subsystems are independent [64]. We then establish in Theorem A5 a simple characterization of information in systems comprised of independent subsystems.

For $i = 1, \ldots, k$, let $\mathcal{A}_i = (A_i, H_{A_i}, \sigma_{A_i})$ be subsystems of $\mathcal{A} = (A, H_A, \sigma_A)$, with the subsets $A_i \subset A$ disjoint from each other. We recall the definition of independent subsystems from Section 5.4.

**Definition A1.** *The subsystems $\mathcal{A}_i$ are* independent *if*

$$H(A_1 \cup \ldots \cup A_k) = H(A_1) + \ldots + H(A_k).$$

We establish the hereditary property of independence first in the case of two subsystems (Lemma A2), using repeated application of the strong subadditivity axiom. We then extend this result in Theorem A4 to arbitrary numbers of subsystems.

**Lemma A2.** *If $\mathcal{A}_1$ and $\mathcal{A}_2$ are independent subsystems of $\mathcal{A}$, then for every pair of subsets $U_1 \subset A_1$, $U_2 \subset A_2$,*
$H(U_1 \cup U_2) = H(U_1) + H(U_2)$.

**Proof.** The strong subadditivity axiom, applied to the sets $A_1$ and $U_1 \cup A_2$, yields

$$H(A_1 \cup A_2) \leq H(A_1) + H(U_1 \cup A_2) - H(U_1).$$

Replacing the left-hand side by $H(A_1) + H(A_2)$ and adding $H(U_1) - H(A_1)$ to both sides yields

$$H(U_1) + H(A_2) \leq H(U_1 \cup A_2). \tag{A9}$$

Now applying strong subadditivity to the sets $U_1 \cup U_2$ and $A_2$ yields

$$H(U_1 \cup A_2) \leq H(U_1 \cup U_2) + H(A_2) - H(U_2).$$

Combining with (A9) via transitivity, we have

$$H(U_1) + H(A_2) \leq H(U_1 \cup U_2) + H(A_2) - H(U_2).$$

Adding $H(U_2) - H(A_2)$ to both sides yields

$$H(U_1) + H(U_2) \leq H(U_1 \cup U_2). \tag{A10}$$

But strong subadditivity applied to $U_1$ and $U_2$ yields

$$H(U_1 \cup U_2) \leq H(U_1) + H(U_2) - H(U_1 \cap U_2) \leq H(U_1) + H(U_2). \tag{A11}$$

We conclude from inequalities (A10) and (A11) that

$$H(U_1 \cup U_2) = H(U_1) + H(U_2). \quad \square$$

We now use an induction argument to extend the hereditary property of independence to any number of subsystems.

**Theorem A4.** *If $\mathcal{A}_1, \ldots, \mathcal{A}_k$ are independent subsystems of $\mathcal{A}$, and $U_i \subset A_i$ for $i = 1, \ldots, k$ then*

$$H(U_1 \cup \ldots \cup U_k) = H(U_1) + \ldots + H(U_k).$$

**Proof.** This follows by induction on $k$. The $k = 1$ case is trivial. Suppose inductively that the statement is true for $k = \tilde{k}$, for some integer $\tilde{k} \geq 1$, and consider the case $k = \tilde{k} + 1$. We have

$$H(U_1) + \ldots + H(U_{\tilde{k}}) + H(U_{\tilde{k}+1}) = H(U_1 \cup \ldots \cup U_{\tilde{k}}) + H(U_{\tilde{k}+1})$$

by the inductive hypothesis, and

$$H(U_1 \cup \ldots \cup U_{\tilde{k}}) + H(U_{\tilde{k}+1}) = H(U_1 \cup \ldots \cup U_{\tilde{k}} \cup U_{\tilde{k}+1})$$

by Lemma A2 (since the subsystem of $\mathcal{A}$ with component set $A_1 \cup \ldots \cup A_{\bar{k}}$ is clearly independent from $A_{\bar{k}+1}$). This completes the proof. $\square$

We now examine the information in dependencies for systems comprised of independent subsystems. For convenience, we introduce a new notion: The *power system* of a system $\mathcal{A}$ is a system $2^{\mathcal{A}} = (2^A, H_{2^{\mathcal{A}}})$, where $2^A$ is the set of all subsets of $A$ (which in set theory is called the *power set* of $A$). In other words, the components of $2^{\mathcal{A}}$ are the subsets of $A$. The information function $H_{2^{\mathcal{A}}}$ on $2^{\mathcal{A}}$ is defined by the relation

$$H_{2^{\mathcal{A}}}(U_1, \ldots, U_k) = H_{\mathcal{A}}(U_1 \cup \ldots \cup U_k). \tag{A12}$$

By identifying the singleton subsets of $2^A$ with the elements of $A$ (that is, identifying each $\{a\} \in 2^A$ with $a \in A$), we can view $\mathcal{A}$ as a subsystem of $2^{\mathcal{A}}$.

This new system allows us to use the following relation: For any integers $k, \ell \geq 0$ and components $a_1, a_2, b_1, \ldots, a_k, c_1, \ldots, c_\ell \in A$,

$$I_{\mathcal{A}}(a_1; a_2; b_1; \ldots; b_k | c_1, \ldots, c_\ell) = I_{\mathcal{A}}(a_1; b_1; \ldots; b_k | c_1, \ldots, c_\ell)$$
$$+ I_{\mathcal{A}}(a_2; b_1; \ldots; b_k | c_1, \ldots, c_\ell) - I_{2^{\mathcal{A}}}(\{a_1, a_2\}; b_1; \ldots; b_k | c_1, \ldots, c_\ell). \tag{A13}$$

This relation generalizes the identity $I(a_1; a_2) = H(a_1) + H(a_2) - H(a_1, a_2)$ to conditional mutual information. It follows directly from the mathematical definition of $I$, Equation (5) of the main text.

We now show that if $\mathcal{B}$ and $\mathcal{C}$ are independent subsystems of $\mathcal{A}$, any conditional mutual information of components $\mathcal{B}$ and components of $\mathcal{C}$ is zero.

**Lemma A3.** *Let* $\mathcal{B} = (B, H_B)$ *and* $\mathcal{C} = (C, H_C)$ *be independent subsystems of* $\mathcal{A}$*. For any components* $b_1, \ldots, b_m, b'_1, \ldots, b'_{m'} \in B$ *and* $c_1, \ldots, c_n, c'_1, \ldots, c'_{n'} \in C$*, with* $m, n \geq 1, m', n' \geq 0$*,*

$$I(b_1; \ldots; b_m; c_1; \ldots; c_n | b'_1, \ldots, b'_{m'}, c'_1, \ldots, c'_{n'}) = 0. \tag{A14}$$

**Proof.** We prove this by induction. As a base case, we take $m = n = 1, m' = n' = 0$. In this case, the statement reduces to $I(b; c) = 0$ for every $b \in B, c \in C$. Since Lemma A2 guarantees that $H(b, c) = H(b) + H(c)$, this claim follows directly from the identity $I(b; c) = H(b) + H(c) - H(b, c)$.

We now inductively assume that the claim is true for all independent subsystems $\mathcal{B}$ and $\mathcal{C}$ of a system $\mathcal{A}$, and all $m \leq \tilde{m}, n \leq \tilde{n}, m' \leq \tilde{m}'$, and $n' \leq \tilde{n}'$, for some integers $\tilde{m}, \tilde{n} \geq 1, \tilde{m}', \tilde{n}' \geq 0$. We show that the truth of the claim is maintained when each of $\tilde{m}, \tilde{n}, \tilde{m}'$, and $\tilde{n}'$ is incremented by one.

We begin by incrementing $m$ to $\tilde{m} + 1$. Applying (A13) yields

$$I_{\mathcal{A}}\left(b_{\tilde{m}}; b_{\tilde{m}+1}; b_1; \ldots; b_{\tilde{m}-1}; c_1; \ldots; c_{\tilde{n}} | b'_1, \ldots, b'_{\tilde{m}'}, c'_1, \ldots, c'_{\tilde{n}'}\right)$$
$$= I_{\mathcal{A}}\left(b_{\tilde{m}}; b_1; \ldots; b_{\tilde{m}-1}; c_1; \ldots; c_{\tilde{n}} | b'_1, \ldots, b'_{\tilde{m}'}, c'_1, \ldots, c'_{\tilde{n}'}\right)$$
$$+ I_{\mathcal{A}}\left(b_{\tilde{m}+1}; b_1; \ldots; b_{\tilde{m}-1}; c_1; \ldots; c_{\tilde{n}} | b'_1, \ldots, b'_{\tilde{m}'}, c'_1, \ldots, c'_{\tilde{n}'}\right)$$
$$- I_{2^{\mathcal{A}}}\left(\{b_{\tilde{m}}; b_{\tilde{m}+1}\}; b_1; \ldots; b_{\tilde{m}-1}; c_1; \ldots; c_{\tilde{n}} | b'_1, \ldots, b'_{\tilde{m}'}, c'_1, \ldots, c'_{\tilde{n}'}\right). \tag{A15}$$

The first two terms of the right-hand side of (A15) are zero by the inductive hypothesis. Furthermore, it is clear from the definition of a power system that $2^{\mathcal{B}}$ and $2^{\mathcal{C}}$ are independent subsystems of $2^{\mathcal{A}}$. Thus the final term on the right-hand size of (A15) is also zero by the inductive hypothesis. In sum, the entire right-hand side of (A15) is zero, and the left-hand side must therefore be zero as well. This proves the claim is true for $m = \tilde{m} + 1$.

We now increment $m'$ to $\tilde{m}' + 1$. From Equation (6) of the main text, we have the relation

$$I_A(b_1; \ldots; b_{\tilde{m}}; c_1; \ldots; c_{\tilde{n}} | b'_1, \ldots, b'_{\tilde{m}'}, c'_1, \ldots, c'_{\tilde{n}'})$$
$$= I_A(b'_{\tilde{m}'+1}; b_1; \ldots; b_{\tilde{m}}; c_1; \ldots; c_{\tilde{n}} | b'_1, \ldots, b'_{\tilde{m}'}, c'_1, \ldots, c'_{\tilde{n}'})$$
$$+ I_A(b_1; \ldots; b_{\tilde{m}}; c_1; \ldots; c_{\tilde{n}} | b'_1, \ldots, b'_{\tilde{m}'}, b'_{\tilde{m}'+1}, c'_1, \ldots, c'_{\tilde{n}'}).$$

The left-hand side above is zero by the inductive hypothesis, and the first term on the right-hand side is zero by the case $m = \tilde{m} + 1$ proven above. Thus the second term on the right-hand side is also zero, which proves the claim is true for $m' = \tilde{m}' + 1$.

Finally, the cases $n = \tilde{n} + 1$ and $n' = \tilde{n}' + 1$ follow by interchanging the roles of $B$ and $C$. The result now follows by induction. $\quad\square$

We next show that for $B$ and $C$ independent subsystems of $A$, the amounts of information in dependencies of $B$ are not affected by additionally conditioning on components of $C$.

**Lemma A4.** *Let $B = (B, H_B, \sigma_B)$ and $C = (C, H_C, \sigma_C)$ be independent subsystems of $A$. For integers $m \geq 1$ and $m', n' \geq 0$, let $b_1, \ldots, b_m \in B$, $c_1, \ldots, c_n, c'_1, \ldots, c'_{n'} \in C$. Then*

$$I(b_1; \ldots; b_m | b'_1, \ldots, b'_{m'}, c'_1, \ldots, c'_{n'}) = I(b_1; \ldots; b_m | b'_1, \ldots, b'_{m'}). \tag{A16}$$

**Proof.** This follows by induction on $n'$. The claim is trivially true for $n' = 0$. Suppose it is true in the case $n' = \tilde{n}'$, for some $\tilde{n}' \geq 0$. By Equation (6) we have

$$I(b_1; \ldots; b_m | b'_1, \ldots, b'_{m'}, c'_1, \ldots, c'_{\tilde{n}'})$$
$$= I(b_1; \ldots; b_m; c'_{\tilde{n}'+1} | b'_1, \ldots, b'_{m'}, c'_1, \ldots, c'_{\tilde{n}'})$$
$$+ I(b_1; \ldots; b_m | b'_1, \ldots, b'_{m'}, c'_1, \ldots, c'_{\tilde{n}'}, c'_{\tilde{n}'+1}). \tag{A17}$$

The left-hand side is equal to $I(b_1; \ldots; b_m | b'_1, \ldots, b'_{m'})$ by the inductive hypothesis, and the first term on the right-hand side is zero by Lemma A3. This completes the proof. $\quad\square$

Finally, it follows from Lemmas A3 and A4 that if $A$ separates into independent subsystems, an irreducible dependency of $A$ has nonzero information only if it includes components from only one of these subsystems. To state this precisely, we introduce a projection mapping from irreducible dependencies of a system $A$ to those of a subsystem $B$ of $A$. This mapping, denoted $\rho_B^A : \mathfrak{D}_A \to \mathfrak{D}_B$, takes an irreducible dependency among the components in $A$, and "forgets" those components that are not in $B$, leaving an irreducible dependency among only the components in $B$. For example, suppose $A = \{a, b, c\}$ and $B = \{b, c\}$. Then

$$\rho_B^A(a; b | c) = b | c$$
$$\rho_B^A(b; c | a) = b; c. \tag{A18}$$

We can now state the following simple characterization of information in systems comprised of independent subsystems:

**Theorem A5.** *Let $A_1, \ldots, A_k$ be independent subsystems of $A$, with $A = A_1 \cup \ldots \cup A_k$. Then for any irreducible dependency $x \in \mathfrak{D}_A$,*

$$I_A(x) = \begin{cases} I_{A_i}(\rho_{A_i}^A(x)), & \text{if } x \text{ includes only components of } A_i \\ & \text{for some } i \in \{1, \ldots, k\}, \\ 0 & \text{otherwise.} \end{cases} \tag{A19}$$

**Proof.** In the case that $x$ includes only components of $\mathcal{A}_i$ for some $i$, the statement follows from Lemma A4. In all other cases, the claim follows from Lemma A3. $\square$

**Appendix D. Additivity of the Complexity Profile**

Here we prove Property 3 of the complexity profile claimed in Section 6: the complexity profile is additive over independent systems.

**Theorem A6.** *Let* $\mathcal{A}_1, \ldots, \mathcal{A}_k$ *be independent subsystems of* $\mathcal{A}$. *Then*

$$C_{\mathcal{A}}(y) = C_{\mathcal{A}_1}(y) + \ldots + C_{\mathcal{A}_k}(y). \tag{A20}$$

**Proof.** We start with the definition

$$C_{\mathcal{A}}(y) = \sum_{\substack{x \in \mathcal{D}_{\mathcal{A}} \\ \sigma(x) \geq y}} I_{\mathcal{A}}(x). \tag{A21}$$

Applying Theorem A5 to each term on the right-hand side yields

$$C_{\mathcal{A}}(y) = \sum_{i=1}^{k} \sum_{\substack{x \in \mathcal{D}_{\mathcal{A}} \\ x \text{ includes only components of } \mathcal{A}_i \\ \sigma(x) \geq y}} I_{\mathcal{A}_i}\left(\rho_{\mathcal{A}_i}^{\mathcal{A}}(x)\right)$$

$$= \sum_{i=1}^{k} \sum_{\substack{x \in \mathcal{D}_{\mathcal{A}_i} \\ \sigma(x) \geq y}} I_{\mathcal{A}_i}(x) = \sum_{i=1}^{k} C_{\mathcal{A}_i}(y). \quad \square$$

**Appendix E. Additivity of Marginal Utility of Information**

Here we prove the additivity property of MUI stated in Section 7. We begin by recalling the mathematical context for this result.

The maximal utility of information, $U(y)$, is defined as the maximal value of the quantity

$$u = \sum_{a \in A} \sigma(a) I(d; a), \tag{A22}$$

as the variables in the set $\{I(d; V)\}_{V \subset A}$ vary subject to the following constraints:

(i)   $0 \leq I(d; V) \leq H(V)$ for all $V \subset A$.
(ii)  For any $W \subset V \subset A$,

$$0 \leq I(d; V) - I(d; W) \leq H(V) - H(W). \tag{A23}$$

(iii) For any $V, W \subset A$,

$$I(d; V) + I(d; W) - I(d; V \cup W) - I(d; V \cap W) \leq H(V) + H(W) - H(V \cup W) - H(V \cap W).$$

(iv)  $I(d; A) \leq y$.

The marginal utility of information, $M(y)$ is defined as the derivative of $U(y)$.

We emphasize for clarity that, while we intuitively regard $I(d; V)$ as the information that a descriptor $d$ imparts about utility $V$, we formally treat the quantities $\{I(d; V)\}_{V \subset A}$ not as functions of two inputs but as variables subject to the above constraints.

Throughout this appendix we consider a system $\mathcal{A} = (A, H_{\mathcal{A}})$ comprising two independent subsystems, $\mathcal{B} = (B, H_{\mathcal{B}})$ and $\mathcal{C} = (C, H_{\mathcal{C}})$. This means that $A$ is the disjoint union of $B$ and $C$, and $H(A) = H(B) + H(C)$. The additivity property of MUI can be stated as

$$M_A(y) = \min_{\substack{y_1+y_2=y \\ y_1,y_2\geq 0}} \max\left\{M_B(y_1), M_C(y_2)\right\}. \tag{A24}$$

Alternatively, this property can be stated in terms of the reflection $\tilde{M}_A(x)$ of $M_A(y)$, with the dependent and independent variables interchanged (see Section 7), as

$$\tilde{M}_A(x) = \tilde{M}_B(x) + \tilde{M}_C(x). \tag{A25}$$

The proof of this property is organized as follows. Our first major goal is Theorem A7, which asserts that $I(d; A) = I(d; B) + I(d; C)$ when $u$ is maximized. Lemmas A5 and A6 are technical relations needed to achieve this result. We then apply the decomposition principle of linear programming to prove an additivity property of $U_A$ (Theorem A8). Theorem A9 then deduces the additivity of $M_A$ from the additivity of $\hat{U}_A$. Finally, in Corollary A1, we demonstrate the additivity of the reflected function $\tilde{M}_A$.

**Lemma A5.** *Suppose the quantities $\{I(d; V)\}_{V \subset A}$ satisfy Constraints (i)–(iv). Then for any subset $V \subset A$,*

$$I(d; V) \geq I(d; V \cap B) + I(d; V \cap C). \tag{A26}$$

**Proof.** Applying Constraint (iii) to the sets $V \cap B$ and $V \cap C$ we have

$$I(d; V \cap B) + I(d; V \cap C) - I(d; V) \leq H(V \cap B) + H(V \cap C) - H(V). \tag{A27}$$

But by Lemma A2, $H(V) = H(V \cap B) + H(V \cap C)$. Thus the right-hand side above is zero, which proves the claim. $\square$

**Lemma A6.** *Suppose the quantities $\{I(d; V)\}_{V \subset A}$ satisfy Constraints (i)–(iv). Suppose further that $W \subset V \subset A$ and $I(d; V) = I(d; V \cap B) + I(d; V \cap C)$. Then $I(d; W) = I(d; W \cap B) + I(d; W \cap C)$.*

**Proof.** Constraint (iii), applied to the sets $V \cap B$ and $W \cup (V \cap C)$, yields

$$I(d; V \cap B) + I\big(d; W \cup (V \cap C)\big) - I(d; V) - I(d; W \cap B)$$
$$\leq H(V \cap B) + H\big(W \cup (V \cap C)\big) - H(V) - H(W \cap B). \tag{A28}$$

By Lemma A2, we have

$$H\big(W \cup (V \cap C)\big) = H(W \cap B) + H(V \cap C) \tag{A29}$$
$$H(V) = H(V \cap B) + H(V \cap C).$$

With these two relations, the right-hand side of (A28) simplifies to zero. Making this simplification and substituting $I(d; V) = I(d; V \cap B) + I(d; V \cap C)$ (as given), we obtain

$$I\big(d; W \cup (V \cap C)\big) - I(d; W \cap B) - I(d; V \cap C) \leq 0. \tag{A30}$$

We next apply Constraint (iii) to $V \cap C$ and $W$, yielding

$$I(d; V \cap C) + I(d; W) - I\big(d; W \cup (V \cap C)\big) - I(d; W \cap C)$$
$$\leq H(V \cap C) + H(W) - H\big(W \cup (V \cap C)\big) - H(W \cap C). \tag{A31}$$

Lemma A2 implies $H(W) = H(W \cap B) + H(W \cap C)$. Combining this relation with (A29), the right-hand side of (A31) simplifies to zero. We then rewrite (A31) as

$$I(d;W) - I(d;W \cap C) \leq I(d;W \cup (V \cap C)) - I(d;V \cap C). \tag{A32}$$

By (A30), the right-hand side above is less than or equal to $I(d;W \cap B)$. Making this substitution and rearranging, we obtain

$$I(d;W) \leq I(d;W \cap B) + I(d;W \cap C). \tag{A33}$$

Combining now with Lemma A5, it follows that $I(d;W) = I(d;W \cap B) + I(d;W \cap C)$ as desired. □

**Theorem A7.** *Suppose the quantities* $\{I(\hat{d};V)\}_{V \subset A}$ *maximixe* $u = \sum_{a \in A} \sigma(a) I(d;a)$ *subject to Constraints (i)–(iv) for some* $0 \leq y \leq H(A)$. *Then*

$$I(\hat{d};A) = I(\hat{d};B) + I(\hat{d};C). \tag{A34}$$

**Proof.** Let $\hat{u} = \sum_{a \in A} \sigma(a) I(\hat{d};a)$ be the maximal value of $u$. By the duality principle of linear programming, the quantities $\{I(\hat{d};V)\}_{V \subset A}$ minimize the value of $I(d;A)$ as $\{I(d;V)\}_{V \subset A}$ varies subject to Constraints (i)–(iii) along with the additional constraint $u \geq \hat{u}$. (Informally, the descriptor $\hat{d}$ achieves utility $\hat{u}$ using minimal information.)

Assume for the sake of contradiction that $I(\hat{d};A) > I(\hat{d};B) + I(\hat{d};C)$. We will obtain a contradiction by showing that there is another set of quantities $\{I(\tilde{d};V)\}_{V \subset A}$, satisfying (i)–(iii) and $\tilde{u} = \hat{u}$, with $I(\tilde{d};A) < I(\hat{d};A)$. Here, $\tilde{u}$ is the utility associated to $\{I(\tilde{d};V)\}_{V \subset A}$; that is, $\tilde{u} = \sum_{a \in A} \sigma(a) I(\tilde{d};a)$. (Informally, we construct a new descriptor $\tilde{d}$ that achieves the same utility as $\hat{d}$ using less information.)

To obtain such quantities $\{I(\tilde{d};V)\}_{V \subset A}$, we first define $S \subset 2^A$ as the set of all subsets $V \subset A$ that satisfy

$$I(\hat{d};V) > I(\hat{d};V \cap B) + I(\hat{d};V \cap C). \tag{A35}$$

We observe that, by Lemma A5, if $V \notin S$, then $I(\hat{d};V) = I(\hat{d};V \cap B) + I(\hat{d};V \cap C)$. It then follows from Lemma A6 that if $W \subset V \subset A$ and $W \in S$, then $V \in S$ as well.

Next we choose $\epsilon > 0$ sufficiently small that, for each $V \in S$, the following two conditions are satisfied:

(1)  $I(\hat{d};V) > I(\hat{d};V \cap B) + I(\hat{d};V \cap C) + \epsilon$,
(2)  $I(\hat{d};V) > I(\hat{d};W) + \epsilon$, for all $W \subset V, W \notin S$.

There is no problem arranging for condition (2) to be satisfied for any particular $V \in S$, since it follows readily from Constraint (ii) on $\hat{d}$ that if $W \subset V$ and $W \notin S$, then $I(\hat{d};V) > I(\hat{d};W)$. We also note that since $A$ is finite, there are only a finite number of conditions to be satisfied as $V$ and $W$ vary, so it is possible to choose an $\epsilon > 0$ satisfying all of them.

Having chosen such an $\epsilon$, we define the quantities $\{I(\tilde{d};V)\}_{V \subset A}$ by

$$I(\tilde{d};V) = \begin{cases} I(\hat{d};V) - \epsilon & V \in S \\ I(\hat{d};V) & \text{otherwise.} \end{cases} \tag{A36}$$

In words, we reduce the amount of information that is imparted about the sets in $S$ by an amount $\epsilon$, while leaving fixed the amount that is imparted about sets not in $S$. Intuitively, one could say that we are exploiting an inefficiency in the amount of information imparted by $\hat{d}$ about sets in $S$, and that the new descriptor $\tilde{d}$ is more efficient in terms of minimizing the information $I(d;A)$ without sacrificing utility.

We will now show that $\tilde{d}$ satisfies Constraints (i)–(iii) and $\tilde{u} = \hat{u}$. First, since $0 \leq I(\tilde{d};V) \leq I(\hat{d};V) \leq H(V)$ for all $V \subset A$, Constraint (i) is clearly satisfied.

For Constraint (ii), consider any $W \subset V \subset A$. If $V$ and $W$ are either both in $S$ or both not in $S$ then $I(\tilde{d};V) - I(\tilde{d};W) = I(\hat{d};V) - I(\hat{d};W)$, and Constraint (ii) is satisfied for $\tilde{d}$ since it is satisfied for $\hat{d}$. It only remains to consider the case that $V \in S$ and $W \notin S$. In this case, we have

$$I(\tilde{d};V) - I(\tilde{d};W) = I(\hat{d};V) - I(\hat{d};W) - \epsilon > 0, \tag{A37}$$

since $V$ and $\epsilon$ satisfy condition (2) above. Furthermore,

$$\begin{aligned} I(\tilde{d};V) - I(\tilde{d};W) &= I(\hat{d};V) - I(\hat{d};W) - \epsilon \\ &\leq H(V) - H(W) - \epsilon \\ &< H(V) - H(W). \end{aligned}$$

Thus Constraint (ii) is satisfied.

To verify Constraint (iii), we must consider a number of cases, only one of which is nontrivial.

- If either

    - none of $V$, $W$, $V \cup W$ and $V \cap W$ belong to $S$,
    - all of $V$, $W$, $V \cup W$ and $V \cap W$ belong to $S$,
    - $V$ and $V \cup W$ belong to $S$ while $W$ and $V \cap W$ do not, or
    - $W$ and $V \cup W$ belong to $S$ while $V$ and $V \cap W$ do not,

    then the difference on the left-hand side of Constraint (iii) has the same value for $d = \hat{d}$ and $d = \tilde{d}$—that is, the changes in each term cancel out in the difference. Thus Constraint (iii) is satisfied for $\tilde{d}$ since it is satisfied for $\hat{d}$.

- If $V$, $W$, and $V \cup W$ belong to $S$ while $V \cap W$ does not, then

    $$\begin{aligned} I(\tilde{d};V) + I(\tilde{d};W) &- I(\tilde{d};V \cup W) - I(\tilde{d};V \cap W) \\ &= I(\hat{d};V) + I(\hat{d};W) - I(\hat{d};V \cup W) - I(\hat{d};V \cap W) - \epsilon. \end{aligned}$$

    The left-hand side of Constraint (iii) therefore decreases when moving from $d = \hat{d}$ to $d = \tilde{d}$. So Constraint (iii) is satisfied for $\tilde{d}$ since it is satisfied for $\hat{d}$.

- The nontrivial case is that $V \cup W$ belongs to $S$ while $V$, $W$ and $V \cap W$ do not. Then

    $$\begin{aligned} I(\tilde{d};V) + I(\tilde{d};W) &- I(\tilde{d};V \cup W) - I(\tilde{d};V \cap W) \\ &= I(\hat{d};V) + I(\hat{d};W) - \left( I(\hat{d};V \cup W) - \epsilon \right) - I(\hat{d};V \cap W). \tag{A38} \end{aligned}$$

By the definition of $S$ and condition (1) on $\epsilon$, we have

$$\begin{aligned} I(\hat{d};V \cup W) - \epsilon &> I\left( \hat{d};(V \cup W) \cap B \right) + I\left( \hat{d};(V \cup W) \cap C \right) \\ I(\hat{d};V) &= I(\hat{d};V \cap B) + I(\hat{d};V \cap C) \\ I(\hat{d};W) &= I(\hat{d};W \cap B) + I(\hat{d};W \cap C) \\ I(\hat{d};V \cap W) &= I\left( \hat{d};(V \cap W) \cap B \right) + I\left( \hat{d};(V \cap W) \cap C \right). \end{aligned}$$

Substituting into (A38) we have

$$\begin{aligned} I(\tilde{d};V) + I(\tilde{d};W) &- I(\tilde{d};V \cup W) - I(\tilde{d};V \cap W) \\ &< I(\hat{d};V \cap B) + I(\hat{d};W \cap B) \\ &\quad - I\left( \hat{d};(V \cup W) \cap B \right) - I\left( \hat{d};(V \cap W) \cap B \right) \\ &\quad + I(\hat{d};V \cap C) + I(\hat{d};W \cap C) \\ &\quad - I\left( \hat{d};(V \cup W) \cap C \right) - I\left( \hat{d};(V \cap W) \cap C \right). \end{aligned}$$

Applying Constraint (iii) on $\hat{d}$ twice to the right-hand side above, we have

$$I(\tilde{d};V) + I(\tilde{d};W) - I(\tilde{d};V\cup W) - I(\tilde{d};V\cap W)$$
$$< H(V\cap B) + H(W\cap B) - H((V\cup W)\cap B) - H((V\cap W)\cap B)$$
$$+ H(V\cap C) + H(W\cap C) - H((V\cup W)\cap C) - H((V\cap W)\cap C).$$

But Lemma A2 implies that $H(Z\cap B) + H(Z\cap C) = H(Z)$ for any subset $Z \subset A$. We apply this to the sets $V$, $W$, $V\cup W$ and $V\cap W$ to simplify the right-hand side above, yielding

$$I(\tilde{d};V) + I(\tilde{d};W) - I(\tilde{d};V\cup W) - I(\tilde{d};V\cap W) < H(V) + H(W) - H(V\cup W) - H(V\cap W),$$

as required.

No other cases are possible, since, as discussed above, any superset of a set in $S$ must also be in $S$.

Finally, it is clear that no singleton subsets of $A$ are in $S$. Thus $I(\tilde{d};a) = I(\hat{d};a)$ for each $a \in A$, and it follows that $\sum_{a\in A} \sigma(a) I(\tilde{d};a) = \hat{u}$.

We have now verified that $\tilde{d}$ satisfies Constraints (i)–(iii) and $U(\tilde{d}) = \hat{u}$. Furthermore, since $A \in S$ by assumption, we have $I(\tilde{d};A) < I(\hat{d};A)$. This contradicts the assertion that $\hat{d}$ minimizes $I(d;A)$ subject to Constraints (i)–(iii) and $U(d) \geq \hat{u}$. Therefore our assumption that $I(\hat{d};A) > I(\hat{d};B) + I(\hat{d};C)$ was incorrect, and we must instead have $I(\hat{d};A) = I(\hat{d};B) + I(\hat{d};C)$. □

**Theorem A8.** *The maximal utility function $U(y)$ is additive over independent subsystems in the sense that*

$$U_A(y) = \max_{\substack{y_1+y_2=y \\ y_1,y_2\geq 0}} (U_B(y_1) + U_C(y_2)). \tag{A39}$$

**Proof.** For a given $y \geq 0$, let $\{I(\hat{d};V)\}_{V\subset A}$ maximixe $u = \sum_{a\in A} \sigma(a) I(d;a)$ subject to Constraints (i)–(iv). Combining Theorem A7 with Lemma A6, it follows that $I(\hat{d};V) = I(\hat{d};V\cap B) + I(\hat{d};V\cap C)$. We may therefore augment our linear program with the additional constraint,

(v)  $I(d;V) = I(d;V\cap B) + I(d;V\cap C)$,

for each $V \subset A$, without altering the optimal solution.

Upon doing so, we can use this new constraint to eliminate the variables $I(d;V)$ for $V$ not a subset of either $B$ or $C$. We thereby reduce the set of variables from $\{I(d;V)\}_{V\subset A}$ to

$$\{I(d;V)\}_{V\subset B} \cup \{I(d;W)\}_{W\subset C}. \tag{A40}$$

We observe that this reduced linear program has the following structure: The variables in the set $\{I(d;V)\}_{V\subset B}$ are restricted by Constraints (i)–(iii) as applied to these variables. Separately, variables in the set $\{I(d;W)\}_{W\subset C}$ are also restricted by Constraints (i)–(iii), as they apply to the variables in this second set. The only constraint that simultaneously involves variables in both sets is Constraint (iv). This constraint can be rewritten as

$$u_B + u_C \leq y, \tag{A41}$$

with

$$u_B = \sum_{b\in B} \sigma(b) I(d;b), \qquad u_C = \sum_{c\in C} \sigma(c) I(d;c). \tag{A42}$$

This structure enables us to apply the decomposition principle for linear programs [112] to decompose the full program into two linear sub-programs, one on the variables $\{I(d;V)\}_{V\subset B}$ and one on $\{I(d;W)\}_{W\subset C}$, together with a coordinating program described by Constraint (iv). The desired result then follows from standard theorems of linear program decomposition [112]. □

.

**Theorem A9.** $M_A$ *is additive over independent subsystems in the sense that*

$$M_A(y) = \min_{\substack{y_1+y_2=y \\ y_1,y_2 \geq 0}} \max \left\{ M_B(y_1), M_C(y_2) \right\}. \tag{A43}$$

**Proof.** We define the function

$$F(y_1; y) = U_B(y_1) + U_C(y - y_1). \tag{A44}$$

The result of Theorem A8 can then be expressed as

$$U(y) = \max_{0 \leq y_1 \leq y} F(y_1; y). \tag{A45}$$

We choose and fix an arbitrary $y$-value $\tilde{y} \geq 0$, and we will prove the desired result for $y = \tilde{y}$.

We observe that $F(y_1; \tilde{y})$ is concave in $y_1$ since $U_B(y_1)$ and $U_C(\tilde{y} - y_1)$ are. It follows that any local maximum of $F(y_1; \tilde{y})$ in $y_1$ is also a global maximum. We assume that the maximum of $F(y_1; \tilde{y})$ in $y_1$ is achieved at a single point $\hat{y}_1$ with $0 < \hat{y}_1 < \tilde{y}$. The remaining cases—that the maximum is achieved at $y_1 = 0$ or $y_1 = \tilde{y}$, or is achieved on a closed interval of $y_1$-values—are trivial extensions of this case.

Assuming we are in the case described above (and again invoking the concavity of $F$ in $y_1$), $\hat{y}_1$ must be the unique point at which the derivative $\frac{\partial F}{\partial y_1}(y_1; \tilde{y})$ changes sign from positive to negative. This derivative can be written

$$\frac{\partial F}{\partial y_1}(y_1; \tilde{y}) = M_B(y_1) - M_C(\tilde{y} - y_1). \tag{A46}$$

It follows that $\hat{y}_1$ is the unique real number in $[0, \tilde{y}]$ satisfying

$$\begin{cases} M_B(y_1) > M_C(\tilde{y} - y_1) & y_1 < \hat{y}_1 \\ M_B(y_1) < M_C(\tilde{y} - y_1) & y_1 > \hat{y}_1. \end{cases} \tag{A47}$$

From inequalities (A47), and using the fact that $M_B(y_1)$ and $M_C(y_2)$ are nonincreasing, piecewise-constant functions, we see that either $M_B(y_1)$ decreases at $y_1 = \hat{y}_1$, or $M_C(y_2)$ decreases at $y_2 = \tilde{y} - \hat{y}_1$, or both. We analyze these cases separately.

**Case A1.** $M_B(y_1)$ *decreases at* $y_1 = \hat{y}_1$, *while* $M_C(y_2)$ *is constant in a neighborhood of* $y_2 = \tilde{y} - \hat{y}_1$.

Pick $\epsilon > 0$ sufficiently small so that $M_C(y_2)$ has constant value for $y_2 \in (\tilde{y} - \hat{y}_1 - \epsilon, \tilde{y} - \hat{y}_1 + \epsilon)$. Then inequalities (A47) remain satisfied with $\tilde{y}$ replaced by any $y \in (\tilde{y} - \epsilon, \tilde{y} + \epsilon)$ and $\hat{y}_1$ fixed. Thus for $y$ in this range, we have

$$U_A(y) = U_B(\hat{y}_1) + U_C(y - \hat{y}_1). \tag{A48}$$

Taking the derivative of both sides in $y$ at $y = \tilde{y}$ yields

$$M_A(\tilde{y}) = M_C(\tilde{y} - \hat{y}_1). \tag{A49}$$

We claim that

$$M_C(\tilde{y} - \hat{y}_1) = \min_{0 \leq y_1 \leq \tilde{y}} \max \left\{ M_B(y_1), M_C(\tilde{y} - y_1) \right\}. \tag{A50}$$

To prove this claim, we first note that, by the inequalities (A47),

$$\max \left\{ M_B(y_1), M_C(\tilde{y} - y_1) \right\} = \begin{cases} M_B(y_1) & y_1 < \hat{y}_1 \\ M_C(\tilde{y} - y_1) & y_1 > \hat{y}_1. \end{cases} \tag{A51}$$

Since both $M_\mathcal{B}$ and $M_\mathcal{C}$ are piecewise-constant and nonincreasing, the minimax in Equation (A50) is achieved for values $y_1$ near $\hat{y}_1$. We therefore can restrict to the range $y_1 \in (\hat{y}_1 - \epsilon, \hat{y}_1 + \epsilon)$. Combining Equation (A51) with the conditions defining Case A1 and the definition of $\epsilon$, we have

$$
\begin{aligned}
\max\left\{M_\mathcal{B}(y_1), M_\mathcal{C}(\tilde{y} - y_1)\right\} = M_\mathcal{B}(y_1) > M_\mathcal{C}(\tilde{y} - \hat{y}_1) \qquad &\text{for } y_1 \in (\hat{y}_1 - \epsilon, \hat{y}_1) \\
\max\left\{M_\mathcal{B}(y_1), M_\mathcal{C}(\tilde{y} - y_1)\right\} = M_\mathcal{C}(\tilde{y} - y_1) = M_\mathcal{C}(\tilde{y} - \hat{y}_1) \qquad &\text{for } y_1 \in (\hat{y}_1, \hat{y}_1 + \epsilon).
\end{aligned}
\tag{A52}
$$

Thus the minimax in Equation (A50) is achieved at a value of $M_\mathcal{C}(\tilde{y} - \hat{y}_1)$ when $y_1 \in (\hat{y}_1, \hat{y}_1 + \epsilon)$, verifying Equation (A50). Combining with Equation (A49), we have

$$
M_\mathcal{A}(y) = \min_{\substack{y_1 + y_2 = y \\ y_1, y_2 \geq 0}} \max\left\{M_\mathcal{B}(y_1), M_\mathcal{C}(y_2)\right\},
\tag{A53}
$$

proving the theorem in this case.

**Case A2.** $M_\mathcal{B}(y_1)$ *is constant in a neighborhood of* $y_1 = \hat{y}_1$*, while* $M_\mathcal{C}(y_2)$ *decreases at* $y_2 = \tilde{y} - \hat{y}_1$*.*

In this case, we define $\hat{y}_2 = \tilde{y} - \hat{y}_1$. The proof then follows exactly as in Case 1, with $\mathcal{B}$ and $\mathcal{C}$ interchanged, and the subscripts 1 and 2 interchanged.

**Case A3.** $M_\mathcal{B}(y_1)$ *decreases at* $y_1 = \hat{y}_1$ *and* $M_\mathcal{C}(y_2)$ *decreases at* $y_2 = \tilde{y} - \hat{y}_1$*.*

This case only occurs at the $y$-values for which $U_\mathcal{A}(y)$ changes slope and $M_\mathcal{A}(y)$ changes value. At these nongeneric points, $M_\mathcal{A}(y)$ (defined as the derivative of $U_\mathcal{A}(y)$) is undefined. We therefore disregard this case. $\square$

We now define $\tilde{M}_\mathcal{A}(x)$ as the reflection of $M_\mathcal{A}(y)$ with the dependent and independent variables interchanged. Since $M_\mathcal{A}$ is positive and nonincreasing, $\tilde{M}_\mathcal{A}$ is a well-defined function given by the formula

$$
\tilde{M}_\mathcal{A}(x) = \max\{y : M_\mathcal{A}(y) \leq x\}.
\tag{A54}
$$

The following corollary gives a simpler expression of the additivity property of MUI.

**Corollary A1.** *If $\mathcal{A}$ consists of independent subsystems $\mathcal{B}$ and $\mathcal{C}$, then $\tilde{M}_\mathcal{A}(x) = \tilde{M}_\mathcal{B}(x) + \tilde{M}_\mathcal{C}(x)$ for all $x \geq 0$.*

**Proof.** Combining the above formula for $\tilde{M}_\mathcal{A}(x)$ with the result of Theorem A9, we write

$$
\begin{aligned}
\tilde{M}_\mathcal{A}(x) &= \max\{y : M_\mathcal{A}(y) \leq x\} \\
&= \max\left\{y : \left(\min_{\substack{y_1 + y_2 = y \\ y_1, y_2 \geq 0}} \max\left\{M_\mathcal{B}(y_1), M_\mathcal{C}(y_2)\right\}\right) \leq x\right\} \\
&= \max\left\{y : \left(\exists y_1, y_2 \geq 0. \, (y_1 + y_2 = y \,\wedge\, \max\left\{M_\mathcal{B}(y_1), M_\mathcal{C}(y_2)\right\} \leq x)\right)\right\} \\
&= \max\left\{(y_1 + y_2) : \left(y_1, y_2 \geq 0 \,\wedge\, M_\mathcal{B}(y_1) \leq x \,\wedge\, M_\mathcal{C}(y_2) \leq x\right)\right\} \\
&= \max\{y_1 : M_\mathcal{B}(y_1) \leq x\} + \max\{y_2 : M_\mathcal{B}(y_2) \leq x\} \\
&= \tilde{M}_\mathcal{B}(x) + \tilde{M}_\mathcal{C}(x). \quad \square
\end{aligned}
$$

## Appendix F. Marginal Utility of Information for Parity Bit Systems

Here we compute the MUI for a family of systems which exhibit exchange symmetry and have a constraint at the largest scale. Systems in this class have $N \geq 3$ components and information function given by

$$H(V) = H_{|V|} = \begin{cases} |V| & |V| \le N - 1 \\ N - 1 & |V| = N. \end{cases} \tag{A55}$$

This includes Example **D** as the case $N = 3$. More generally, this family includes systems of $N - 1$ independent random bits together with one parity bit.

Since these systems have exchange symmetry, we can apply the argument of Section 9 to obtain the reduced set of constraints:

(i)  $0 \le I_n \le H_n$ for all $n \in \{1, \ldots, N\}$,
(ii)  $0 \le I_n - I_{n-1} \le H_n - H_{n-1}$ for all $n \in \{1, \ldots, N\}$,
(iii)  $I_n + I_m - I_{n+m-\ell} - I_\ell \le H_n + H_m - H_{n+m-\ell} - H_\ell$ for all $n, m, \ell \in \{1, \ldots, N\}$,
(iv)  $I_N \le y$.

Above, $I_n$ denotes information that a descriptor imparts about any set of $n$ components, $0 \le n \le N$. Constraint (ii), in the case $n = N$, yields

$$0 \le I_N - I_{N-1} \le H_N - H_{N-1} = 0, \tag{A56}$$

and therefore,

$$I_{N-1} = I_N. \tag{A57}$$

Constraint (iii), in the case $m + n \le N - 1$, yields

$$I_m + I_n - I_{m+n} \le H_m + H_n - H_{m+n} = 0, \tag{A58}$$

and therefore

$$I_m + I_n \le I_{m+n} \qquad \text{for all } m, n \ge 0 \text{ with } m + n \le N - 1. \tag{A59}$$

By iteratively applying Equation (A59) we arrive at the inequality

$$(N - 1)I_1 \le I_{N-1}. \tag{A60}$$

Combining (A57), (A60) and Constraint (iv), we have

$$I_1 \le \frac{I_{N-1}}{N - 1} = \frac{I_N}{N - 1} \le \frac{y}{N - 1}. \tag{A61}$$

By definition, the utility of a descriptor is

$$u = NI_1. \tag{A62}$$

Combining with (A61) yields the inequality

$$u \le \frac{Ny}{N - 1}. \tag{A63}$$

Inequality (A63) places a limit on the utility of any descriptor of this system. To complete the argument, we exhibit a descriptor for which equality holds in (A63). This descriptor is defined by

$$I_m = \begin{cases} \frac{m}{N-1} \min\{y, N - 1\} & 0 \le m \le N - 1 \\ \min\{y, N - 1\} & m = N. \end{cases} \tag{A64}$$

It is straightforward to verify that Constraints (i)–(iv) are satisfied by this descriptor. Combining Equations (A62) and (A64), the utility of this descriptor is

$$u = \min \left\{ \frac{Ny}{N-1}, N \right\}. \tag{A65}$$

By inequality (A63), this descriptor achieves optimal utility. Taking the derivative with respect to $y$, we obtain the marginal utility of information

$$M(y) = \begin{cases} \frac{N}{N-1} & 0 \le y \le N-1 \\ 0 & y > N-1. \end{cases} \tag{A66}$$

Setting $N = 3$, we recover the MUI for example **D** as stated in the main text, Equation (25).

**References**

1. Bar-Yam, Y. *Dynamics of Complex Systems*; Westview Press: Boulder, CO, USA, 2003.
2. Haken, H. *Information and Self-Organization: A Macroscopic Approach to Complex Systems*; Springer: New York, NY, USA, 2006.
3. Miller, J.H.; Page, S.E. *Complex Adaptive Systems: An Introduction to Computational Models of Social Life*; Princeton University Press: Princeton, NJ, USA, 2007.
4. Boccara, N. *Modeling Complex Systems*; Springer: New York, NY, USA, 2010.
5. Newman, M.E.J. Complex Systems: A Survey. *Am. J. Phys.* **2011**, *79*, 800–810.
6. Kwapień, J.; Drożdż, S. Physical approach to complex systems. *Phys. Rep.* **2012**, *515*, 115–226.
7. Sayama, H. *Introduction to the Modeling and Analysis of Complex Systems*; Open SUNY: Bunghamton, NY, USA, 2015.
8. Sethna, J.P. *Statistical Mechanics: Entropy, Order Parameters, and Complexity*; Oxford University Press: Oxford, UK, 2006.
9. Shannon, C. A mathematical theory of communication. *Bell Syst. Tech. J.* **1948**, *27*, 379–423.
10. Cover, T.M.; Thomas, J.A. *Elements of Information Theory*; Wiley: Hoboken, NJ, USA, 1991.
11. Prokopenko, M.; Boschetti, F.; Ryan, A.J. An information-theoretic primer on complexity, self-organization, and emergence. *Complexity* **2009**, *15*, 11–28.
12. Gallagher, R.G. *Information Theory and Reliable Communication*; Wiley: Hoboken, NJ, USA, 1968.
13. Bar-Yam, Y. Multiscale complexity/entropy. *Adv. Complex Syst.* **2004**, *7*, 47–63.
14. Bar-Yam, Y. Multiscale variety in complex systems. *Complexity* **2004**, *9*, 37–45.
15. Bar-Yam, Y.; Harmon, D.; Bar-Yam, Y. Computationally tractable pairwise complexity profile. *Complexity* **2013**, *18*, 20–27..
16. Metzler, R.; Bar-Yam, Y. Multiscale complexity of correlated Gaussians. *Phys. Rev. E* **2005**, *71*, 046114.
17. Gheorghiu-Svirschevski, S.; Bar-Yam, Y. Multiscale analysis of information correlations in an infinite-range, ferromagnetic Ising system. *Phys. Rev. E* **2004**, *70*, 066115.
18. Stacey, B.C.; Allen, B.; Bar-Yam, Y. Multiscale Information Theory for Complex Systems: Theory and Applications. In *Information and Complexity*; Burgin, M., Calude, C.S., Eds.; World Scientific: Singapore, 2017; pp. 176–199.
19. Grassberger, P. Toward a quantitative theory of self-generated complexity. *Int. J. Theor. Phys.* **1986**, *25*, 907–938.
20. Crutchfield, J.P.; Young, K. Inferring statistical complexity. *Phys. Rev. Lett.* **1989**, *63*, 105–108.
21. Crutchfield, J.P. The calculi of emergence: Computation, dynamics and induction. *Phys. D Nonlinear Phenom.* **1994**, *75*, 11–54.
22. Misra, V.; Lagi, M.; Bar-Yam, Y. *Evidence of Market Manipulation in the Financial Crisis*; Technical Report 2011-12-01; NECSI: Cambridge, MA, USA, 2011.
23. Harmon, D.; Lagi, M.; de Aguiar, M.A.; Chinellato, D.D.; Braha, D.; Epstein, I.R.; Bar-Yam, Y. Anticipating Economic Market Crises Using Measures of Collective Panic. *PLoS ONE* **2015**, *10*, e0131871.
24. Green, H. *The Molecular Theory of Fluids*; North–Holland: Amsterdam, The Netherlands, 1952.
25. Nettleton, R.E.; Green, M.S. Expression in terms of molecular distribution functions for the entropy density in an infinite system. *J. Chem. Phys.* **1958**, *29*, 1365–1370.

26. Wolf, D.R. Information and Correlation in Statistical Mechanical Systems. Ph.D. Thesis, University of Texas, Austin, TX, USA, 1996.
27. Kardar, M. *Statistical Physics of Particles*; Cambridge University Press: Cambridge, UK, 2007.
28. Kadanoff, L.P. Scaling laws for Ising models near $T_c$. *Physics* **1966**, *2*, 263.
29. Wilson, K.G. The renormalization group: Critical phenomena and the Kondo problem. *Rev. Mod. Phys.* **1975**, *47*, 773.
30. McGill, W.J. Multivariate information transmission. *Psychometrika* **1954**, *46*, 26–45.
31. Han, T.S. Multiple mutual information and multiple interactions in frequency data. *Inf. Control* **1980**, *46*, 26–45.
32. Yeung, R.W. A new outlook on Shannon's information measures. *IEEE Trans. Inf. Theory* **1991**, *37*, 466–474.
33. Jakulin, A.; Bratko, I. Quantifying and visualizing attribute interactions. *arXiv* **2003**, arXiv:cs.AI/0308002.
34. Bell, A.J. The co-information lattice. In Proceedings of the Fifth International Workshop on Independent Component Analysis and Blind Signal Separation (ICA), Nara, Japan, 1–4 April 2003; Volume 2003.
35. Bar-Yam, Y. A mathematical theory of strong emergence using multiscale variety. *Complexity* **2004**, *9*, 15–24.
36. Krippendorff, K. Information of interactions in complex systems. *Int. J. Gen. Syst.* **2009**, *38*, 669–680.
37. Leydesdorff, L. Redundancy in systems which entertain a model of themselves: Interaction information and the self-organization of anticipation. *Entropy* **2010**, *12*, 63–79.
38. Kolchinsky, A.; Rocha, L.M. Prediction and modularity in dynamical systems. *arXiv* **2011**, arXiv:1106.3703.
39. James, R.G.; Ellison, C.J.; Crutchfield, J.P. Anatomy of a bit: Information in a time series observation. *Chaos Interdiscip. J. Nonlinear Sci.* **2011**, *21*, 037109.
40. Tononi, G.; Sporns, O.; Edelman, G.M. A measure for brain complexity: Relating functional segregation and integration in the nervous system. *Proc. Natl. Acad. Sci. USA* **1994**, *91*, 5033–5037.
41. Ay, N.; Olbrich, E.; Bertschinger, N.; Jost, J. A unifying framework for complexity measures of finite systems. In Proceedings of the European Complex Systems Society (ECCS06), Oxford, UK, 25 September 2006.
42. Bar-Yam, Y. *Complexity of Military Conflict: Multiscale Complex Systems Analysis of Littoral Warfare*; Technical Report; NECSI: Cambridge, MA, USA, 2003.
43. Granovsky, B.L.; Madras, N. The noisy voter model. *Stoch. Process. Appl.* **1995**, *55*, 23–43.
44. Faddeev, D.K. On the concept of entropy of a finite probabilistic scheme. *Uspekhi Mat. Nauk* **1956**, *11*, 227–231.
45. Khinchin, A.I. *Mathematical Foundations of Information Theory*; Dover: New York, NY, USA, 1957.
46. Lee, P. On the axioms of information theory. *Ann. Math. Stat.* **1964**, *35*, 415–418.
47. Rényi, A. *Probability Theory*; Akadémiai Kiadó: Budapest, Hungary, 1970.
48. Daróczy, Z. Generalized information functions. *Inf. Control* **1970**, *16*, 36–51.
49. Dos Santos, R.J. Generalization of Shannon's theorem for Tsallis entropy. *J. Math. Phys.* **1997**, *38*, 4104–4107.
50. Abe, S. Axioms and uniqueness theorem for Tsallis entropy. *Phys. Lett. A* **2000**, *271*, 74–79.
51. Tsallis, C. Possible generalization of Boltzmann-Gibbs statistics. *J. Stat. Phys.* **1988**, *52*, 479–487.
52. Gell-Mann, M.; Tsallis, C. *Nonextensive Entropy: Interdisciplinary Applications*; Oxford University Press: Oxford, UK, 2004.
53. Furuichi, S. Information theoretical properties of Tsallis entropies. *J. Math. Phys.* **2006**, *47*, 023302.
54. Steudel, B.; Janzing, D.; Schölkopf, B. Causal Markov condition for submodular information measures. *arXiv* **2010**, arXiv:1002.4020.
55. Dougherty, R.; Freiling, C.; Zeger, K. Networks, matroids, and non-Shannon information inequalities. *IEEE Trans. Inf. Theory* **2007**, *53*, 1949–1969.
56. Li, M.; Vitányi, P. *An Introduction to Kolmogorov Complexity and Its Applications*; Springer Science & Business Media: New York, NY, USA, 2009.
57. Chaitin, G.J. A theory of program size formally identical to information theory. *J. ACM* **1975**, *22*, 329–340.
58. May, R.M.; Arinaminpathy, N. Systemic risk: The dynamics of model banking systems. *J. R. Soc. Interface* **2010**, *7*, 823–838.
59. Haldane, A.G.; May, R.M. Systemic risk in banking ecosystems. *Nature* **2011**, *469*, 351–355.
60. Beale, N.; Rand, D.G.; Battey, H.; Croxson, K.; May, R.M.; Nowak, M.A. Individual versus systemic risk and the Regulator's Dilemma. *Proc. Natl. Acad. Sci. USA* **2011**, *108*, 12647–12652.
61. Erickson, M.J. *Introduction to Combinatorics*; Wiley: Hoboken, NJ, USA, 1996.
62. Williams, P.L.; Beer, R.D. Nonnegative decomposition of multivariate information. *arXiv* **2010**, arXiv:1004.2515.

63. James, R.G.; Crutchfield, J.P. Multivariate Dependence Beyond Shannon Information. *arXiv* **2016**, arXiv:1609.01233.
64. Perfect, H. Independence theory and matroids. *Math. Gaz.* **1981**, *65*, 103–111.
65. Studený, M.; Vejnarová, J. The multiinformation function as a tool for measuring stochastic dependence. In *Learning in Graphical Models*; Springer: Dodrecht, The Netherlands, 1998; pp. 261–297.
66. Schneidman, E.; Still, S.; Berry, M.J.; Bialek, W. Network information and connected correlations. *Phys. Rev. Lett.* **2003**, *91*, 238701.
67. Polani, D. Foundations and formalizations of self-organization. In *Advances in Applied Self-Organizing Systems*; Springer: New York, NY, USA, 2008; pp. 19–37.
68. Wets, R.J.B. Programming Under Uncertainty: The Equivalent Convex Program. *SIAM J. Appl. Math.* **1966**, *14*, 89–105.
69. James, R.G. Python Package for Information Theory. *Zenodo* **2017**, doi:10.5281/zenodo.235071.
70. Slonim, N.; Tishby, L. Agglomerative information bottleneck. *Adv. Neural Inf. Process. Syst. NIPS* **1999**, *12*, 617–623.
71. Shalizi, C.R.; Crutchfield, J.P. Information bottlenecks, causal states, and statistical relevance bases: How to represent relevant information in memoryless transduction. *Adv. Complex Syst.* **2002**, *5*, 91–95,
72. Tishby, N.; Pereira, F.C.; Bialek, W. The information bottleneck method. *arXiv* **2000**, arXiv:physics/0004057.
73. Ziv, E.; Middendorf, M.; Wiggins, C. An information-theoretic approach to network modularity. *Phys. Rev. E* **2005**, *71*, 046117.
74. Peng, C.K.; Buldyrev, S.V.; Havlin, S.; Simons, M.; Stanley, H.E.; Goldberger, A.L. Mosaic organization of DNA nucleotides. *Phys. Rev. E* **1994**, *49*, 1685–1689.
75. Gell-Mann, M.; Lloyd, S. Information measures, effective complexity, and total information. *Complexity* **1996**, *2*, 44–52.
76. Hu, K.; Ivanov, P.; Chen, Z.; Carpena, P.; Stanley, H.E. Effect of Trends on Detrended Fluctuation Analysis. *Phys. Rev. E* **2002**, *64*, 011114.
77. Vereshchagin, N.; Vitányi, P. Kolmogorov's structure functions and model selection. *IEEE Trans. Inf. Theory* **2004**, *50*, 3265–3290.
78. Grünwald, P.; Vitányi, P. Shannon information and Kolmogorov complexity. *arXiv* **2004**, arXiv:cs.IT/0410002.
79. Vitányi, P. Meaningful information. *IEEE Trans. Inf. Theory* **2006**, *52*, 4617–4626.
80. Moran, P.A.P. Random processes in genetics. *Math. Proc. Camb. Philos. Soc.* **1958**, *54*, 60–71.
81. Harmon, D.; Stacey, B.C.; Bar-Yam, Y. *Networks of Economic Market Independence and Systemic Risk*; Technical Report 2009-03-01 (updated); NECSI: Cambridge, MA, USA, 2010.
82. Stacey, B.C. Multiscale Structure in Eco-Evolutionary Dynamics. Ph.D. Thesis, Brandeis University, Waltham, MA, USA, 2015.
83. Domb, C.; Green, M.S. (Eds.) *Phase Transitions and Critical Phenomena*; Academic Press: New York, NY, USA, 1972.
84. Kardar, M. *Statistical Physics of Fields*; Cambridge University Press: Cambridge, UK, 2007.
85. Jacob, F.; Monod, J. Genetic regulatory mechanisms in the synthesis of proteins. *J. Mol. Biol.* **1961**, *3*, 318–356.
86. Britten, R.J.; Davidson, E.H. Gene regulation for higher cells: A theory. *Science* **1969**, *165*, 349–357.
87. Carey, M.; Smale, S. *Transcriptional Regulation in Eukaryotes: Concepts, Strategies, and Techniques*; Cold Spring Harbor Laboratory Press: Cold Spring Harbor, NY, USA, 2001.
88. Elowitz, M.B.; Levine, A.J.; Siggia, E.D.; Swain, P.S. Stochastic gene expression in a single cell. *Science* **2002**, *297*, 1183–1186.
89. Lee, T.I.; Rinaldi, N.J.; Robert, F.; Odom, D.T.; Bar-Joseph, Z.; Gerber, G.K.; Hannett, N.M.; Harbison, C.T.; Thompson, C.M.; Simon, I.; et al. Transcriptional Regulatory Networks in *Saccharomyces cerevisiae*. *Science* **2002**, *298*, 799–804.
90. Boyer, L.A.; Lee, T.I.; Cole, M.F.; Johnstone, S.E.; Levine, S.S.; Zucker, J.P.; Guenther, M.G.; Kumar, R.M.; Murray, H.L.; Jenner, R.G.; et al. Core Transcriptional Regulatory Circuitry in Human Embryonic Stem Cells. *Cell* **2005**, *122*, 947–956.
91. Chowdhury, S.; Lloyd-Price, J.; Smolander, O.P.; Baici, W.C.; Hughes, T.R.; Yli-Harja, O.; Chua, G.; Ribeiro, A.S. Information propagation within the Genetic Network of *Saccharomyces cerevisiae*. *BMC Syst. Biol.* **2010**, *4*, 143.
92. Hopfield, J.J. Neural networks and physical systems with emergent collective computational abilities. *Proc. Natl. Acad. Sci. USA* **1982**, *79*, 2554–2558.

93. Rabinovich, M.I.; Varona, P.; Selverston, A.I.; Abarbanel, H.D.I. Dynamical principles in neuroscience. *Rev. Mod. Phys.* **2006**, *78*, 1213–1265.

94. Schneidman, E.; Berry, M.J.; Segev, R.; Bialek, W. Weak pairwise correlations imply strongly correlated network states in a neural population. *Nature* **2006**, *440*, 1007–1012.

95. Bonabeau, E.; Dorigo, M.; Theraulaz, G. *Swarm Intelligence: From Natural to Artificial Systems*; Oxford University Press: Oxford, UK, 1999.

96. Vicsek, T.; Zafeiris, A. Collective motion. *Phys. Rep.* **2012**, *517*, 71–140.

97. Berdahl, A.; Torney, C.J.; Ioannou, C.C.; Faria, J.J.; Couzin, I.D. Emergent sensing of complex environments by mobile animal groups. *Science* **2013**, *339*, 574–576.

98. Ohtsuki, H.; Hauert, C.; Lieberman, E.; Nowak, M.A. A simple rule for the evolution of cooperation on graphs and social networks. *Nature* **2006**, *441*, 502–505.

99. Allen, B.; Lippner, G.; Chen, Y.T.; Fotouhi, B.; Momeni, N.; Yau, S.T.; Nowak, M.A. Evolutionary dynamics on any population structure. *Nature* **2017**, *544*, 227–230.

100. Mandelbrot, B.; Taylor, H. On the distribution of stock price differences. *Oper. Res.* **1967**, *15*, 1057–1062.

101. Mantegna, R.N. Hierarchical structure in financial markets. *Eur. Phys. J. B Condens. Matter Complex Syst.* **1999**, *11*, 193–197.

102. Sornette, D. *Why Stock Markets Crash: Critical Events in Complex Financial Systems*; Princeton University Press: Princeton, NJ, USA, 2004.

103. May, R.M.; Levin, S.A.; Sugihara, G. Complex systems: Ecology for bankers. *Nature* **2008**, *451*, 893–895.

104. Schweitzer, F.; Fagiolo, G.; Sornette, D.; Vega-Redondo, F.; Vespignani, A.; White, D.R. Economic Networks: The New Challenges. *Science* **2009**, *325*, 422–425.

105. Harmon, D.; De Aguiar, M.; Chinellato, D.; Braha, D.; Epstein, I.; Bar-Yam, Y. Predicting economic market crises using measures of collective panic. *arXiv* **2011**, arXiv:1102.2620.

106. Schrödinger, E. *What Is Life? The Physical Aspect of the Living Cell and Mind*; Cambridge University Press: Cambridge, UK, 1944.

107. Brillouin, L. The negentropy principle of information. *J. Appl. Phys.* **1953**, *24*, 1152–1163.

108. Stacey, B.C. Multiscale Structure of More-than-Binary Variables. *arXiv* **2017**, arXiv:1705.03927.

109. Ashby, W.R. *An Introduction to Cybernetics*; Chapman & Hall: London, UK, 1956.

110. Stacey, B.C.; Bar-Yam, Y. *Principles of Security: Human, Cyber, and Biological*; Technical Report 2008-06-01; NECSI: Cambridge, MA, USA, 2008.

111. Dorogovtsev, S.N. *Lectures on Complex Networks*; Oxford University Press: Oxford, UK, 2010.

112. Dantzig, G.B.; Wolfe, P. Decomposition principle for linear programs. *Oper. Res.* **1960**, *8*, 101–111.

*entropy*

MDPI

*Article*

# Synergy and Redundancy in Dual Decompositions of Mutual Information Gain and Information Loss

Daniel Chicharro [1,2,]* and Stefano Panzeri [2]

[1]    Department of Neurobiology, Harvard Medical School, Boston, MA 02115, USA
[2]    Neural Computation Laboratory, Center for Neuroscience and Cognitive Systems@UniTn, Istituto Italiano di Tecnologia, Rovereto (TN) 38068, Italy; stefano.panzeri@iit.it
*    Correspondence: daniel.chicharro@iit.it or Daniel_Chicharro@hms.harvard.edu; Tel.: +39-0464-808696

Academic Editor: Mikhail Prokopenko
Received: 30 December 2016; Accepted: 13 February 2017; Published: 16 February 2017

**Abstract:** Williams and Beer (2010) proposed a nonnegative mutual information decomposition, based on the construction of information gain lattices, which allows separating the information that a set of variables contains about another variable into components, interpretable as the unique information of one variable, or redundant and synergy components. In this work, we extend this framework focusing on the lattices that underpin the decomposition. We generalize the type of constructible lattices and examine the relations between different lattices, for example, relating bivariate and trivariate decompositions. We point out that, in information gain lattices, redundancy components are invariant across decompositions, but unique and synergy components are decomposition-dependent. Exploiting the connection between different lattices, we propose a procedure to construct, in the general multivariate case, information gain decompositions from measures of synergy or unique information. We then introduce an alternative type of lattices, information loss lattices, with the role and invariance properties of redundancy and synergy components reversed with respect to gain lattices, and which provide an alternative procedure to build multivariate decompositions. We finally show how information gain and information loss dual lattices lead to a self-consistent unique decomposition, which allows a deeper understanding of the origin and meaning of synergy and redundancy.

**Keywords:** information theory; mutual information decomposition; synergy; redundancy

**MSC:** 94A15, 94A17

## 1. Introduction

The aim to determine the mechanisms producing dependencies in a multivariate system, and to characterize these dependencies, has motivated several proposals to breakdown the contributions to the mutual information between sets of variables [1]. This problem is interesting from a theoretical perspective in information theory, but it is also crucial from an empirical point of view in many fields of systems and computational biology, e.g., [2–6]. For example, in neuroscience breaking down the contributions to mutual information between sets of variables is fundamental to make any kind of progress in understanding neural population coding of sensory information. This breakdown is, in fact, necessary to identify the unique contributions of individual classes of neurons, and of interactions among them, to the sensory information carried by neural populations [7,8], is necessary to understand how information in populations of neurons contributes to behavioural decisions [9,10], and to understand how information is transmitted and further processed across areas [11].

Consider the mutual information $I(\mathbf{S}; \mathbf{R})$ between two possibly multivariate sets of variables $\mathbf{S}$ and $\mathbf{R}$, here thought, for the sake of example, as a set of sensory stimuli, $\mathbf{S}$, and neural responses

**R**, but generally any sets of variables. An aspect that has been widely studied is how dependencies within each set contribute to the information. For example, the mutual information breakdown of [12,13] quantifies the global contribution to the information of conditional dependencies between the variables in **R**, and has been applied to study how interactions among neurons shape population coding of sensory information. Subsequent decompositions, based on a maximum entropy approach, have proposed to subdivide this global contribution, separating the influence of dependencies of different orders [14,15]. However, these types of decompositions do not ensure that all terms in the decomposition are nonnegative and hence should be better interpreted as a comparison of the mutual information across different alternative system's configurations [16,17]. Two concepts tightly related to these types of decompositions are those of redundancy and synergy, e.g., [18]. Redundancy refers to the existence of common information about **S** that could be retrieved from different variables contained in **R** used separately. Conversely, synergy refers to the existence of information that can only be retrieved when jointly using the variables in **R**. Traditionally, synergy and redundancy had been quantified together, with the measure called interaction information [19] or co-information [20]. A positive value of this measure is considered as a signature of redundancy being present in the system, while a negative value is associated with synergy, so that redundancy and synergy have traditionally been considered as mutually exclusive.

The seminal work of [21] introduced a new approach to decompose the mutual information into a set of nonnegative contributions. Let us consider first the bivariate case. Without loss of generality, from now on we assume $S$ to be a univariate variable, if not stated otherwise. It was shown in [21] that the mutual information can be decomposed into four terms:

$$I(S;12) = I(S;1.2) + I(S;1\backslash2) + I(S;2\backslash1) + I(S;12\backslash1,2). \qquad (1)$$

The term $I(S;1.2)$ refers to a redundancy component between variables 1 and 2. The terms $I(S;1\backslash2)$ and $I(S;2\backslash1)$ quantify a component of the information that is unique of 1 and of 2, respectively, that is, some information that can be obtained from one of the variables alone but that cannot be obtained from the other alone. The term $I(S;12\backslash1,2)$ refers to the synergy between the two variables, the information that is unique for the joint source 12 with respect to the variables alone. Note that in this decomposition, a redundancy and a synergy component can exist simultaneously. In fact, [21] showed that the measure of co-information is equivalent to the difference between the redundancy and the synergy terms of Equation (1). Generally, [21] defined this type of decomposition for any multivariate set of variables $\{\mathbf{R}\}$. The key ingredients for this general formulation were the definition of a general measure of redundancy and the association of each decomposition comprising $n$ variables to a lattice structure, constructed with different combinations of groups of variables ordered by defining an ordering relation. We will review this general formulation linking decompositions and lattices in great detail below.

Different parts of the framework introduced by [21] have generated different levels of consensus. The conceptual framework of nonnegative decompositions of mutual information, with distinguishable redundancy and synergy contributions and with lattices underpinning the decompositions, has been adopted by many others, e.g., [22–24]. Conversely, it has been argued that the specific measure $I_{min}$ originally used to determine the terms of the decomposition does not properly quantify redundancy, e.g., [22,23]. Accordingly, much of the subsequent efforts have focused in finding the right measures to define the components of the decomposition. From these alternative proposals, some take as the basic component to derive the terms in the decomposition another measure of redundancy [22,25], but also a measure of synergy [23], or of unique information [24]. In contrast to $I_{min}$, these measures fulfill the identity axiom [22], introduced to prevent that for **S** composed by two independent variables, a redundancy component is obtained for **R** being a copy of **S**. Indeed, apart from proposing other specific measures, subsequent studies have proposed a set of axioms which state desirable properties of these measures [22,23,26–28]. However, there is no full consensus on which are the axioms that should be imposed. Furthermore, it has been shown that some of these axioms are incompatible with each other [26]. In particular, [26] provided a counterexample illustrating that nonnegativity is not ensured for the components of the decomposition in the multivariate case if assuming the identity axiom. Some contributions have also studied the relation between the measures that contain different number of variables [26,29]. For some specific type of variables, namely multivariate Gaussians with a univariate $S$, the equivalence between some of the proposed measure has been proven [30].

To our knowledge, perhaps because of these difficulties in finding a proper measure to construct the decompositions, less attention has been paid to study the properties of the lattices associated with the decompositions. We here focus on examining these properties and the basic constituents that are used to construct the decompositions from the lattices. We generalize the type of lattices introduced by [21] and we examine the relation between the information-theoretic quantities associated with different lattices (Section 3.1), discussing when a certain lattice is valid (Section 3.2). We show that the connection between the components of different lattices can be used to extend to the multivariate case decompositions for which, to our knowledge, there was currently no available method to determine their components in the multivariate case, e.g., [23,24]. In particular, we introduce an iterative hierarchical procedure that allows building decompositions when using as a basic component a measure of synergy or unique information (Section 3.3). Motivated by this analysis, we introduce a new type of lattices, namely information loss lattices in contrast to the information gain lattices described in [21]. We show that these loss lattices are more naturally related to synergy measures, as opposed to gain lattices more naturally related to redundancy measures (Section 4). The information loss lattices provide an alternative and more direct procedure to construct the mutual information decompositions from a synergy measure. This procedure is equivalent to the one used in the information gain lattices with a redundancy measure [21], and does not require considering the connection between different lattices, oppositely to the iterative hierarchical procedure. Given these alternative options to build mutual information decompositions, we ask how consistent the decompositions obtained from each procedure are. This lead us to identify the existence of dual information gain and loss lattices which, independently of the procedure used, allow constructing a unique mutual information decomposition, compatible with the existence of unique notions of redundancy, synergy, and unique information (Section 5). Other open questions related to the selection of the measures and the axioms are out of the scope of this work.

## 2. A Brief Review of Lattice-Based Mutual Information Decompositions

We first review some basic facts regarding the existing decompositions of [21] as a first step for the extensions we propose in this work. In relation to the bivariate decomposition of Equation (1), it was also shown in [21] that

$$I(S;1) = I(S;1.2) + I(S;1\backslash 2), \tag{2}$$

and a similar relation holds for $I(S;2)$. Accordingly, given the standard information-theoretic equalities [31]

$$I(S;12) = I(S;1) + I(S;2|1) \tag{3a}$$
$$= I(S;2) + I(S;1|2), \tag{3b}$$

also the conditional mutual information is decomposed as

$$I(S;2|1) = I(S;2\backslash 1) + I(S;12\backslash 1,2), \tag{4}$$

and analogously for $I(S;1|2)$. That is, each variable can contain some information that is redundant to the other and some part that is unique. Conditioning one variable on the other removes the redundant component of the information but adds the synergistic component, resulting in the conditional information being the sum of the unique and synergistic terms.

We now review the construction of the lattices and their relation to the decompositions. A lattice is composed by a set of *collections*. This set is defined as

$$\mathcal{A}(\mathbf{R}) = \{\alpha \in \mathcal{P}(\mathbf{R})\backslash\{\varnothing\} : \forall\, \mathbf{A}_i, \mathbf{A}_j \in \alpha, \mathbf{A}_i \not\subseteq \mathbf{A}_j\}, \tag{5}$$

where $\mathcal{P}(\mathbf{R})\backslash\{\varnothing\}$ is the set of all nonempty subsets of the set of nonempty *sources* that can be formed from $\{\mathbf{R}\}$, where a source $\mathbf{A}$ is a subset of the variables $\{\mathbf{R}\}$. That is, each collection $\alpha$ is itself a set of sources, and each source $\mathbf{A}$ is a set of variables. The domain of the collections included in the lattice is established by the constraint that a collection cannot contain sources that are a superset of another source in the collection. This restriction is justified in detail in [21], based on the idea that the redundancy between a source and any superset of it is equal to the information of that source. Given the set of collections $\mathcal{A}(\mathbf{R})$, the lattice is constructed defining an ordering relation between the collections. In particular:

$$\forall\, \alpha, \beta \in \mathcal{A}(\mathbf{R}), (\alpha \preceq \beta \Leftrightarrow \forall \mathbf{B} \in \beta, \exists \mathbf{A} \in \alpha, \mathbf{A} \subseteq \mathbf{B}), \tag{6}$$

that is, for two collections $\alpha$ and $\beta$, $\alpha \preceq \beta$ if for each source in $\beta$ there is a source in $\alpha$ that is a subset of that source. This ordering relation is reflexive, transitive, and antisymmetric. The lattices constructed for the case of $n = 2$ and $n = 3$ using this ordering relation are shown in Figure 1A,B. The order is partial because an order does not exist between all pairs of collections, for example between the collections at the same level of the lattice. In this work, we use a different notation than in [21], which allows us to shorten the expressions a bit. For example, instead of writing $\{1\}\{23\}$ for the collection composed by the source containing variable 1 and the source containing variables 2 and 3, we write 1.23, that is, we save the curly brackets that indicate for each source the set of variables and we use instead a dot to separate the sources.

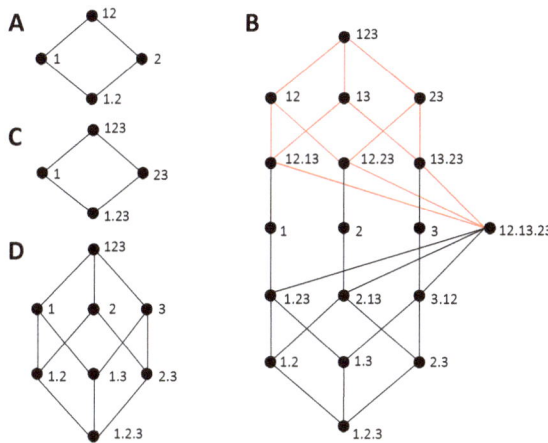

**Figure 1.** Information gain decompositions of different orders and for different subsets of collections of sources. (**A,B**) Lattices constructed from the complete domain of collections as defined by Equation (5) for $n = 2$ and $n = 3$, respectively. Pale red edges in (**B**) identify the embedded lattice formed by collections that do not contain univariate sources. (**C**) Alternative decomposition based only on sources 1 and 23. (**D**) Alternative decomposition that does not contain bivariate sources.

Each collection in the lattice is associated with a measure of the redundancy between the sources composing the collection. Reference [21] defined a measure of redundancy, called $I_{min}$, that is well defined for any collection. In this work, we do not need to consider the specific definition of $I_{min}$. What is relevant for us is that, when ascending the lattice, $I_{min}$ monotonically increases, being a cumulative measure of information and reaching the total amount of information at the top of the lattice. Based on this accumulation of information, we will from now on refer to the type of lattices introduced by [21] as *information gain lattices*. Furthermore, we will generically refer to the terms quantifying the information accumulated in each collection as *cumulative terms* and denote the cumulative term of a collection $\alpha$ by $I(S, \alpha)$. The reason for this change of terminology will become evident when we introduce the information loss lattices, since redundancy is not specific to the information gain lattices and it appears also in the loss lattices but not associated with cumulative terms, and thus we need to disentangle it nominally from the cumulative terms, even if in the formulation of [21] they are inherently associated.

Independently of which measure is used to define the cumulative terms $I(S, \alpha)$, two axioms originally required by [21] ensure that these terms and their relations are compatible with the lattice. First, the symmetry axiom requires that $I(S, \alpha)$ is invariant to the order of the sources in the collection, in the same way that the domain $\mathcal{A}(\mathbf{R})$ of collections does not distinguish the order of the sources. Second, the monotonicity axiom requires that $I(S, \alpha) \leq I(S, \beta)$ if $\alpha \preceq \beta$. Another axiom from [21] ensures that the cumulative terms are linked to the actual mutual information measures of the variables. In particular, the self-redundancy axiom requires that, when the collection is formed by a single source $\alpha = \mathbf{A}$, then $I(S; \alpha = \mathbf{A}) = I(S; \mathbf{A})$, that is, the cumulative term is equal to the directly calculable mutual information of the variables in $\mathbf{A}$. The other axioms that have been proposed are not related to the construction of the information gain lattice itself. Conversely, they are motivated by desirable properties of a measure of redundancy or desirable properties of the terms in the decomposition, such as the nonnegativity axiom. The fulfillment of a proper set of additional axioms is important to endow this type of mutual information decompositions with a meaning, regarding how to interpret each component of the decomposition. However, these additional axioms do not determine, nor are they determined, by the properties of the lattice. We will not examine these other axioms in detail in this work, but we refer to Appendix C for a discussion of the requirements to obtain nonnegative terms in the decomposition.

The mutual information decomposition was constructed in [21] by implicitly defining partial information measures associated with each node, such that the cumulative terms are obtained from the sum of partial information measures:

$$I(S, \alpha) = \sum_{\beta \in \downarrow \alpha} \Delta_C(S; \beta). \tag{7}$$

In particular, $\downarrow \alpha$ refers to the set of collections lower than or equal to $\alpha$, given the ordering relation (see Appendix A for details). Again, here we will adopt a different terminology and we will refer to $\Delta_C(S; \beta)$ as the *incremental term* of the collection $\beta$ in lattice $C$, instead of as the partial information measure. This is because, as we will see, it is convenient to consider incremental terms as increments that can equally be of information gain or information loss. Given the link between the cumulative terms and the mutual information of the variables imposed by the self-redundancy axiom, the decomposition of the total mutual information results from applying Equation (7) to the collection $\alpha = \{\mathbf{R}\}$. As proved in (Theorem 3, [21]), Equation (7) can be inverted to:

$$\Delta_C(S; \alpha) = I(S; \alpha) - \sum_{k=1}^{|\alpha^-|} (-1)^{k-1} \sum_{\substack{\mathcal{B} \subseteq \alpha^- \\ |\mathcal{B}| = k}} \sum_{\beta \in \bigcap_{\gamma \in \mathcal{B}} \downarrow \gamma} \Delta_C(S; \beta) \tag{8a}$$

$$= I(S, \alpha) - \sum_{k=1}^{|\alpha^-|} (-1)^{k-1} \sum_{\substack{\mathcal{B} \subseteq \alpha^- \\ |\mathcal{B}| = k}} I(S; \bigwedge \mathcal{B}), \tag{8b}$$

where $\alpha^-$ is the cover set of $\alpha$ and $\bigwedge \mathcal{B}$ is the infimum of the set $\mathcal{B}$. The cover set of $\alpha$ is composed by the collections that are immediate descendants from $\alpha$ in the lattice. The infimum of a set of collections is the upper collection that can be reached descending from all collections of the set. We will also refer to the set formed by all the collections for which cumulative terms determine the incremental term $\Delta_C(S; \alpha)$ as the increment sublattice $\Diamond\alpha$. This sublattice is formed by $\alpha$ and by all the collections that appear as infimums in the summands of Equation (8b). See Appendix A for details on the definition of these concepts and other properties of the lattices.

## 3. Extended Information Gain Decompositions from Redundancy, Uniqueness or Synergy Measures

In this section, we still focus on the information gain decompositions introduced by [21]. We first extend their approach to comprise a more general set of lattices, built based on subsets of the domain of collections determined in Equation (5). We study the relation between the terms from different lattices, showing how the incremental terms of the full lattice are mapped to the incremental terms of smaller lattices. We also indicate that the existence of deterministic relations between the variables imposes constraints to the range of valid lattices. We then show that considering the connection between different lattices is not only useful to better interpret the decompositions based on a subset of collections, but also in practice provides a way to build multivariate decompositions. In particular, we introduce an iterative hierarchical procedure to build multivariate decompositions from synergy or unique information measures, something that was not possible from the full lattices alone.

### 3.1. Relations between Information Gain Decompositions with Different Subsets of Sources Collections

Reference [21] studied how to decompose the mutual information in decompositions composed by all the collections of sources in $\mathcal{A}(\mathbf{R})$. Figure 1A,B show the corresponding lattices for $n = 2$ and $n = 3$, respectively. However, the number of collections in these decompositions rapidly increases when the number of variables increases (e.g., 7579 collections for $n = 5$ [21]), which may render the decompositions difficult to handle in practice. Here, we generalize their approach in a straightforward way, considering decompositions composed by any subset $\mathcal{C} \subseteq \mathcal{A}(\mathbf{R})$ whose elements still form a lattice (see Appendix C for a discussion of more general decompositions based on subsets that do not form a lattice). Given that the ordering relation of Equation (6) is a pairwise relation that does not depend on the set of collections considered, any lattice is built following the same rule as in the full lattice: each collection is connected by an edge to the collections in its cover set (see Appendix A), which depends on the subset of collections used to construct the lattice. For example, Figure 1C shows the decomposition formed by the collections that combine the sources 1 and 23. In Figure 1B, the red edges indicate the decomposition based on collections combining the sources 12, 13, 23, without further decomposing the contribution of single variables separately. Oppositely, Figure 1D shows the decomposition based on the sources 1, 2, and 3, which does not include bivariate sources resulting from merging these univariate sources. A certain decomposition can be embedded within a bigger one, as indicated in Figure 1B, but generally considering more collections alters the structure of the lattice, by modifying the cover relations between the nodes. For example, the decomposition of Figure 1D is not embedded in Figure 1B. Similarly, the cover relations in the bivariate decomposition of $\mathcal{A}(\{1,2\})$ in Figure 1A change in the trivariate decomposition of $\mathcal{A}(\{1,2,3\})$ in Figure 1B since nodes 12.13 and 12.23 appear between 12 and 1 and 2, respectively. The same occurs between 1.2 and 1 and 2, with nodes 1.23 and 2.13, respectively. Furthermore, the down set of 12, in comparison to the bivariate lattice, comprises others nodes because of the presence of 12.13.23.

When studying multivariate systems, the nature and relation between the variables may provide some a priori information in favor of a certain decomposition. For example, in the case of Figure 1C, variables 2 and 3 can correspond to two signals recorded from the same subsystem, while 1 is a signal from a different subsystem. This may render a bivariate decomposition more adequate, even if having three variables. For example, this is a common scenario when recording brain signals from different brain areas and the analysis of interactions can be carried out at different spatial scales [8]. Similarly, in the case of Figure 1D, one may prefer to simplify the analysis without explicitly considering all synergistic contributions of bivariate sources. Another possibility is that, even if it is known that a system is composed by a certain number of variables, only a subset is available for the analysis, and it is thus important to understand how the influence of the missing variables is reflected in each term of the decomposition. For example, if in a trivariate system only 1 and 2 are observed, we would like to understand how the terms in the full decomposition for $n = 3$ are reflected in the full decomposition for $n = 2$ restricted to 1 and 2. Again, this is a common scenario when studying neural population coding of sensory stimuli, since usually only simultaneous recordings from a subset of the neural population, or from one of the brain regions involved, is available. In any case, in order to better choose the most useful decomposition given a certain set of concrete variables, and to understand how the different decompositions are related, we need to consider how the terms from one decomposition are mapped to another. Furthermore, as we will see in Section 3.3, the relationship between lattices has also practical applications to build multivariate decompositions from synergy or unique information measures.

The connection between the terms in two different decompositions is qualitatively different for the cumulative terms, $I(S, \alpha)$, and the incremental terms $\Delta_C(S; \alpha)$. A cumulative term $I(S, \alpha)$ quantifies the information about $S$ that is redundant within a certain collection of sources $\alpha$. This information is well defined without considering which is the set $C$ of collections that has been selected, that is, it depends only on $S$ and $\alpha$. Accordingly, the cumulative terms of information gain $I(S, \alpha)$ are invariant across decompositions. Oppositely, as we here explicitly indicate in our notation, the incremental terms $\Delta_C(S; \alpha)$ are, in general, decomposition-dependent. Although Equation (8) was derived in [21] only to express the incremental terms as a function of cumulative terms in the full lattices, it is straightforward to check that, by construction, the relation of Equation (7) also can be inverted to Equation (8) in lattices formed by subsets of $\mathcal{A}(\mathbf{R})$ (See Appendix C). From Equation (8), it can be seen that the cumulative terms used to calculate $\Delta_C(S; \alpha)$ depend on the specific structure of the lattice, in particular on which is the increment sublattice $\Diamond \alpha$. This is summarized indicating that:

$$I(S; \alpha) = I_C(S; \alpha) = I_{C'}(S; \alpha), \forall C, C', \tag{9}$$

while for the incremental terms, only a sufficient condition for equality across decompositions can be formulated:

$$\Diamond_C \alpha = \Diamond_{C'} \alpha \Rightarrow \Delta_C(S; \alpha) = \Delta_{C'}(S; \alpha), \tag{10}$$

which is a direct consequence of Equation (9) given the dependence of the incremental terms on the cumulative terms (Equation (8)).

The invariance of cumulative terms implies that each cumulative term that is present in two decompositions provides an equation that relates the incremental terms in those decompositions, since in each lattice cumulative terms result from the accumulation of increments according to Equation (7). In particular, for two decompositions $C$ and $C'$ with a common collection $\alpha$

$$I(S; \alpha) = \sum_{\beta \in \downarrow_C \alpha} \Delta_C(S; \beta) = \sum_{\beta \in \downarrow_{C'} \alpha} \Delta_{C'}(S; \beta). \tag{11}$$

In general, these type of relations impose some constraints that involve several incremental terms from each decomposition. For example, when connecting the incremental terms of the decompositions of Figure 1A,C, we get only the constraint $\Delta_A(S; 1) + \Delta_A(S; 1.2) = \Delta_C(S; 1) + \Delta_C(S; 1.23)$, given the only common node $I(S; 1)$. An important type of comparison is between full lattices of different order. In this case, the constraints allow breaking down each of the incremental terms of the lower order lattice as a sum of incremental terms from the higher order lattice. Figure 2 shows the case for $n = 2, 3$ when the set $\mathcal{A}(\{1, 2\})$ of the full decomposition for $n = 2$ is a subset of $\mathcal{A}(\{1, 2, 3\})$, corresponding to the full decomposition for $n = 3$. This determination of each incremental term of one lattice as a function of the incremental terms of another lattice holds, in general, when the set of collections of the smaller lattice is a subset of the collections used in the bigger lattice. This happens not only when comparing a full lattice with a full lattice of lower order (Figure 2), but also for $n = 3$ in Figure 1B, when comparing a full lattice with any lattice composed by a subset of the collections (Figure 1A,C,D).

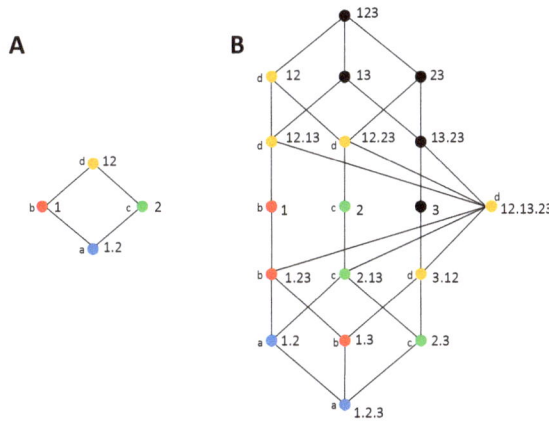

**Figure 2.** Mapping between the incremental terms of the bivariate lattice for 1, 2 and the full trivariate lattice for 1, 2, 3. (**A**) The bivariate lattice with each node marked with a different color (and also, redundantly, with a different lower case letter, for no-color printing). (**B**) The trivariate lattice with the nodes coloured consistently with the mapping to the bivariate lattice. In more detail, the incremental term of each node of the bivariate lattice is obtained as the sum of the incremental terms of the nodes of the trivariate lattice with the same color.

Figure 3A,B further shows how the incremental terms of the sublattice of Figure 1D can be broken down as a sum of the incremental terms of the full lattice. However, in general, the fact that each incremental term of the smaller lattice can be expressed as a combination of the incremental terms of the bigger lattice, does not mean that each incremental term of the latter is related to a single incremental term of the former. This is illustrated in Figure 3C,D. The set of collections of the lattice in Figure 3C is a subset of the set of the lattice in Figure 3D. However, now each incremental term of Figure 3D contributes to more than one incremental term of Figure 3C. For example, $\Delta_C(S;1) = \Delta_D(S;2) + \Delta_D(S;1.2) + \Delta_D(S;2.3)$, but also $\Delta_C(S;1) = \Delta_D(S;1) + \Delta_D(S;1.2) + \Delta_D(S;1.3)$, with $\Delta_D(S;1.2)$ contributing to both. Furthermore, since the incremental terms of the full lattice can only contribute once to the decomposition of $I(S;123)$, the fact that an incremental term of Figure 3D contributes twice positively, is balanced by a negative contribution to another incremental term of Figure 3C. Combining the correspondences of Figure 3C,D and Figure 3A,B, we can see that the incremental terms of the full lattice of Figure 3B contribute in an overlapping way to the incremental terms of Figure 3C, that is, there is no function assigning a single term in Figure 3C to the terms of the full lattice. To understand which lattices produce overlapping projections, let us examine in more detail the constraint of Equation (11) for $I(S;1)$ and $I(S;2)$ in the lattices of Figure 3C,D. Each cumulative term is obtained as the sum of all incremental terms from nodes reached descending from the node of the cumulative term (Equation (7)). In Figure 3D, 1.2 is the infimum of 1 and 2, that is, the first node common to the descending paths from 1 and 2. This means that the incremental terms from this infimum and nodes reached descending from it contribute to both $I(S;1)$ and $I(S;2)$. If these nodes are not present in Figure 3C, then some upper incremental term will have to account for them. In this case, the node 1.2.3 is present in Figure 3C, but 1.2 is not and thus has to be accounted for both in the incremental terms of 1 and 2. This type of loss of the infimum node does not occur when the full lattice is reduced to the one of Figure 3A. For example, the infimum of 1 and 2 is 1.2 and this node is preserved, even if the intermediate nodes 1.23 and 2.13 are not. Therefore, in general, preserving the structure of the infimums when taking a subset of $\mathcal{A}(\mathbf{R})$ to construct a lattice is what determines if a non-overlapping projection exists or not. If for two collections in the subset their infimum collection is not included, then Equation (7) can only be fulfilled for both cumulative terms by a multiple contribution of the incremental term of the infimum in the full lattice to the incremental terms of the smaller lattice.

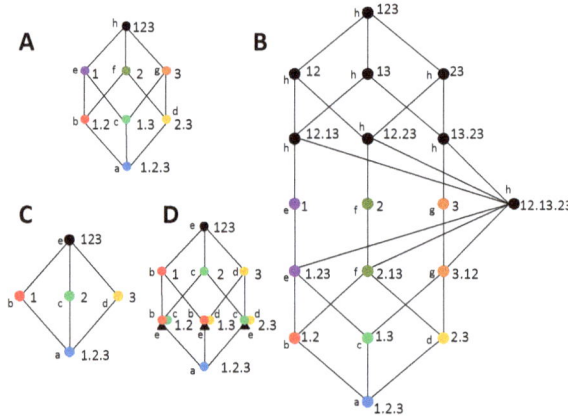

**Figure 3.** Mapping of the incremental terms of the full lattice to lattices formed by subsets of collections (**A**) The lattice of Figure 1D with each node marked with a different color (and also a different lower case letter, for no-color printing). (**B**) The trivariate lattice with the nodes coloured consistently with the mapping to the lattice of (A). In more detail, the incremental term of each node of the smaller lattice is obtained as the sum of the incremental terms of the nodes of the trivariate lattice with the same color. (**C**) Another lattice obtained from a subset of the collections, with each node marked with a different color (and lower case letter). (**D**) The lattice of (A) now with its nodes coloured consistently with the mapping to the lattice of (C). In contrast to the mapping between (B) and (A), here each incremental term of (D) can contribute to more than one incremental term of (C), with a positive (circle) or negative (triangles) contribution.

### 3.2. Checking the Validity of Lattices in the Presence of Deterministic Relations between the Variables

As a last remark regarding the selection of different lattices we indicate that, when deterministic relationships exist between the variables, the definition of the domain of collections (Equation (5)) and the ordering relation (Equation 6) can impose some limitations on the decompositions that are possible. In particular, Equation (5) excludes any collection in which a source is a superset of any other. Consider, for example, the case of three variables $1, 2, 3$ such that 12 completely determine 3. Accordingly, in the decomposition of Figure 1B, several collections are altered, since the source 12 could be replaced by 123. This leads to the presence of invalid collections in the set, such as 123.13 instead of 12.13, since 13 is a subset of 123. Similarly, given the deterministic relation, one could reduce 123 to 12, duplicating this last collection and affecting the ordering relation of the top element with 13 and 23. In general, this means that a certain lattice cannot be taken as valid a priori. Conversely, it should be verified, for each specific set of variables, if the collections that compose it are valid once the properties of the variables are taken into account.

The exclusion of certain lattices in the presence of deterministic relations can be seen as a limitation of the decomposition framework, but on the other hand this verification turns out to be important to avoid problematic cases. In particular, it allows avoiding the counterexample provided in [26] to show that it is not always possible when $n > 2$, independently of how the terms of the decomposition are defined, to obtain a nonnegative decomposition. This counterexample is based on three variables such that any pair deterministically determines the third. Without using any specific definition of the measures associated with the nodes, the authors proved that at least a certain incremental term of the lattice of Figure 1B is negative in this case. However, given the deterministic relations between the variables, all the collections comprising a bivariate source need to be excluded from the set, since these bivariate sources are equivalent to the source 123 and thus any other source in the collection is a subset of this one. Similarly, 123 can be reduced to any collection containing a single source composed by a pair of the variables, which duplicates these collections and affects the ordering relations. In Appendix B, we reconsider in more detail the counterexample of [26]. We show that taking into account the deterministic relations leads to the construction of a smaller lattice that complies the constraints of Equations (5) and (6) and for which nonnegativity is preserved. Therefore, we can reinterpret the violation of the nonnegativity axiom as a more specific non-compliance of the constraints for a certain set of variables. However, note that the possibility to deal with cases such as the one raised in [26] by adapting the lattice does not preclude from the potential existence of negative incremental terms. As we reviewed above, the definition of the proper measure for the decompositions is an open question, and finding a measure that ensures generally the nonnegativity of the incremental terms, or identifying the properties of the variables or the decompositions that ensures this nonnegativity, is out of the scope of this work. Only in Appendix C we review the requirements to obtain nonnegative incremental terms. For this purpose, we reexamine from a more general perspective several of the Theorems of [21], identifying the key ingredients of the proofs that are sustained by lattice properties, by general properties captured in the axioms that have been proposed [22], or by properties specific of their measure $I_{min}$.

### 3.3. An Iterative Hierarchical Procedure to Determine Information Gain Cumulative Terms from Synergy or Unique Information Measures by Combining Information Gain Decompositions of Different Orders

Above, we have examined possible alternative decompositions of the mutual information and the relations among them from a generic perspective, based only on the structure of the decompositions and the general properties of the cumulative and incremental terms. Apart from the general advantage of understanding these relations in order to select and interpret the decompositions, we now show that these relations also have a more immediate practical implication and provide a way to determine the components of multivariate decompositions when the cumulative terms are not predefined but only a measure that can be associated with certain incremental terms is used as the basis to construct the decomposition. If a measure of accumulated mutual information gain $I(S, \alpha)$ is defined, it is straightforward to calculate all the terms in the decomposition. The cumulative terms are already defined based on this measure and the incremental terms can be calculated using Equation (8). This was the case in the seminal work of [21], where the cumulative terms were defined as the redundancy measures $I_{min}$. However, the calculation of all terms is not straightforward if the measure defined as the basis to construct the decomposition does not define the cumulative terms. In fact, in the different proposals that exist so far, the basic component chosen to calculate the other terms has alternatively been a redundancy measure [21,22,25,32], a synergy measure [23], or a unique information measure [24]. When the measure of redundancy is sufficiently general to be taken as a general definition of all cumulative terms, the multivariate decompositions can be built in a direct way. Conversely, when a measure of synergy or unique information is taken as the basis, only bivariate decompositions can be directly built and only bivariate decompositions have been studied [23,24]. To our knowledge, there is currently no general procedure to construct the decomposition associated with an information gain lattice in the multivariate case from the definition of a measure of synergy or unique information. We here describe an iterative hierarchical procedure to do so, which relies on the relations between lattices of different order.

We start by considering first the bivariate case, to understand why, in this case, the lattice can be constructed from synergy or unique information measures in a simple way and why this procedure does not apply directly for the multivariate case. In the bivariate case, Equations (1) and (2) provide

$m = 3$ equations to relate the $K = 4$ terms of the bivariate decomposition with $I(S;1)$, $I(S;2)$, and $I(S;12)$, so that defining one of the four terms is enough to identify the other three. However, this direct procedure cannot similarly be applied for $n > 2$. This can be understood, already for $n = 3$, considering the number of cumulative terms which are directly calculable as a mutual information thanks to the self-redundancy axiom. These terms are the ones related to the collections formed by a single source, 1, 2, 3, 12, 13, 23, and 123. This means that only $m = 7$ equations analogous to Equations (1) and (2) are available to calculate the $K = 18$ cumulative terms. If the measure taken as basis of the decomposition is defined generally for each node (as $I_{min}$ in [21]) this is not a problem, and these seven equations are directly fulfilled as special cases of Equation (7). However, if the measure taken as the basis is a measure of synergy or uniqueness, then it does not define directly the cumulative terms, but only certain incremental terms. This difference is clear already for $n = 2$. The redundancy $I(S;1.2)$ is a cumulative term in the decomposition, in particular it corresponds to the bottom element of the lattice. Conversely, the unique information terms $I(S;1\backslash 2)$ and $I(S;2\backslash 1)$, as well as the synergy $I(S;12\backslash 1,2)$ correspond to incremental terms. That is, the particularity of redundancy measures such as $I_{min}$ is that they provide a definition for all the cumulative terms of the mutual information gain decomposition, while the measures of unique information or synergy, for $n > 2$, do not provide a definition applicable to all the incremental terms of the lattice.

We will now describe a new iterative hierarchical procedure to build information gain multivariate decompositions from measures of synergy or unique information. For simplicity, we will focus on the case of $n = 3$, but the logic of the procedure can be extrapolated to the general multivariate case. As we will show in Section 4 by introducing the alternative information loss lattices, this is not the only way to build these multivariate decompositions. In fact, this procedure can lead to some inconsistencies if it is applied to any lattice without a careful examination of the correspondence between incremental terms and synergy or unique information measures. However, if the lattices and measures are properly chosen, as we will discuss after introducing the dual decompositions of information gain and information loss in Section 5, the different procedures to build multivariate decompositions are consistent and the same cumulative terms and incremental terms are obtained independently of how they are calculated.

The key ingredient for this procedure is the invariance of the cumulative terms across decompositions, as indicated in Equation (9). Based on this invariance, we can resort to the bivariate decompositions in order to calculate many of the cumulative terms of the trivariate decomposition of Figure 1B. Indeed, from the 18 minus 7 terms that do not correspond directly to the mutual information of a single source, all except the ones of the collections 12.13.23 and 1.2.3 appear also in a bivariate decomposition. For example, 1.2 is part of the decomposition in Figure 1A, and 1.23 is part of the one in Figure 1C. Analogous bivariate decompositions exist for 1.3, 2.3, 2.13, 3.12, 12.13, 12.23, and 13.23. For each of these bivariate decompositions, if a definition of bivariate synergy is defined, it can be used to determine the corresponding bivariate redundancy, which, being a cumulative term, is invariant and can be used equally in the trivariate decomposition. Accordingly, it is the connection between different decompositions what allows us to calculate most of the terms. This same procedure of using the bivariate decompositions could be used if instead of a definition of synergy, we used a definition of unique information. Finally, to calculate 1.2.3 and 12.13.23, we can use the smaller trivariate decompositions of Figure 1D and the one composed by the red edges of Figure 1B, respectively. In these two smaller trivariate decompositions, after using the bivariate ones to calculate the corresponding cumulative terms, the situation becomes the same as for the bivariate case: all cumulative terms are already calculated except one, which means that it suffices to define a single measure, either a synergy or unique information, in order to be able to retrieve the complete set of cumulative and incremental terms.

This procedure provides a way to construct multivariate decompositions, simply by recurrently using decompositions of a lower order to calculate the cumulative terms. Since the cumulative terms are invariant, at each order, most of them are not specific of that order and appear also in lattices of lower orders. However, this approach leads to some inconsistencies if it is applied to any lattice without carefully considering how a synergy or unique information measure can be associated with an incremental term. In particular, consider that a measure of synergy is provided, which should allow identifying the top incremental term of any of the decompositions used in the iterative procedure. For example, a measure of synergy should determine the incremental term $\Delta(S; 123\backslash 12.13.23)$ of Figure 1B and $\Delta(S; 123\backslash 1.2.3)$ of Figure 1D. However, the same measure of synergy cannot be taken to determine the same top incremental term $\Delta(S; 123\backslash 1.2.3)$ of Figure 1D for other lattices, since the incremental terms are decomposition-specific. For example, consider the alternative decompositions presented in Figure 4A,B, and Figure 4C,D, respectively. In Figure 4A,B, if the same definition of $\Delta(S; 123\backslash 1.2.3)$ is used based on the synergy measure, this results in a different form for $I(S; 1.2.3)$ in the two lattices, because the increment sublattices $\Diamond_A 123 \neq \Diamond_B 123$. This contradicts the invariance of the cumulative terms across lattices. Figure 4C,D presents another contradiction resulting from directly using the same synergy definition: since collections 123, 1, and 23 are common to both decompositions, the same expression would be obtained for the redundancy $I(S; 1.23)$ and $I(S; 1.2.3)$ depending on the lattice used. For both examples, the problem is that a definition of synergy is expected to depend only on $S$ and on the sources among which synergy is quantified, but cannot be context-dependent, in opposition to the incremental terms, which are always context-dependent in the sense that they are decomposition-specific.

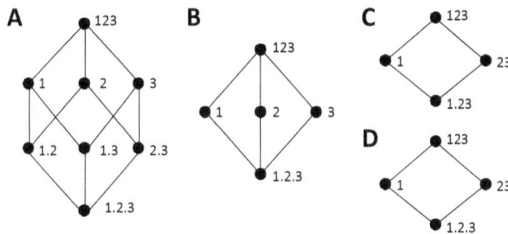

**Figure 4.** Examples of information gain lattices that result in inconsistencies when trying to derive redundancy terms from a synergy definition, as explained in Section 3.3.

In more detail, given the conceptual meaning of synergy, when a synergy measure is to be associated with an incremental term, the precise synergy contribution to be quantified is only determined by the collection $\alpha$ and the collections on its cover $\alpha^-$. For example, in Figure 4A, $\Delta(S; 123\backslash 1.2.3)$ quantifies the overall synergistic increase in information that can arise from combining any of the single variables. However, the incremental term is not only determined by $\alpha$ and $\alpha^-$, conversely, it depends on the whole increment sublattice $\Diamond\alpha$. We have seen in Figure 3C,D that $\Delta(S; 123\backslash 1.2.3)$ is a different incremental term for these lattices, with a different expression as a function of the incremental terms of the full lattice. Accordingly, it is not straightforward to interpret the top incremental term as the one quantifying the synergistic component of the mutual information, since different possible decompositions result in different terms. This issue does not arise for the bivariate decomposition because a single decomposition involving a synergistic component is possible. In general, for a given definition of synergy or unique information, or more generally for any measure which is conceptually defined in order to capture a certain portion of the total mutual information (e.g., the union-information in [23]), the incremental terms to which the measure can be associated is not directly determined by the local structure of the lattice related to the variables involved in the definition of the measure, but by the whole incremental sublattice. The meaning of the measure assigned to an incremental term has to be consistent with its expression as a function of the incremental terms of the full lattice. Only with this careful examination of the correspondence between measures and incremental terms, the iterative procedure can be used to build multivariate decompositions.

## 4. Decompositions of Mutual Information Loss

Although the iterative hierarchical procedure allows determining the information gain decompositions from a synergy measure, the construction has to proceed in a reversed way, determining the cumulative terms from the incremental terms using lower order lattices. This means that the full lattice at a given order cannot be determined separately, but always by jointly constructing a range of lower order lattices. We now examine if this asymmetry between redundancy measures, corresponding to cumulative terms, and synergy measures, corresponding to incremental terms, is intrinsic to the notions of redundancy and synergy, or if conversely this correspondence can be inverted. Indeed, we introduce the type of information loss lattices in which synergy is associated with cumulative terms and redundancy with incremental terms. Apart from the implications of the possibility of this inversion in the understanding of the nature synergy and redundant contributions, a practical application of information loss lattices is that, using them as the basis of the mutual information decomposition, the decomposition can be directly determined from a synergy measure in the same way that it can be directly determined from a redundancy measure when using gain lattices. In Section 5, we will study in which cases information gain and loss lattices are dual, providing the same decomposition of the mutual information.

As a first step to obtain a mutual information decomposition associated with an information loss lattice, we define a new ordering relation between the collections. In the lattices associated with the decompositions of mutual information gain, the ordering relation is defined such that upper nodes correspond to collections which cumulative terms have more information about $S$ than each of the cumulative terms in their down set. Oppositely, in the lattice associated with a decomposition of mutual information loss, an upper node corresponds to a higher loss of the total information contained in the whole set of variables about $S$. The domain of the collections valid for the information loss decomposition can be defined analogously to the case of information gain:

$$\mathcal{A}^*(\mathbf{R}) = \{\alpha \in \mathcal{P}(\mathbf{R}) \backslash \{\mathbf{R}\} : \forall\, \mathbf{A}_i, \mathbf{A}_j \in \alpha, \mathbf{A}_i \not\subseteq \mathbf{A}_j\}. \tag{12}$$

Note that this domain is equivalent to the one of the information gain decompositions (Equation (5)), except that the collection corresponding to the source containing all variables $\{\mathbf{R}\}$ is excluded instead of the empty collection. This is because, in the same way that no information gain can be accumulated with no variables, no loss can be accumulated with all variables. Furthermore, $\mathcal{A}^*(\mathbf{R})$ excludes collections that contain sources that are supersets of other sources of the same collection, equally to $\mathcal{A}(\mathbf{R})$. An ordering relation is introduced analogously to Equation (6):

$$\forall\, \alpha, \beta \in \mathcal{A}^*(\mathbf{R}), (\alpha \preceq \beta \Leftrightarrow \forall \mathbf{B} \in \beta, \exists \mathbf{A} \in \alpha, \mathbf{B} \subseteq \mathbf{A}). \tag{13}$$

This ordering relation differs from the one of lattices associated with information gain decompositions in that now upper collections should contain subset sources and not the opposite. Figure 5 shows several information loss decompositions analogous to the gain decompositions of Figure 1. For the lattices of Figure 5A,C,D, the only difference with respect to Figure 1 is the top node, where the collection containing all variables is replaced by the empty set. Indeed, the empty set results in the highest information loss. For the full trivariate decomposition of Figure 5B, there are many more changes in the structure of the lattice with respect to Figure 1B. In particular, now the smaller embedded lattice indicated with the red edges corresponds to the one of Figure 5D, while the lattice of Figure 1D is not embedded in Figure 1B. An intuitive way to interpret the mutual information loss decomposition is in terms of the marginal probability distributions from which information can be obtained for each collection of sources. Each source in a collection indicates a certain probability distribution that is available. For example, the collection 12.13, composed by the sources 12 and 13, is associated with the preservation of the information contained in the marginal distributions $p(S, 1, 2)$ and $p(S, 1, 3)$. Note that all distributions are joint distributions of the sources and $S$. In this view, the extra information contained in $p(S; \mathbf{R})$ that cannot be obtained from the marginals preserved, corresponds to the accumulated information loss. Accordingly, the information loss decompositions can be connected to hierarchical decompositions of the mutual information [33,34]. Furthermore, information loss associated with the preservation of only certain marginal distributions can be formulated in terms of maximum entropy [24], which renders loss lattices suitable to extend previous work studying neural population coding with the maximum entropy framework [15].

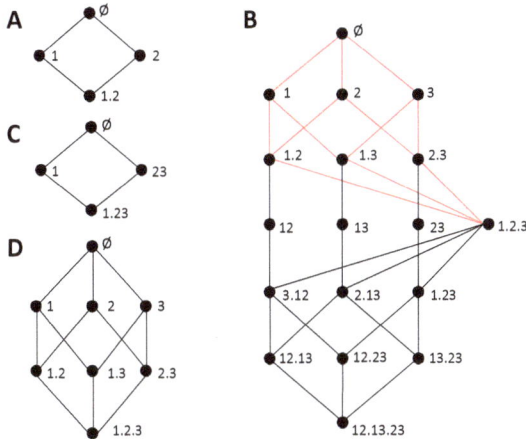

**Figure 5.** Information loss decompositions of different orders and for different subsets of collections of sources. (**A–D**) The lattices are analogous to the information gain lattices of Figure 1. Note that the lattice embedded in (**B**), indicated with the pale red edges, corresponds to the one shown in (**D**), differently than in Figure 1.

We will use the notation $L(S; \alpha)$ to refer to the cumulative terms of the information loss decomposition, in comparison to the cumulative terms of information gain $I(S; \alpha)$. For the incremental terms, since they also correspond to a difference of information (in this case lost information) we will use the same notation. This will be further justified below when examining the dual relationship between certain information gain and loss lattices. However, when we want to explicitly indicate the type of lattice to which an incremental term belongs, we will explicitly distinguish $\Delta I$ and $\Delta L$. Importantly, the role of synergy measures and redundancy measures is exchanged in the information loss lattice with respect to the information gain lattice. In particular, in the information loss lattices, the bottom element of the lattice corresponds to the synergistic term that in the information gain lattices is located at the top element. This represents a qualitative difference because now it is the synergy measure which is associated with cumulative terms, and redundancy is quantified by an incremental term. For example, in Figure 5B, $L(S; 12.13.23)$ quantifies the information loss of considering only the sources 12.13.23 instead of the joint source 123, which is a synergistic component. On the other hand, the incremental term $\Delta(S; \varnothing \backslash 1, 2, 3)$ quantifies the information loss of either removing the source 1, or removing 2, or removing 3. Since the information loss quantified is associated with removing any of these sources, it means that the loss corresponds to information which was redundant to these three sources. This reasoning applies also to identify the unique nature of other incremental terms of the information loss lattice. For example, $\Delta(S; 12.13 \backslash 23)$ can readily be interpreted as the unique information contained in 23 that is lost when having only sources 12.13.

Analogously to the information gain lattices, we require that the measures used for the cumulative terms fulfill the symmetry and monotonicity axioms, so that the cumulative terms inherit the structure of the lattice. We also require the fulfilment of the self-redundancy axiom, so that the cumulative terms are linked to mutual information measures of the variables when the collections have a single source. In particular, for $\alpha = \mathbf{A}$, the loss due to only using $\mathbf{A}$ corresponds to $I(S; \{\mathbf{R}\}) - I(S; \mathbf{A})$, which is the conditional mutual information $I(S; \{\mathbf{R}\}|\mathbf{A})$. That is, in the same way that the cumulative terms $I(S; \alpha = \mathbf{A})$ should be directly calculable as the mutual information of the variables, the cumulative terms $L(S; \alpha = \mathbf{A})$ should be directly calculable as conditional mutual information. Apart from these axioms directly related to the construction of the information loss lattice, we do not assume any further constraint to the measures used as cumulative terms $L(S; \alpha)$, and the selection of the concrete measures will determine the interpretability of the decomposition. Since, like in the information gain lattice, the top cumulative term is equal to the total information ($L(S; \varnothing) = I(S; \{\mathbf{R}\})$), the lattice is a decomposition of the mutual information for a certain set of variables $\{\mathbf{R}\}$. In more detail, the relations between cumulative and incremental terms are defined totally equivalent to the ones of the information gain lattices

$$L(S, \alpha) = \sum_{\beta \in \downarrow \alpha} \Delta_C L(S; \beta), \tag{14}$$

and

$$\Delta_C L(S; \alpha) = L(S; \alpha) - \sum_{k=1}^{|\alpha^-|} (-1)^{k-1} \sum_{\substack{\mathcal{B} \subseteq \alpha^- \\ |\mathcal{B}| = k}} \sum_{\beta \in \bigcap_{\gamma \in \mathcal{B}} \downarrow \gamma} \Delta_C L(S; \beta) \tag{15a}$$

$$= L(S, \alpha) - \sum_{k=1}^{|\alpha^-|} (-1)^{k-1} \sum_{\substack{\mathcal{B} \subseteq \alpha^- \\ |\mathcal{B}| = k}} L(S; \bigwedge \mathcal{B}). \tag{15b}$$

The definition of information loss lattices simplifies the construction of mutual information decompositions from a synergy measure. If a synergy measure can generically be used to define the cumulative terms of the loss lattice analogously to how a redundancy measure, for example, $I_{min}$ defines the cumulative terms of the gain lattice, then these equations relating cumulative and incremental terms can be applied to identify all the remaining terms. Accordingly, the introduction of information loss lattices solves the problem of the ambiguity of the synergy terms derived from information gain lattices, which was caused by the identification of synergistic contributions with incremental terms, which are decomposition-specific by construction. In the information loss decomposition, the synergy contributions are identified with cumulative terms, and thus are not decomposition-specific. That is, if a conceptually proper measure of synergy is proposed, it can be used to construct the decomposition straightforwardly, without the need to examine to which terms the measure can be assigned, as it happens with the correspondence with incremental terms in the gain lattices. Note however, that there is still a difference between the degree of invariance of the cumulative terms in the information gain decompositions and in the information loss decompositions. The loss is per se relative to a maximum amount of information that can be achieved. This means that the cumulative terms of the information loss decomposition are only invariant across decompositions that have in common the set of variables from which the collections are constructed. Taking this into account, the relations between different lattices with different subsets of collections are analogous to the ones examined in Section 3.1 for the gain lattices. Similarly, an analogous iterative hierarchical procedure can be used with the loss lattices to build multivariate decompositions by associating redundancy or unique information measures to the incremental terms of the loss lattice.

## 5. Dual Decompositions of Information Gain and Information Loss

The existence of alternative decompositions, associated with information gain and information loss lattices, respectively, raises the question of to which degree these decompositions are consistent. A different quantification of synergy and redundancy for each lattice type would not be compatible with the decomposition being meaningful with regards to unique notions of synergy and redundancy. Comparing the information gain lattices and the information loss lattices, we see that the former seem adequate to quantify unambiguously redundancy and the latter to quantify unambiguously synergy, in the sense that there is a reverse in which measures correspond to the decomposition-invariant cumulative terms. However, we would like to understand in more detail, how the two types of lattices are connected, i.e., which relations exist between the cumulative or incremental terms of each other, and how to quantify synergy and redundancy together. We now study how information gain and information loss components can be mapped between the information gain and information loss lattices, and we define lattice duality as a set of conditions that impose a symmetry in the structure of the lattices such that for dual lattices the set of incremental terms is the same, leading to a unique total mutual information decomposition.

Considering how the total mutual information decomposition is obtained by applying Equations (7) and (14) to $I(S; \{\mathbf{R}\})$ and to $L(S; \varnothing)$, respectively in the gain and loss lattices, it is clear that both types of lattices can be used to track the accumulation of information gain or the accumulation of information loss. Each node $\alpha$ partitions the total information in two parts: the accumulated gain (loss) and the rest of the information, which is hence a loss (gain), respectively, for each type of lattice. In particular:

$$I(S; \alpha) = \sum_{\beta \in \downarrow \alpha} \Delta I(S; \beta) \tag{16a}$$

$$I(S; \{\mathbf{R}\}) - I(S; \alpha) = \sum_{\beta \in (\downarrow \alpha)^C} \Delta I(S; \beta) \tag{16b}$$

$$L(S; \alpha') = \sum_{\beta \in \downarrow \alpha'} \Delta L(S; \beta) \tag{16c}$$

$$I(S; \{\mathbf{R}\}) - L(S; \alpha') = \sum_{\beta \in (\downarrow \alpha')^C} \Delta L(S; \beta), \tag{16d}$$

where $(\downarrow \alpha)^C = \mathcal{C} \setminus \downarrow \alpha$ is the complementary set to the down set of $\alpha$ given the particular set of collections $\mathcal{C}$ used to build a lattice. These equations indicate that in the information gain lattice all nodes (collections) out of the down set of $\alpha$ correspond to the information not gained by $\alpha$, or equivalently, to the information loss by using $\alpha$ instead of the whole set of variables. Analogously, in the information loss lattice, all nodes out of the down set of $\alpha'$ contain the information not lost by $\alpha'$, i.e., the information gained by using $\alpha'$. Accordingly, in both types of lattices, we can say that each collection $\alpha$ partitions the lattice into an accumulation of gained and lost information. This means that the terms $I(S; \{\mathbf{R}\}) - I(S; \alpha)$, when descending the gain lattice instead of ascending it, follow a monotonic accumulation of loss, in the same way that the terms $L(S; \alpha')$ follow a monotonic accumulation of loss when ascending the loss lattice. Vice-versa, the terms $I(S; \{\mathbf{R}\}) - L(S; \alpha')$, when descending the loss lattice instead of ascending it, follow a monotonic accumulation of gain, in the same way that the terms $I(S; \alpha)$ follow a monotonic accumulation of gain when ascending the gain lattice. However, these equations do not establish any mapping between the gain and loss lattices. In particular, they do not determine any correspondence between $I(S; \{\mathbf{R}\}) - I(S; \alpha)$ and any cumulative term of the loss lattice and $I(S; \{\mathbf{R}\}) - L(S; \alpha')$ and any term of the gain lattice. They only describe how information gain and loss are accumulated within each type of lattice separately. For the mutual information decomposition to be unique, any accumulation of information loss or gain, ascending or descending a lattice, has to rely on the same incremental terms.

To understand when this is possible, we study how cumulative terms of information loss or gain are mapped from one type of lattice to the other. While in some cases it seems possible to establish a connection between the components of a pair composed by an information gain and an information loss lattice, in other cases the lack of a match is immediately evident. Consider the examples of Figure 6. In Figure 6A,C we reconsider the information gain lattices of Figure 4C,D, which we examined in Section 3.3 to illustrate that we arrive to an inconsistency when trying to extract the bottom cumulative term from the directly calculable mutual informations of 1 and 23 and the same definition of synergy. Figure 4B,D show information loss lattices candidates to be the dual lattice of these gain lattices, respectively, based on the correspondence of the bottom and top collections. While the two information gain lattices only differ from each other in the bottom collection, the information loss lattices are substantially more different, with a different number of nodes. This occurs because, as we discussed above, the concept of a redundancy 1.2.3 is associated with a loss that is common to removing any of the three variables, considered as the only source of information, and thus, in any candidate dual lattice, a separation of 2 and 3 from the source 23 is required to quantify this redundancy.

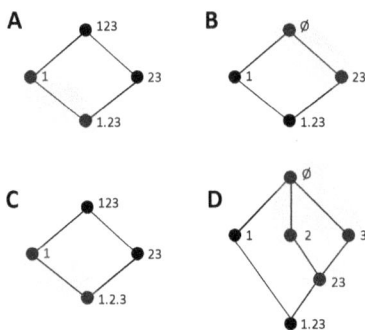

**Figure 6.** The correspondence between information gain and information loss lattices. (**A,C**) Examples of information gain lattices. (**B,D**) Information loss lattices candidates to be their dual lattices, respectively. The shaded areas comprise the collections corresponding to incremental terms that contribute to $I(S; 1)$ in each lattice.

The fact that the information gain lattice of Figure 6C and the information loss lattice of Figure 6D have a different number of nodes already indicates that a complete match between their components is not possible. For example, consider the decomposition of $I(S; 1)$ in the information gain lattice, as indicated by the nodes comprised in the shaded area in Figure 6C. $I(S; 1)$ is decomposed into two incremental terms. To understand which nodes are associated with $I(S; 1)$ in the information loss lattice we argue, based on Equation (16)d, that since the node 1 is related to the accumulated loss $L(S; 1) = I(S; 123) - I(S; 1)$, and $L(S; \emptyset) = I(S; 123)$, this means that the sum of all the incremental terms which are not in the down set of 1 must correspond to $I(S; 1)$. These nodes are indicated by the shaded area in Figure 6D. Clearly, there is no match between the incremental terms of the information gain lattice and of the information loss lattice, since in the former $I(S; 1)$ is decomposed into two incremental terms and in the latter is decomposed into four incremental terms. Conversely, for the lattices of Figure 6A,B, the number of incremental terms is the same, which does not preclude from a match.

As another example to gain some intuition about the degree to which gain and loss lattices can be connected, we now reexamine the other two lattices of Figure 4. The blue shaded area of Figure 7A indicates the down set of 1, containing all the incremental terms accumulated in $I(S; 1)$. The complementary set $(\downarrow 1)^C$, indicated by the pink shaded area in Figure 7A, by construction accumulates the remaining information (Equation (16b)), which in this case is $I(S; 23|1)$. These two complementary sets of the information gain lattice are mapped to two dual sets in the information loss lattice, as shown in Figure 7B. In Figure 7C, we analogously indicate the sets formed by partitioning the gain lattice given the collection 1, and in Figure 7D the corresponding sets in the information loss lattice. In comparison to the example of Figure 6C,D, for which we have already indicated that there is no correspondence between the gain and loss lattices; here, in neither of the two examples, is this correspondence precluded by the difference in the total number of nodes of the gain and loss lattices. However, in Figure 7C,D, the number of nodes is not preserved in the mapping of the partition sets corresponding to collection 1 from the gain to the loss lattice, which means that the incremental terms cannot be mapped one-to-one from one lattice to the other.

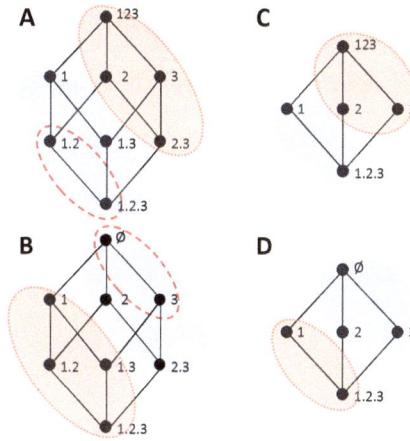

**Figure 7.** Correspondence between information gain and information loss lattices. (**A,C**) Examples of information gain lattices. (**B,D**) Information loss lattices candidates to be their dual lattice, respectively. The blue shaded areas comprise the collections corresponding to incremental terms that contribute to $I(S; 1)$ in each lattice. The pink shaded areas surrounded with a dotted line comprise the collections corresponding to incremental terms that contribute to the complementary information $I(S; 23|1)$ in each lattice. In (**A,B**), the dashed red lines encircle the incremental terms contributing to $I(S; 1.2)$.

So far, we have examined the correspondence of how collection 1 partitions the total mutual information in different lattices (Figures 6 and 7). This collection is representative of collections $\alpha$ containing a single source, and hence associated with a directly calculable mutual information, e.g., $I(S; 1)$. Their corresponding loss in the partition is associated as well with a directly calculable conditional mutual information, e.g., $I(S; 23|1)$. Accordingly, we can extend Equation (16b,d) to:

$$L(S; \mathbf{A}) = I(S; \{\mathbf{R}\}) - I(S; \mathbf{A}) = \sum_{\beta \in (\downarrow \mathbf{A})^C} \Delta I(S; \beta) \tag{17a}$$

$$I(S; \mathbf{A}') = I(S; \{\mathbf{R}\}) - L(S; \mathbf{A}') = \sum_{\beta \in (\downarrow \mathbf{A}')^C} \Delta L(S; \beta). \tag{17b}$$

While in Equation (16b), $I(S; \{\mathbf{R}\}) - I(S; \alpha)$ is a quantification of information loss only within the information gain lattice, the equality in Equation (17a) maps this information loss to the correspondent cumulative term in the information loss lattice. Analogously, Equation (17b) allows mapping a quantification of gain only within the information loss lattice to the cumulative term in the information gain lattice. That is, while Equation (16a,b) only regard the representation of information gain and loss in the gain lattice, and Equation (16c,d) the representation of information gain and loss in the loss lattice, Equation (17) indicates the mapping of the gain and loss representations across lattices, for the single source collections. This mapping was possible for all the pairs of lattices examined, including the ones of Figures 6C,D and 7C,D, which we have shown cannot be dual. This is because Equation (17) only relies on the self-redundancy axiom and the definition of how cumulative and incremental terms are related (Equations (7) and (14)), and hence this mapping can be done between any arbitrary pair of lattices. However, this direct mapping between the two types of lattices does not hold for collections composed by more than one source. For example, consider the mapping of the cumulative term $I(S; 1.2)$, composed by the incremental terms indicated by the dashed red ellipse in Figure 7A. Now in the information loss lattice we cannot take the collection 1.2 to find the corresponding partition, because the role of the collection 1.2 in the gain and in the loss lattice is different. The collection 1.2 indicates the redundant information gain with sources 1, 2, and the loss of ignoring other sources apart from 1, 2, respectively.

For dual lattices, since the set of incremental terms has to be the same—so that the mutual information decomposition is unique—this mapping cannot be limited to collections composed by single sources. This means that any collection $\alpha$ of the gain lattice should determine a partition of the incremental terms in the loss lattice that allows retrieving $I(S; \alpha)$, and analogously in the gain lattice to retrieve $L(S; \alpha)$. For example, for the cumulative term $I(S; 1.2)$, to identify the appropriate partition in the information loss lattice, we argue that the redundant information between 1 and 2 cannot be contained in the accumulated loss of preserving only 1 or only 2. Accordingly, $I(S; 1.2)$ corresponds to the sum of the incremental terms outside the union of the down sets of 1 and 2 in the loss lattice. Following this reasoning, in general:

$$I(S; \alpha) = \sum_{\beta \in (\bigcup_{\mathbf{B} \in \alpha} \downarrow \mathbf{B})^C} \Delta L(S; \beta), \tag{18a}$$

$$L(S; \alpha') = \sum_{\beta \in (\bigcup_{\mathbf{B} \in \alpha'} \downarrow \mathbf{B})^C} \Delta I(S; \beta), \tag{18b}$$

where the same argument led to relate $L(S; \alpha')$ to gain incremental terms. These equations reduce to Equation (17) for collections with a single source. In Figure 7B, we indicate with the dashed red ellipse the mapping determined by Equation (18a) for $I(S; 1.2)$. We can now compare how the cumulative terms $I(S; \alpha)$ are obtained as a sum of incremental terms in the gain and loss lattice, respectively. In the gain lattice, incremental terms are accumulated in $\downarrow \alpha$, descending from $\alpha$ (Equation (7)). In the loss lattice, they are accumulated in a set defined by complementarity to the union of descending sets, which means that these terms can be reached ascending the loss lattice (Equation (18a)). However, this does not imply that all incremental terms decomposing $I(S; \alpha)$ can be obtained ascending from a single node. This can be seen comparing the set of collections related to $I(S; 1)$ (blue shaded area) in the information loss lattices of Figure 7B,D. In Figure 7B, this set corresponds to $\uparrow 2.3$, while in Figure 7D there is no $\beta$ such that the set corresponds to $\uparrow \beta$, and the incremental terms can only be reached ascending from 2 and 3 separately. However, in order for the decomposition to be unique, the set of incremental terms has to be equal in both dual lattices, and thus Equations (7) and (18a) should be equivalent. The same holds for the cumulative terms $L(S; \alpha)$ and Equation (14) and (18b). Furthermore, plugging Equation (18a,b) in Equations (8b) and (15b), respectively, we obtain equations relating the incremental terms of the two lattices:

$$\Delta I(S;\alpha) = \sum_{\substack{\beta \in (\bigcup_{B \in \alpha} \downarrow \mathbf{B})^C}} \Delta L(S;\beta) - \sum_{k=1}^{|\alpha^-|} (-1)^{k-1} \sum_{\substack{\mathcal{B} \subseteq \alpha^- \\ |\mathcal{B}| = k}} \sum_{\substack{\beta \in (\bigcup_{B \in \wedge \mathcal{B}} \downarrow \mathbf{B})^C}} \Delta L(S;\beta) \qquad (19a)$$

$$\Delta L(S;\alpha') = \sum_{\substack{\beta \in (\bigcup_{B \in \alpha'} \downarrow \mathbf{B})^C}} \Delta I(S;\beta) - \sum_{k=1}^{|\alpha'^-|} (-1)^{k-1} \sum_{\substack{\mathcal{B} \subseteq \alpha'^- \\ |\mathcal{B}| = k}} \sum_{\substack{\beta \in (\bigcup_{B \in \wedge \mathcal{B}} \downarrow \mathbf{B})^C}} \Delta I(S;\beta). \qquad (19b)$$

If the lattices paired are dual, the right hand side of Equation (19a) has to simplify to a single incremental term $\Delta L(S;\beta)$, and similarly the right hand side of Equation (19b) has to simplify to a single incremental term $\Delta I(S;\beta)$. Taking these constraints into account, we define duality between information gain and loss lattices imposing this one-to-one mapping of the incremental terms:

**Lattice duality**: An information gain lattice associated with a set $\mathcal{C}$ and an information loss lattice associated with a set $\mathcal{C}'$, built according to the ordering relations of Equations (6) and (13), and fulfilling the constraints of Equations (7), (8), (14) and (15), are dual if and only if

$$\forall \alpha \in \mathcal{C} \; \exists \beta \in \mathcal{C}' : \Delta I(S;\alpha) = \Delta L(S;\beta), \qquad (20a)$$

$$\forall \alpha \in \mathcal{C} \; \exists \beta \in \mathcal{C}' : I(S;\alpha) = \sum_{\gamma \in \downarrow \alpha} \Delta I(S;\gamma) = \sum_{\gamma \in \uparrow \beta} \Delta L(S;\gamma), \qquad (20b)$$

$$\forall \alpha' \in \mathcal{C}' \; \exists \beta' \in \mathcal{C} : \Delta L(S;\alpha') = \Delta I(S;\beta'), \qquad (20c)$$

$$\forall \alpha' \in \mathcal{C}' \; \exists \beta' \in \mathcal{C} : L(S;\alpha') = \sum_{\gamma \in \downarrow \alpha'} \Delta L(S;\gamma) = \sum_{\gamma \in \uparrow \beta'} \Delta I(S;\gamma). \qquad (20d)$$

Equation (20a,c) ensure that the set of incremental terms is the same for both lattices, so that the mutual information decomposition is unique. Equation (20b,d) ensure that the mapping between lattices of Equation (18) is consistent with the intrinsic relation of cumulative and incremental terms within each lattice, introducing a symmetry between the descending and ascending paths of the lattices. This definition does not provide a procedure to construct the dual information loss lattice from an information gain lattice, or vice-versa. However, we have found and we here conjecture that a necessary condition for two lattices to be dual is that they contain the same collections except {**R**} at the top of the gain lattice being replaced by ∅ at the loss lattice. This is not a sufficient condition, as can be seen from the counterexample of Figure 7C,D. Importantly, the lattices constructed from the full domain of collections, $\mathcal{A}\{\mathbf{R}\}$ for the gain and $\mathcal{A}^*\{\mathbf{R}\}$ for the loss, are dual. This is because the gain and loss full lattices both contain all possible incremental terms differentiating the different contributions for a certain number of variables $n$. For lattices constructed from a subset of the collections, the way their incremental terms can be expressed as a sum of different incremental terms of the full lattice (Section 3.1) gives us an intuition of why not all the lattices have their dual pair. In particular, the combination of incremental terms of the full lattice in the incremental terms of a smaller lattice can be specific for each type of lattice, and this causes that, in general, the resulting incremental terms of the smaller lattice can no longer fulfill the constraint connecting incremental and cumulative terms in the other type of lattice. However, duality is not restricted to full lattices. In Figure 8, we show an example of dual lattices, the pair already discussed in Figure 7A,B. We detail all the cumulative and incremental terms in these lattices. While the cumulative terms are specific to each lattice, the incremental terms, in agreement with Equation (20a,c), are common to both. In more detail, the incremental terms are mapped from one lattice to the other by an up/down and right/left reversal of the lattice. From these two reversals, the right/left is purely circumstantial, a consequence of our choice to locate the collections common to both lattices in the same location (for example, to have the collections ordered 1, 2, 3 in both lattices instead of 3, 2, 1 for one of them). Oppositely, the up/down reversal is inherent to the duality between the lattices and reflects the relation between the summation in down sets or up sets in the summands of Equation (20b,d).

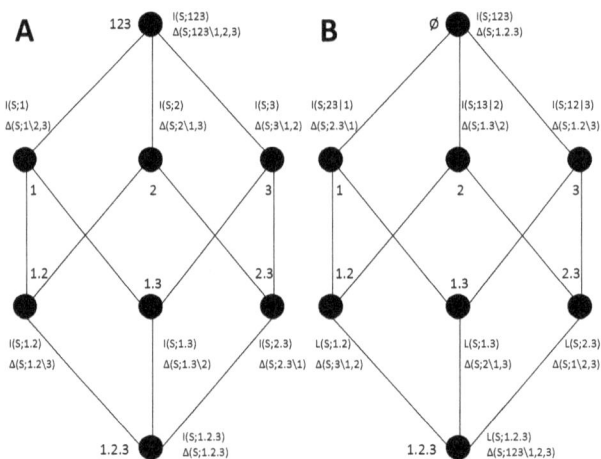

**Figure 8.** Dual trivariate decompositions for the sets of collections that do not contain bivariate sources. (**A**) Information gain lattice. (**B**) Information loss lattice. In each node together with the collection, the corresponding cumulative and incremental terms are indicated. Note that the incremental terms are common to both lattices and can be mapped by reversing the lattice up/down and right/left. In the information loss lattice, the cumulative terms of collections containing single sources, $L(S; i)$, $i = 1, 2, 3$, are directly expressed as the corresponding conditional information.

To provide a concrete example of information gain and information loss dual decompositions, we here adopted and extended to the multivariate case the bivariate synergy measure defined in [24]. Table 1 lists all the resulting expressions when the extension of this measure is used to determine all the terms in both decompositions. This measure is extended in a straightforward way to the multivariate case, and in particular for the trivariate case defines $L(S; i.j)$ and $L(S; i.j.k)$. The bivariate redundancy measure also already used in [24] corresponds to $I(S; i.j)$. The rest of incremental terms can be obtained from the information loss lattice using Equation (15). Note that we could have proceeded in a similar way starting from a definition of the cumulative terms in the gain lattice, such as $I_{min}$, and then determining the terms of the loss lattice. Here, we use this concrete decomposition only as an example and it is out of the scope of this work to characterize the properties of the resulting terms. Alternatively, we focus on discussing the properties related with the duality of the decompositions.

Most importantly, the dual lattices provide a self-consistent quantification of synergy and redundancy. Equation (20a,c), together with the fact that the bottom incremental terms of lattices are also cumulative terms, ensure that, combining different dual lattices of different order $n$ and composed by different subsets, as studied in Section 3.1, all incremental terms correspond to a bottom cumulative term of a certain lattice. For example, for the lattices of Figure 8, the bottom cumulative term in the information gain lattice, the redundancy $I(S; i.j.k)$, is equal to the top incremental term of the loss lattice, $\Delta(S; i.j.k)$. Similarly, in the bottom cumulative term of the loss lattice, the synergy $L(S; i.j.k)$, is equal to the top incremental term of the gain lattice $\Delta(S; ijk \backslash i, j, k)$.

**Table 1.** Components of the mutual information dual decompositions of Figure 8 based on the synergy measure defined in [24]. Note that we here build this decomposition for exemplary purpose and we do not study the degree to which these measures fulfill other axioms proposed for a proper decomposition.

| Term | Measure |
|------|---------|
| $\Delta(S; i.j.k) = I(S; i.j.k)$ | $\min\limits_{i.j.k} I(S; ijk) - \min\limits_{i.j} I(S; j\|i) - \min\limits_{i.k} I(S; i\|k) - \min\limits_{j.k} I(S; k\|j)$ |
| $\Delta(S; i.j\backslash k)$ | $\min\limits_{i.k} I(S; ijk) + \min\limits_{j.k} I(S; ijk) - \min\limits_{i.j.k} I(S; ijk) - I(S; k)$ |
| $\Delta(S; i\backslash j, k)$ | $\min\limits_{i.j.k} I(S; ijk) - \min\limits_{j.k} I(S; ijk)$ |
| $\Delta(S; ijk\backslash i, j, k) = L(S; i.j.k)$ | $I(S; ijk) - \min\limits_{i.j.k} I(S; ijk)$ |
| $I(S; i.j)$ | $I(S; j) - \min\limits_{i.j} I(S; j\|i)$ |
| $L(S; i.j)$ | $I(S; ijk) - \min\limits_{i.j} I(S; ijk)$ |
| $L(S; i)$ | $I(S; jk\|i)$ |

For dual lattices, the iterative procedure of Section 3.3 can be applied to recover the components of the information gain lattice from a definition of synergy and the components calculated in this way are equal to the ones obtained from the mapping of incremental terms from one lattice to the other. In more detail, let us refer to the bottom and top terms by $\bot$ and $\top$, respectively, and distinguish between generic terms such as $I(S; \alpha)$ and a specific measure assigned to it, $\bar{I}(S; \alpha)$. One can define the synergistic top incremental term of the gain lattice using the measure assigned to the bottom cumulative term of the loss lattice, imposing $\Delta I(S; \top) \equiv \bar{L}(S; \bot)$ and self-consistency ensures that the measures obtained with the iterative procedure fulfill $\bar{I}(S; \bot) = \Delta \bar{L}(S; \top)$. Similarly, self-consistency ensures that, if one takes as a definition of redundancy for the cumulative terms of the gain lattice the measure assigned to the incremental terms of the loss lattice based on a definition of synergy, consistent incremental terms are obtained in the gain lattice. That is, $I(S; \bot) \equiv \Delta \bar{L}(S; \top)$ results in $\Delta \bar{I}(S; \top) = \bar{L}(S; \bot)$. It can be checked that these self-consistency properties do not hold in general, for example for the lattices of Figure 7C,D. The properties of dual lattices guarantee that, within the class of dual lattices connected by the decomposition-invariance of cumulative terms, inconsistencies of the type discussed in Section 3.3 do not occur because all lattices share the same correspondence between incremental terms and measures, and all the terms in the decompositions are not decomposition-dependent.

As a last point, the existence of dual information gain and loss lattices, with redundancy measures and synergy measures playing an interchanged role, also indicates, by contrast, that unique information has a qualitatively different role in this type of mutual information decompositions. The measures of unique information always correspond to incremental terms and cannot be taken as the cumulative terms to build the decomposition because they are intrinsically asymmetric in the way different sources determine the measure, which contradicts the symmetry axiom required to connect collections in the lattice to cumulative terms. Despite this difference, the iterative hierarchical procedure of Section 3.3 provides a way to build the mutual information decomposition from a unique information measure, and duality ensures that the decomposition is self-consistent with alternatively having built the decomposition using the resulting redundancy or synergy measures as the cumulative terms of the gain and loss lattice, respectively.

## 6. Discussion

In this work, we extended the framework of [21] focusing on the lattices that underpin the mutual information decompositions. We started generalizing the type of information gain lattices introduced by [21]. By considering more generally which information gain lattices can be constructed (Section 3.1), we reexamined the constraints that [21] identified for the lattices' components (Equation (5)) and ordering relation (Equation (6)). These constraints were motivated by the link of each node in the lattice with a measure of accumulated information. We argued that it is necessary to check the validity of each specific lattice given each specific set of variables and that this checking can overcome the problems found by [26] with the original lattices described in [21] for the multivariate case. In particular, we showed that the existence of nonnegative components in the presence of deterministic relations between the variables is directly a consequence of the non-compliance of the validity constraints. This indicates that valid multivariate lattices are not a priori incompatible with a mutual information nonnegative decomposition.

For our generalized set of information gain lattices, we examined the relations between the terms in different lattices (Section 3.1). We pointed out that the two types of information-theoretic quantities associated with the lattices have different invariance properties: The cumulative terms of the information gain lattices are invariant across decompositions, while the incremental terms are decomposition-dependent and are only connected across lattices through the relations resulting from the invariance of the cumulative terms. This produces a qualitative difference in the properties of the redundancy components of the decompositions, which are associated with cumulative terms in the information gain lattices, and the unique or synergy components, which correspond to incremental terms. This difference has practical consequences when trying to construct a mutual information decomposition from a measure of redundancy or a measure of synergy or unique information, respectively. In the former case, as described in [21], the terms of the decomposition can be derived straightforwardly given that the redundancy measure identifies the cumulative terms. In the latter, for the multivariate case, it is not straightforward to construct the decomposition because the synergy or uniqueness measures only allow identifying specific incremental terms. Exploiting the connection between different lattices that results from the invariance of the cumulative terms, we proposed an iterative hierarchical procedure to generally construct information gain multivariate decompositions from a measure of synergy or unique information (Section 3.3). To our knowledge, there was currently no method to build multivariate decompositions from these types of measures and thus this procedure allows application to the multivariate case measures of synergy [23] or unique information [24] for which associated decompositions had only been constructed for the bivariate case. However, the application of this procedure led us to recognize inconsistencies in the determination of decompositions components across lattices. We argued that these inconsistencies are a consequence of the intrinsic decomposition-dependence of incremental terms, to which synergy and unique information components are associated in the information gain lattices. We explained these inconsistencies based on how the components of the full lattice are mapped to components of smaller lattices and indicated that this mapping should be considered to determine if the conceptual meaning of a synergy or unique information measure is compatible with the assignment of the measure to a certain incremental term. With a compatible assignment of measures to incremental terms, the iterative hierarchical procedure provides a consistent way to build multivariate decompositions.

We then introduced an alternative decomposition of the mutual information based on information loss lattices (Section 4). The role of redundancy and synergy components is exchanged in the loss lattices with respect to the gain lattices, with the ones associated with the cumulative terms now being the synergy components. We defined the information loss lattices analogously to the gain lattices, determining validity constraints for the components and introducing an ordering relation to construct the lattices. Cumulative and incremental terms are related in the same way as in the gain lattices, establishing the connection between the lattice and the mutual information decomposition. This type of lattices allows the ready determination of the information decomposition from a definition of synergy. Furthermore, the information loss lattices can be useful in relation to other alternative information decompositions [33–35].

The existence of different procedures to construct mutual information decompositions, using a redundancy or synergy measure to directly define the cumulative terms of the information gain or loss lattice, respectively, or using the iterative hierarchical procedure to indirectly determine cumulative terms, raised the question of how consistent are the decompositions obtained from these different methods. Therefore, we studied, in general, the correspondence between information gain and information loss lattices. The final contribution of this work was the definition of dual gain and loss lattices (Section 5). Within a dual pair, the gain and loss lattices share the incremental terms, which can be mapped one-to-one from the nodes of one lattice to the other. Duality ensures self-consistency, so that the redundancy components obtained from a synergy definition are the same as the synergy components obtained from the corresponding redundancy definition. Accordingly, for dual lattices, any of the procedures can be equivalently used, leading to a unique mutual information decomposition compatible with the existence of unique notions of redundancy, synergy, and unique information. Nonetheless, each type of lattice expresses in a more transparent way different aspects of the decomposition, and allows different components to be extracted more easily, and thus may be preferable depending on the analysis.

As in the original work of [21] that we aimed to extend, we have here considered generic variables, without making any assumption about their nature and relations. However, a case which deserves special attention is that of variables associated with time-series, so that information decompositions allow the study of the dynamic dependencies in the system [36,37]. Practical examples include the study of multiple-site recordings of the time course of neural activity at different brain locations, with the aim of understanding how information is processed across neural systems [38]. In such cases of time-series variables, a widely-used type of mutual information decomposition aims to separate the contribution to the information of different causal interactions between the subsystems, e.g., [39,40]. Consideration of synergistic effects is also important when trying to characterize the causal relations [41]. In fact, when causality is analyzed by quantifying statistical predictability using conditional mutual information, a link between these other decompositions and the one of [21] can be readily established [42,43].

The proposal of [21] has proven to be a fruitful conceptual framework and connections to other approaches to study information in multivariate systems have been explored [44–46]. However, despite subsequent attempts, e.g., [22–25], it is still an open question how to decompose in multivariate systems the mutual information into nonnegative contributions that can be interpreted as synergy, redundancy, or unique components. This issue constitutes the main challenge that limits so far the practical applicability of the framework. Another challenge for this type of decomposition is to be able to further relate the terms in the decomposition with a functional description of the parts composing the system [10]. In this direction, an attractive extension could be to adapt the decompositions to an interventional approach [10,47], instead of one based on statistical predictability. This could allow one to better understand how the different components of the mutual information decomposition are determined by the mechanisms producing dependencies in the system. In practice, this would help, for example, to dissect information transmission in neural circuits during behavior, which can be done by combining the analysis of time-series recordings of neural activity using information decompositions with space-time resolved interventional approaches based on brain perturbation techniques such as optogenetics [10,48,49]. This interventional approach could be incorporated to the framework by adopting interventional information-theoretic measures suited to quantify causal effects [50–52].

Overall, this work provides a wider perspective to the ground constituents of the mutual information decompositions introduced by [21], introduces new types of lattices, and helps to clarify the relation between synergy and redundancy measures with the lattice's components. The consolidation of this theoretical framework is expected to foster future applications. An advance of this work of practical importance is that it describes how to build mutual information multivariate decompositions from redundancy, synergy, or unique information measures and shows that different procedures are consistent, leading to a unique decomposition when dual information gain and information loss lattices are used.

**Acknowledgments:** This work was supported by the Fondation Bertarelli and by the Autonomous Province of Trento, Call "Grandi Progetti 2012", project "Characterizing and improving brain mechanisms of attention-ATTEND". We are grateful to Eugenio Piasini, Houman Safaai, Giuseppe Pica, Jan Bim, Robin Ince, and Vito de Feo for useful discussions on these topics.

**Author Contributions:** All authors contributed to the design of the research. The research was carried out by Daniel Chicharro. The manuscript was written by Daniel Chicharro with the contribution of Stefano Panzeri. Both authors have read and approved the final manuscript.

**Conflicts of Interest:** The authors declare no conflict of interest.

## Appendix A. Lattice Theory Definitions

We here review some concepts of lattice theory and of the construction of information decompositions based on lattices. For further review and references to specialized textbooks see [21].

**Definition A1.** *A pair $\langle X, \leq \rangle$ is a partially ordered set or poset if $\leq$ is a binary relation on X that is reflexive, transitive and antisymmetric.*

**Definition A2.** *Let $\langle X, \leq \rangle$ be a poset, and let $Y \subseteq X$. An element $x \in X$ is a lower bound for Y if $\forall y \in Y, y \geq x$. An upper bound for Y is defined dually.*

**Definition A3.** *An element $x \in X$ is the greatest lower bound or infimum for Y , denoted inf Y, if x is a lower bound of Y and $\forall y \in Y$ and $\forall z \in X; y \geq z$ implies $x \geq z$. The least upper bound or supremum for Y, denoted sup Y, is defined dually.*

**Definition A4.** *A poset $\langle X, \leq \rangle$ is a lattice if, and only if, $\forall x, y \in X$ both inf $\{x, y\}$ and sup $\{x, y\}$ exist in X. For $Y \subseteq X$, we use $\bigwedge Y$ and $\bigvee Y$ to denote the infimum and supremum of all elements in Y, respectively.*

**Definition A5.** *For $a, b \in X$, we say that a is covered by b if $a < b$ and $a \leq c < b \Rightarrow a = c$. The set of elements that are covered by b is denoted by $b^-$.*

**Definition A6.** *For any $x \in X$, the down-set of x is the set $\downarrow x = \{y \in X : y \leq x\}$. The up-set $\uparrow x$ of x is defined analogously.*

Apart from these definitions from lattice theory, we here introduce, as a concept more specific of the information decompositions, the concept of increment sublattice:

**Definition A7.** *For a lattice built with the collections set $\mathcal{C}$, for any $\alpha \in \mathcal{C}$, the increment sublattice is $\Diamond \alpha = \{\bigwedge \mathcal{B} : \mathcal{B} \subseteq \alpha^-, |\mathcal{B}| = 1, ..., |\alpha^-|\}$.*

## Appendix B. Validity Checking to Overcome the Nonnegativity Counterexample of [26]

We here examine in more detail the nonnegativity counterexample studied in [26] that we mentioned in Section 3.1. In this example, two variables $Y_1 Y_2$ are independently uniformly distributed binary variables, and a third is generated as $Y_3 = Y_1$ *XOR* $Y_2$. Furthermore, $\mathbf{S} = (Y_1, Y_2, Y_3)$. The variables have deterministic relations, such that any pair $\{Y_i, Y_j\}, i \neq j$ determines the third. We start by reviewing their arguments. The identity axiom proposed by [22] imposes that $I(Y_i Y_j; Y_i.Y_j) = I(Y_i; Y_j)$, and in this case $I(Y_i; Y_j) = 0$ bit, $i \neq j$. Given the deterministic relation between the variables, this implies that $I(\mathbf{S}; Y_i.Y_j) = 0$ bit, $i \neq j$. By monotonicity ascending the lattice of Figure 1B, also $I(\mathbf{S}; Y_1.Y_2.Y_3) = 0$ bit. Accordingly, also the incremental terms of the corresponding nodes vanish. In the next level of the gain lattice, $I(\mathbf{S}; Y_i.Y_j Y_k) = I(Y_1 Y_2 Y_3; Y_i.Y_j Y_k)$ and hence applying again the identity axiom, $I(\mathbf{S}; Y_i.Y_j Y_k) = I(Y_i; Y_j Y_k) = 1$ bit. This also leads to $\Delta I(\mathbf{S}; Y_i.Y_j Y_k \backslash Y_j, Y_k) = 1$ bit. Furthermore, by monotonicity, $I(\mathbf{S}; Y_1 Y_2.Y_1 Y_3.Y_2 Y_3) \leq I(\mathbf{S}; Y_1 Y_2 Y_3) = 2$ bit. This leads to $\Delta I(\mathbf{S}; Y_1 Y_2.Y_1 Y_3.Y_2 Y_3 \backslash Y_1, Y_2, Y_3) \leq 2 - 3$ bit $= -1$ bit. Since this derivation is based on the axioms and not on the specific properties of the measures used, this proves that, for the lattice of Figure 1B and for this specific set of variables, there is no measure that can be used to define the terms in the decomposition so that nonnegativity is respected.

We completely agree with the derivation of [26]. What we argue is that, in this case, the non-compliance of nonnegativity is a direct consequence of how the deterministic relations between the variables render some of the collections that form part of the lattice of Figure 1B invalid according to the constraints that define the domain of collections (Equation (5)), and render some ordering relations invalid according to the ordering rule of Equation (6). Therefore, adopting the generalized framework that we have proposed, this counterexample can be reinterpreted by saying that the full lattice is not valid for these variables, but that still other lattices are possible. In particular, for the lattice of Figure 1B, one can use the deterministic relations between the variables to substitute each bivariate source $Y_i Y_j$ by $Y_1 Y_2 Y_3$, and then check which collections are invalid. After removing these invalid collections and rebuilding the edges between the remaining collections according to the ordering relation, the lattice of Figure 1D is obtained.

However, it can be checked, following a derivation analogous to the one of [26], that also for the lattice of Figure 1D, nonnegativity is not accomplished, in particular by $I(\mathbf{S}; Y_1 Y_2 Y_3 \backslash Y_1, Y_2, Y_3)$. This is because, still by the deterministic relations, the top collection could be reduced to any collection $Y_i Y_j$. In contrast to the lattice of Figure 1B, in Figure 1D this reduction would not lead to a duplication of a collection, since no bivariate sources are present in other nodes, but it still invalidates the ordering relations in the lattice. In particular, if $Y_1 Y_2 Y_3$ is replaced by $Y_i Y_j$, the edge between $Y_i Y_j$ and $Y_k$ has to be removed. The remaining structure is not a lattice anymore, given the Definition 4 in Appendix A. In Appendix C, we briefly discuss more general information decompositions for structures that are not lattices, but here we still restrict ourselves to lattices. Within the set of lattices, it is now clear that, in this case, the deterministic relations render invalid any lattice containing the three variables, and thus only lattices analogous to the one of Figure 1A can be built. For these lattices with two variables, $I(\mathbf{S}; Y_i . Y_j) = 0$ bit, $I(\mathbf{S}; Y_i) = 1$ bit, and $I(\mathbf{S}; Y_i Y_j) = 2$ bit lead to all incremental terms being nonnegative. Instead of a counterexample of the nonnegativity of the incremental terms, we can interpret this case as an example in which the relations between the variables invalidate certain lattices. The possibility to generally construct multivariate nonnegative decompositions, even after these validity checking, remains an open question.

## Appendix C. The Requirements for the Nonnegativity of the Decomposition Incremental Terms

We here review the proofs of Theorems 3–5 of [21] from a general perspective, identifying their key ingredients. The aim is to recognize which constraints exist to further generalize the type of structures that can be used to build mutual information decompositions while preserving the same relation between the structures and the information-theoretic terms. Furthermore, we want to identify the properties required to ensure nonnegativity for the incremental terms, and assess the degree to which these properties can be shared by other measures or are mainly specific of the form of the measure $I_{min}$ proposed in [21]. This is important because the proposal of [21] is the only one in which nonnegativity of the decomposition components has been proven for the multivariate case. This appendix does not aim to be fully autonomous and assumes the previous reading of the proofs in [21].

We start by discussing Theorem 3 of [21]. The theorem states the expression for the incremental terms of the information gain lattices that we indicated in Equation (8). The expression of Equation (8a) results directly from the implicit definition of the incremental terms in Equation (7) and does not require that the structure formed by the collections given the ordering relation is a lattice. Conversely, Equation (8b) requires that, at least for the elements in $\Diamond \alpha$, the structure forms a lattice, namely the increment sublattice. Although [21] formulated this theorem specifically for $I_{min}$, it does not depend on the properties of the measure and relies only on the lattice properties and the connection between the lattice and the information decomposition given by Equation (7). This is why we can use the expressions of Equation (8) without any specification about the form of the mutual information measures used to build the decomposition. Furthermore, the relations in Equation (20b,d) involving the up-sets can be similarly inverted as an extension of Theorem 3.

We now consider Theorem 4 of [21]. For the proof of this theorem, not only lattice properties but also the properties of $I_{min}$ were used. We are interested in separating which of these properties correspond to the axioms generically required for any measure of redundancy, e.g., [28], and which are specific of the form of $I_{min}$. First, the proof uses Theorem 3 and Lemma 2 of [21], which do not depend on the specific properties of $I_{min}$, nor in any generic axiom for redundancy measures. Note however that the proof uses Equation (8b) and not only Equation (8a) to express the incremental terms as a function of cumulative terms, and thus, for a certain $\alpha$, only holds if the structure is compatible with a lattice for $\Diamond\alpha$. Second, the proof relies on a very specific property of the form of $I_{min}$: For a given collection, this measure is defined based on a minimum operation acting on a set of values, each value associated with one of the sources contained in the collection. In more detail, each value corresponds to the Specific Information for the corresponding source, and thus it is nonnegative and monotonicity holds between sources with more variables. This means that, when considering each summand in $I_{min}$ for $S = s$, a cumulative term $I(S = s; \alpha)$ is a function of the cumulative terms associated with the collections formed by each of the sources in $\alpha$ alone. This is relevant because it allows the relation of the measures in each node of the lattice beyond the generic relations characteristic of the decomposition. In more detail, in the proof, this allows the substitution of a minimum operation acting on the sources contained in the infimum of a set of collections by two minimum operations, acting on the collections in that set and on the sources in each of these collections, respectively. Furthermore, this substitution is only valid for lattices keeping the infimums structure of the full lattices.

Finally, Theorem 5, which proves the nonnegativity of the incremental terms, relies on Theorem 4, on the nonnegativity of cumulative terms $I(S; \alpha)$, and on the monotonicity of the Specific Information. Overall, we see that the specific closed form expression of the incremental terms stated in Theorem 4 is fundamental to prove the nonnegativity of the incremental terms. The key property of $I_{min}$ to prove Theorem 4 does not follow from the generic axioms proposed for redundancy measures, and is not shared by other measures that have been proposed, e.g., [22–24]. This renders the proof of Theorem 4 and 5 specific to $I_{min}$, in contrast to the proof of Theorem 3. Accordingly, our reexamination of the proofs of [21] helps to point out that any attempt to prove the nonnegativity of the mutual information decomposition based on an alternative measure cannot, in general, follow the same procedure.

## Appendix D. Another Example of Dual Decompositions

As a second example of a pair of dual decompositions, we show in Figure A1, also for the case of three variables, the decompositions for the sets of collections that do not contain univariate sources.

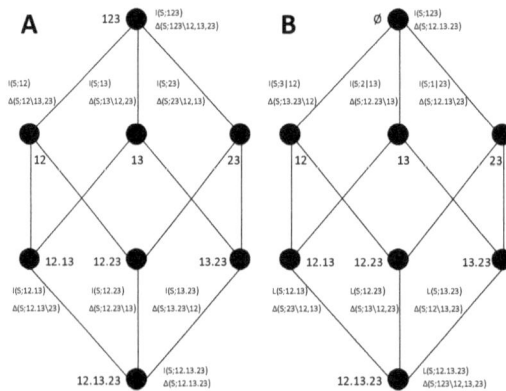

**Figure A1.** Analogous to Figure 8 but for the trivariate decomposition based only on collections that do not contain univariate sources.

# References

1. Timme, N.; Alford, W.; Flecker, B.; Beggs, J.M. Synergy, redundancy, and multivariate information measures: An experimentalist's perspective. *J. Comput. Neurosci.* **2014**, *36*, 119–140.

2. Anastassiou, D. Computational analysis of the synergy among multiple interacting genes. *Mol. Syst. Biol.* **2007**, *3*, 83, doi:10.1038/msb4100124.

3. Lüdtke, N.; Panzeri, S.; Brown, M.; Broomhead, D.; Montemurro, M.; Kell, D. Information-theoretic Sensitivity Analysis: A general method for credit assignment in complex networks. *J. R. Soc. Interface* **2008**, *19*, 223–235.

4. Watkinson, J.; Liang, K.; Wang, X.; Zheng, T.; Anastassiou, D. Inference of regulatory gene interactions from expression data using three-way mutual information. *Ann. N. Y. Acad. Sci.* **2009**, *1158*, 302–313.

5. Oizumi, M.; Albantakis, L.; Tononi, G. From the phenomenology to the mechanisms of consciousness: Integrated information theory 3.0. *PLoS Comput. Biol.* **2014**, *10*, e1003588.

6. Faes, L.; Marinazzo, D.; Nollo, G.; Porta, A. An Information-Theoretic Framework to Map the Spatiotemporal Dynamics of the Scalp Electroencephalogram. *IEEE Trans. Biomed. Eng.* **2016**, *63*, 2488–2496.

7. Averbeck, B.B.; Latham, P.E.; Pouget, A. Neural correlations, population coding and computation. *Nat. Rev. Neurosci.* **2006**, *7*, 358–366.

8. Panzeri, S.; Macke, J.; Gross, J.; Kayser, C. Neural population coding: Combining insights from microscopic and mass signals. *Trends Cogn. Sci.* **2015**, *19*, 162–172.

9. Haefner, R.; Gerwinn, S.; Macke, J.; Bethge, M. Inferring decoding strategies from choice probabilities in the presence of correlated variability. *Nat. Neurosci.* **2013**, *16*, 235–242.

10. Panzeri, S.; Harvey, C.D.; Piasini, E.; Latham, P.E.; Fellin, T. Cracking the neural code for sensory perception by combining statistics, intervention, and behavior. *Neuron* **2017**, *93*, 491–507.

11. Wibral, M.; Vicente, R.; Lizier, J.T. *Directed Information Measures in Neuroscience*; Springer: Berlin/Heidelberg, Germany, 2014.

12. Panzeri, S.; Schultz, S.; Treves, A.; Rolls, E.T. Correlations and the encoding of information in the nervous system. *Proc. Biol. Sci.* **1999**, *266*, 1001–1012.

13. Pola, G.; Thiele, A.; Hoffmann, K.P.; Panzeri, S. An exact method to quantify the information transmitted by different mechanisms of correlational coding. *Netw. Comput. Neural Syst.* **2003**, *14*, 35–60.

14. Amari, S. Information geometry on hierarchy of probability distributions. *IEEE Trans. Inf. Theory* **2001**, *47*, 1701–1711.

15. Ince, R.A.A.; Senatore, R.; Arabzadeh, E.; Montani, F.; Diamond, M.E.; Panzeri, S. Information-theoretic methods for studying population codes. *Neural Netw.* **2010**, *23*, 713–727.

16. Latham, P.E.; Nirenberg, S. Synergy, Redundancy, and Independence in Population Codes, Revisited. *J. Neurosci.* **2005**, *25*, 5195–5206.

17. Chicharro, D. A Causal Perspective on the Analysis of Signal and Noise Correlations and Their Role in Population Coding. *Neural Comput.* **2014**, *26*, 999–1054.

18. Schneidman, E.; Bialek, W.; Berry, M.J. Synergy, redundancy, and independence in population codes. *J. Neurosci.* **2003**, *23*, 11539–11553.

19. McGill, W.J. Multivariate information transmission. *Psychometrika* **1954**, *19*, 97–116.

20. Bell, A.J. The co-information lattice. In Proceedings of the 4th international Symposium on Independent Component Analysis and Blind Source Separation, Nara, Japan, 1–4 April 2003; pp. 921–926.

21. Williams, P.L.; Beer, R.D. Nonnegative Decomposition of Multivariate Information. *arXiv* **2010**, arXiv:1004.2515.

22. Harder, M.; Salge, C.; Polani, D. Bivariate measure of redundant information. *Phys. Rev. E* **2013**, *87*, 012130.

23. Griffith, V.; Koch, C. Quantifying synergistic mutual information. *arXiv* **2013**, arXiv:1205.4265v6.

24. Bertschinger, N.; Rauh, J.; Olbrich, E.; Jost, J.; Ay, N. Quantifying unique information. *Entropy* **2014**, *16*, 2161–2183.

25. Ince, R.A.A. Measuring multivariate redundant information with pointwise common change in surprisal. *arxiv* **2016**, arxiv:1602.05063.

26. Rauh, J.; Bertschinger, N.; Olbrich, E.; Jost, J. Reconsidering unique information: Towards a multivariate information decomposition. In Proceedings of the 2014 IEEE International Symposium on Information Theory, Honolulu, HI, USA, 29 June–4 July 2014; pp. 2232–2236.

27. Williams, P.L. *Information Dynamics: Its Theory and Application to Embodied Cognitive Systems*. Ph.D. Thesis, Indiana University, Bloomington, IN, USA, 2011.
28. Griffith, V.; Chong, E.K.P.; James, R.G.; Ellison, C.J.; Crutchfield, J.P. Intersection Information based on Common Randomness. *Entropy* **2014**, *16*, 1985–2000.
29. Bertschinger, N.; Rauh, J.; Olbrich, E.; Jost, J. Shared Information—New Insights and Problems in Decomposing Information in Complex Systems. In *Proceedings of the European Conference on Complex Systems 2012*; Springer: Cham, Switzerland, 2013; pp. 251–269.
30. Barrett, A. Exploration of synergistic and redundant information sharing in static and dynamical Gaussian systems. *Phys. Rev. E* **2015**, *91*, 052802.
31. Cover, T.M.; Thomas, J.A. *Elements of Information Theory*, 2nd ed.; Wiley: Hoboken, NJ, USA, 2006.
32. Ince, R.A.A. The Partial Entropy Decomposition: Decomposing multivariate entropy and mutual information via pointwise common surprisal. *arxiv* **2017**, arxiv:1702.01591v1.
33. Olbrich, E.; Bertschinger, N.; Rauh, J. Information decomposition and synergy. *Entropy* **2015**, *17*, 3501–3517.
34. Perrone, P.; Ay, N. Hierarchical quantification of synergy in channels. *arXiv* **2016**, arXiv:1512.03614
35. Schneidman, E.; Still, S.; Berry, M.J.; Bialek, W. Network information and connected correlations. *Phys. Rev. Lett.* **2003**, *91*, 238701.
36. Chicharro, D.; Ledberg, A. Framework to study dynamic dependencies in networks of interacting processes. *Phys. Rev. E* **2012**, *86*, 041901.
37. Faes, L.; Kugiumtzis, D.; Nollo, G.; Jurysta, F.; Marinazzo, D. Estimating the decomposition of predictive information in multivariate systems. *Phys. Rev. E* **2015**, *91*, 032904.
38. Valdes-Sosa, P.; Roebroeck, A.; Daunizeau, J.; Friston, K. Effective connectivity: Influence, causality and biophysical modeling. *Neuroimage* **2011**, *58*, 339–361.
39. Solo, V. On causality and Mutual information. In Proceedings of the 47th IEEE Conference on Decision and Control, Cancun, Mexico, 9–11 December 2008; pp. 4639–4944.
40. Chicharro, D. On the spectral formulation of Granger causality. *Biol. Cybern.* **2011**, *105*, 331–347.
41. Stramaglia, S.; Cortes, J.M.; Marinazzo, D. Synergy and redundancy in the Granger causal analysis of dynamical networks. *New J. Phys.* **2014**, *16*, 105003.
42. Williams, P.L.; Beer, R.D. Generalized Measures of Information Transfer. *arXiv* **2011**, arXiv:1102.1507v1.
43. Lizier, J.; Flecker, B.; Williams, P. Towards a synergy-based approach to measuring information modification. In Proceedings of the IEEE Symposium on Artificial Life, Singapore, 16–19 April 2013; pp. 43–51.
44. Wibral, M.; Priesemann, V.; Kay, J.W.; Lizier, J.T.; Phillips, W.A. Partial information decomposition as a unified approach to the specification of neural goal functions. *Brain Cogn.* **2017**, *112*, 25–38.
45. Banerjee, P.K.; Griffith, V. Synergy, redundancy, and common information. *arXiv* **2015**, arXiv:1509.03706v1.
46. James, R.G.; Crutchfield, J.P. Multivariate Dependence Beyond Shannon Information. *arXiv* **2016**, arXiv:1609.01233.
47. Chicharro, D.; Panzeri, S. Algorithms of causal inference for the analysis of effective connectivity among brain regions. *Front. Neuroinform.* **2014**, *8*, 64, doi:10.3389/fninf.2014.00064.
48. O'Connor, D.H.; Hires, S.A.; Guo, Z.; Li, N.; Yu, J.; Sun, Q.Q.; Huber, D.; Svoboda, K. Neural coding during active somatosensation revealed using illusory touch. *Nat. Neurosci.* **2013**, *16*, 958–965.
49. Otchy, T.; Wolff, S.; Rhee, J.; Pehlevan, C.; Kawai, R.; Kempf, A.; Gobes, S.; Olveczky, B. Acute off-target effects of neural circuit manipulations. *Nature* **2015**, *528*, 358–363.
50. Ay, N.; Polani, D. Information flows in causal networks. *Adv. Complex Syst.* **2008**, *11*, 17–41.
51. Lizier, J.T.; Prokopenko, M. Differentiating information transfer and causal effect. *Eur. Phys. J. B.* **2010**, *73*, 605–615.
52. Chicharro, D.; Ledberg, A. When Two Become One: The Limits of Causality Analysis of Brain Dynamics. *PLoS ONE* **2012**, *7*, e32466.

*Article*

# Specific and Complete Local Integration of Patterns in Bayesian Networks

**Martin Biehl [1,2,\*], Takashi Ikegami [3] and Daniel Polani [2]**

[1]   Araya Incorporation, 2F Mori 15 Building, 2-8-10 Toranomon, Minato-ku, Tokyo 105-0001, Japan
[2]   School of Computer Science, University of Hertfordshire, Hatfield AL10 9AB, UK; d.polani@herts.ac.uk
[3]   Department of General Systems Studies, University of Tokyo, 3-8-1 Komaba, Meguro-ku,
      Tokyo 153-8902, Japan; ikeg@sacral.c.u-tokyo.ac.jp
\*   Correspondence: martin@araya.org; Tel.: +81-3-6550-9977

Academic Editor: Mikhail Prokopenko
Received: 19 March 2017; Accepted: 12 May 2017; Published: 18 May 2017

**Abstract:** We present a first formal analysis of specific and complete local integration. Complete local integration was previously proposed as a criterion for detecting entities or wholes in distributed dynamical systems. Such entities in turn were conceived to form the basis of a theory of emergence of agents within dynamical systems. Here, we give a more thorough account of the underlying formal measures. The main contribution is the disintegration theorem which reveals a special role of completely locally integrated patterns (what we call *ι-entities*) within the trajectories they occur in. Apart from proving this theorem we introduce the disintegration hierarchy and its refinement-free version as a way to structure the patterns in a trajectory. Furthermore, we construct the least upper bound and provide a candidate for the greatest lower bound of specific local integration. Finally, we calculate the *ι*-entities in small example systems as a first sanity check and find that *ι*-entities largely fulfil simple expectations.

**Keywords:** identity over time; Bayesian networks; multi-information; entity; persistence; integration; emergence; naturalising agency

## 1. Introduction

This paper investigates a formal measure and a corresponding criterion we developed in order to capture the notion of *wholes* or *entities* within Bayesian networks in general and multivariate Markov chains in particular. The main focus of this paper is to establish some formal properties of this criterion.

The main intuition behind wholes or entities is that combinations of some events/phenomena in space(-time) can be considered as more of a *single* or coherent "thing" than combinations of other events in space(-time). For example, the two halves of a soap bubble (The authors thank Eric Smith for pointing out the example of a soap bubble.) together seem to form more of a single thing than one half of a floating soap bubble together with a piece of rock on the ground. Similarly, the soap bubble at time $t_1$ and the "same" soap bubble at $t_2$ seem more like temporal parts of the *same* thing than the soap bubble at $t_1$ and the piece of rock at $t_2$. We are trying to formally define and quantify what it is that makes some spatially and temporally extended combinations of parts entities but not others.

We envisage spatiotemporal entities as a way to establish not only the problem of *spatial identity* but also that of *temporal identity* (also called *identity over time* [1]). In other words, in addition to determining which events in "space" (e.g., which values of different degrees of freedom) belong to the same structure spatiotemporal entities should allow the identification of the structure at a time $t_2$ that is the future (or past if $t_2 < t_1$) of a structure at time $t_1$. Given a notion of identity over time, it becomes possible to capture which things persist and in what way they persist. Without a notion of identity over time, it seems persistence is not defined. The problem is how to decide whether something persisted from $t_1$ to $t_2$ if we cannot tell what at $t_2$ would count as the future of the original thing.

In everyday experience problems concerning identity over time are not of great concern. Humans routinely and unconsciously connect perceived events to spatially and temporally extended entities. Nonetheless, the problem has been known since ancient times, in particular with respect to artefacts that exchange their parts over time. A famous example is the Ship of Theseus which has all of its planks exchanged over time. This leads to the question whether it is still the *same* ship. From the point of view of physics and chemistry living organisms also exchange their parts (e.g., the constituting atoms or molecules) over time. In the long term we hope our theory can help to understand identity over time for these cases. For the moment, we are particularly interested in identity over time in formal settings like cellular automata, multivariate Markov chains, and more generally dynamical Bayesian networks. In these cases a formal notion of spatiotemporal entities (i.e., one defining spatial and temporal identity) would allow us to investigate persistence of entities/individuals formally. The persistence (and disappearance) of individuals are in turn fundamental to Darwinian evolution [2,3]. This suggests that spatiotemporal entities may be important for the understanding of the emergence of Darwinian evolution in dynamical systems.

Another area in which a formal solution to the problem of identity over time, and thereby entities (In the following, if not stated otherwise, we always mean *spatiotemporal* entities when we refer to entities.), might become important is a theory of intelligent agents that are space-time embedded as described by Orseau and Ring [4]. Agents are examples of entities fulfilling further properties e.g., exhibition of actions, and goal-directedness (cf. e.g., [5]). Using the formalism of reinforcement learning Legg and Hutter [6] proposes a definition of intelligence. Orseau and Ring [4] argue that this definition is insufficient. They dismiss the usual assumption that the environment of the reinforcement agent cannot overwrite the agent's memory (which in this case is seen as the memory/tape of a Turing machine). They conclude that in the most realistic case there only ever is one memory that the agent's (and the environment's) data is embedded in. They note that the difference between agent and environment then disappears. Furthermore, that the policy of the agent cannot be freely chosen anymore, only the initial condition. In order to measure intelligence according to Legg and Hutter [6] we must be able to define reward functions. This seems difficult without the capability to distinguish the agent according to some criterion. Towards the end of their publication Orseau and Ring [4] propose to define a "heart" pattern and use the duration of its existence as a reward. This seems a too specific approach to us since it basically defines identity over time (of the heart pattern) as invariance. In more general settings a pattern that maintains a more general criterion of identity over time would be desirable. Ideally, this criterion would also not need a specifically designed heart pattern. Another advantage would be that reward functions different from lifetime could be used if the agent were identifiable. An entity criterion in the sense of this paper would be a step in this direction.

## 1.1. Illustration

In order to introduce the contributions of this paper we illustrate the setting of our work further.

This illustration should only be taken as a motivation for what follows and not be confused with a result. The reason we don't use a concrete example is simply that we lack the necessary computational means (which are considerable as we will discuss in Section 5).

Let us assume we are given the entire time-evolution (what we will call a *trajectory*) of some known multivariate dynamical system or stochastic process. For example, a trajectory of a one-dimensional elementary cellular automaton showing a glider collision like Figure 1a (This is produced by the rule 62 elementary cellular automaton with time increasing from left to right. However, this does not matter here. For more about this system see e.g., Boccara et al. [7]).

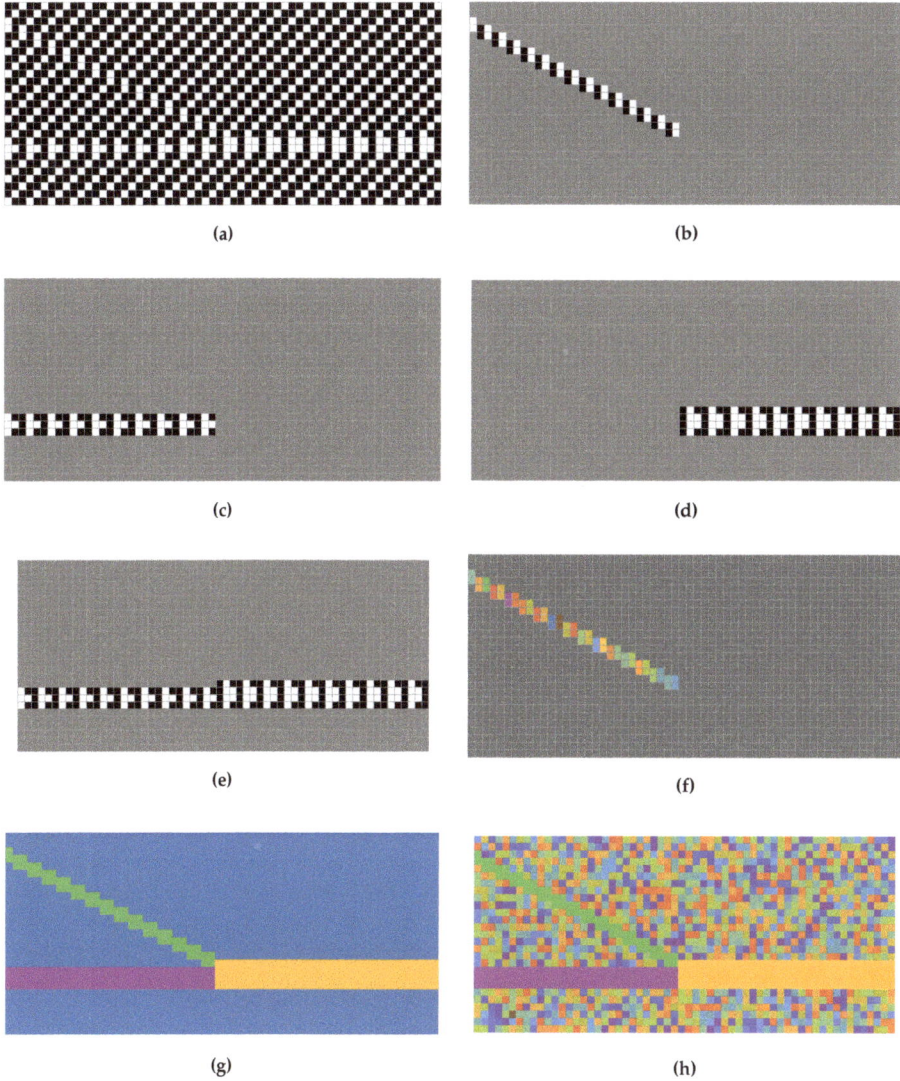

**Figure 1.** Illustration of concepts from this paper on the time-evolution (trajectory) of a one-dimensional elementary cellular automaton. Time-steps increase from left to right. None of the shown structures are derived from principles. They are manually constructed for illustrative purposes. In (**a**) we show the complete (finite) trajectory. Naively, two gliders can be seen to collide and give rise to a third glider; In (**b**–**d**) we show (spatiotemporal) patterns fixing the variables (allegedly) pertaining to a first, second, and a third glider; In (**e**) we show a pattern fixing the variables of what could be a glider that absorbs the first glider from before and maintains its identity; In (**f**) we show a partition into the time-slices of the pattern of the first glider; In (**g**) we show a partition of the trajectory with three parts coinciding with the gliders and one part encompassing the rest; In (**h**) we show again a partition with three parts coinciding with the gliders but now all other variables are considered as individual parts.

We take the point of view here argued for in previous work [8] that entities are phenomena that occur within trajectories and that they can be represented by (spatiotemporal) patterns. Patterns in this sense fix a part of the variables in a trajectory to definite values and leave the rest undetermined. In Figure 1b–d we show such patterns that occur in Figure 1a with the undetermined variables coloured grey and the determined ones taking on those of the trajectory. Visually speaking, a pattern is a snippet from a trajectory that it occurs in.

From Figure 1a we would probably expect that what we are seeing are two gliders colliding and forming a third. However, it may also be that one of the gliders absorbs the other, maintains some form of identity, and only changes its appearance (e.g., it "grows"). This highlights the problem of identity over time. While the spatial identity of such patterns has been treated multiple times in the literature their identity of over time is rarely dealt with.

Our approach evaluates the "integration" of spatiotemporally extended patterns at once. According to our proposed entity-criterion a pattern is an $\iota$-entity if, due to the dynamics of the system, every part of this pattern (which is again a pattern) makes all other parts more probable. Identity over time is then included since future parts have to make past parts more probable and vice versa. In principle this would allow us to detect if one of the gliders absorbs another one without loosing its identity. For example, this could result in an entity as in Figure 1e.

In order to detect entities the straightforward approach is to evaluate the entity-criterion for every spatiotemporal pattern in a given trajectory. Evaluating our entity-criterion of positive complete local integration (CLI) for a given pattern corresponds to splitting the pattern into parts in every possible way and calculating whether all the resulting parts make each other more probable. This means evaluating the specific local integration (SLI) with respect to all *partitions* of the set of variables occupied by the pattern.

## 1.2. Contributions

This paper contains four contributions.

We first give a more formal definition of patterns. Since each pattern uniquely specifies a set of trajectories (those trajectories that the pattern occurs in) one might be tempted to reduce the analysis to that of sets of trajectories. We show that this is not possible since not all sets of trajectories have a pattern that specifies them.

Second, we try to get a general intuition for the patterns whose parts make all other parts more probable. For this we show how to construct patterns that, for given probability of the whole pattern, achieve the least upper bound of specific local integration (SLI). These turn out to be patterns for which each part only occurs if and only if the whole pattern occurs. We also construct a pattern that, again for given probability of the whole pattern, has *negative* SLI. These pattern (which may achieve the greatest lower bound of SLI) occur if either the whole pattern occurs or the pattern occurs up to exactly one part of it, which does not occur.

Third, we prove the disintegration theorem. This is the main contribution. We saw that patterns are snippets of trajectories. We can also look at the whole trajectory as a single pattern. Like all patterns the trajectory can be split up into parts, i.e., partitioned, resulting in a set of patterns. Among the partitions we find examples such as those in Figure 1g,h. These are very particular partitions picking out the gliders among all possible parts. This suggests that finding such special partitions provides a (possibly different) notion of entities.

One intuition we might have is that entities are the most "independent" parts of a trajectory. In other words we could look for the partition whose parts make the other parts *less* probable. The disintegration theorem then shows that this approach again leads to the $\iota$-entities. This shows that $\iota$-entities do not only have an intuitive motivation but also play a particular role in the structure of probabilities of entire trajectories.

It is not directly the parts of the partitions that minimise SLI for a trajectory which are $\iota$-entities. To get $\iota$-entities we first classify all partitions of the trajectory according to their SLI value. Then within each such class we choose the partitions for which no refining partition (A refining partition is one that further partitions any of the parts of the original partition.) achieves an even lower level of SLI.

So according to the disintegration theorem a $\iota$-entity is not only a pattern that is integrated with respect to every possible partition of the pattern but also a pattern that occurs in partitions that minimise (in a certain sense) the integration of trajectories.

A side effect of the disintegration theorem is that we naturally get a kind of hierarchy of $\iota$-entities called the disintegration hierarchy. For each trajectory and its different levels of SLI we find different decompositions of the trajectory into $\iota$-entities.

Fourth, we calculate the $\iota$-entities and disintegration hierarchy for two simple example systems. Our example systems show that in general the partitions at a particular disintegration level are not unique. This means that there are overlapping $\iota$-entities at those levels. Furthermore, the same $\iota$-entity can occur on multiple levels of the disintegration.

We do not thoroughly discuss the disintegration hierarchies in this paper and postpone this to future publications. Here we only note that many entities in the real world occur within hierarchies as well. For example, animals are entities that are composed of cells which are themselves entities.

### 1.3. Related Work

We now give a quick overview of related work. More in depth discussions will be provided after we formally introduce our definitions.

To our knowledge the measure of CLI has been proposed for the first time by us in [8]. However, this publication contained none of the formal or numerical results in the present paper. From a formal perspective the measures of SLI and CLI are a combination of existing concepts. SLI localises multi-information [9,10] in the way proposed by Lizier [11] for other information theoretic measures. In order to get the CLI we apply the weakest-link approach proposed by Tononi and Sporns [12], Balduzzi and Tononi [13] to SLI.

Conceptually, our work is most closely related to Beer [14]. The notion of spatiotemporal patterns used there to capture blocks, blinkers, and gliders is equivalent to the *patterns* we define more formally here. This work also contains an informal entity-criterion that directly deals with identity over time (not only space). It differs significantly from our proposal as it depends on the re-occurrence of certain transitions at later times in a pattern whereas our criterion only depends on the probabilities of parts of the patterns without the need for any re-occurrences.

The *organisations* of chemical organisation theory [15] may also be interpreted as entity-criteria. In Fontana and Buss [15] these are defined in the following way:

> The observer will conclude that the system is an organisation to the extent that there is a compressed description of its objects and of their relations.

The direct intuition is different from ours and it is not clear to us in how far our entity-criterion is equivalent to this. This will be further investigated in the future.

It is worth noting that viewing entities/objects/individuals as patterns occurring within a trajectory is in contrast to an approach that models them as sets of random variables/stochastic processes (e.g., a set of cells in a CA in contrast to a set of specific values of a set of cells). An example of the latter approach are the information theoretic individuals of Krakauer et al. [16]. These individuals are identified using an information theoretic notion of autonomy due to Bertschinger et al. [17]. The latter notion of autonomy is also somewhat related to the idea of integration here. Autonomy contains a term that measures the degree to which a random variable representing an individual at timestep $t$ determines the random variable representing it at $t + 1$. Similarly, CLI requires that every part of an entity pattern makes every other part more probable, in the extreme case this means that every part determines that every other part of the pattern also occurs. However, formally autonomy evaluates random variables and not patterns directly.

At the most basic level the intuition behind entities is that some spatiotemporal patterns are more special than others. Defining (and usually finding) more important spatiotemporal patterns or structures (also called coherent structures) has a long history in the theory of cellular automata and distributed dynamical systems. As Shalizi et al. [18] have argued most of the earlier definitions and methods [19–22] require previous knowledge about the patterns being looked for. They are therefore not suitable for a general definition of entities. More recent definitions based on information theory [18,23,24] do not have this limitation anymore. The difference to our entity-criterion is that they do not treat identity over time. They are well suited to identify gliders at each time-step for example, but if two gliders collide and give rise to a third glider as in Figure 1a these methods (by design) say nothing about the identity of the third glider. i.e., they cannot make a difference between a glider absorbing another one and two gliders producing a new one. While we have not been able to show that our approach actually makes such distinctions for gliders, it could do so in principle.

We note here that the approach of identifying individuals by Friston [25] using Markov blankets has the same shortcoming as the spatiotemporal filters. For each individual time-step it returns a partition of all degrees of freedom into internal, sensory, active, and external degrees. However, it does not provide a way to resolve ambiguities in the case of multiple such partitions colliding.

Among research related to integrated information theory (IIT) there are approaches (a first one by Balduzzi [26] and a more recently by Hoel et al. [27]) that can be used to determine specific spatiotemporal patterns in a trajectory. They can therefore be interpreted to define a notion of entities even if that is not their main goal. These approaches are aimed at establishing the optimal spatiotemporal coarse-graining to describe the dynamics of a system. For a given trajectory we can then identify the patterns that instantiate a macro-state/coarse-grain that is optimal according to their criterion.

In contrast to our approach the spatiotemporal grains are determined by their interactions with other grains. In our case the entities are determined first and foremost by their internal relations.

The consequence seems to be that a pattern can be an entity in one trajectory and not an entity in another even if it occurs in both. In our conception a pattern is an entity in all trajectories it occurs in.

## 2. Notation and Background

In this section we briefly introduce our notation for sets of random variables (Since every set of jointly distributed random variables can be seen as a Bayesian network and vice versa we use these terms interchangeably.) and their partition lattices.

In general, we use the convention that upper-case letters $X, Y, Z$ are random variables, lower-case letters $x, y, z$ are specific values/outcomes of random variables, and calligraphic letters $\mathcal{X}, \mathcal{Y}, \mathcal{Z}$ are state spaces that random variables take values in. Furthermore:

**Definition 1.** *Let $\{X_i\}_{i \in V}$ be a set of random variables with totally ordered finite index set $V$ and state spaces $\{\mathcal{X}_i\}_{i \in V}$ respectively. Then for $A, B \subseteq V$ define:*

1. $X_A := (X_i)_{i \in A}$ *as the joint random variable composed of the random variables indexed by $A$, where $A$ is ordered according to the total order of $V$,*
2. $\mathcal{X}_A := \prod_{i \in A} \mathcal{X}_i$ *as the state space of $X_A$,*
3. $x_A := (x_i)_{i \in A} \in \mathcal{X}_A$ *as a value of $X_A$,*
4. $p_A : \mathcal{X}_A \to [0, 1]$ *as the probability distribution (or more precisely probability mass function) of $X_A$ which is the joint probability distribution over the random variables indexed by $A$. If $A = \{i\}$ i.e., a singleton set, we drop the parentheses and just write $p_A = p_i$,*
5. $p_{A,B} : \mathcal{X}_A \times \mathcal{X}_B \to [0, 1]$ *as the probability distribution over $\mathcal{X}_A \times \mathcal{X}_B$. Note that in general for arbitrary $A, B \subseteq V, x_A \in \mathcal{X}_A$, and $y_B \in \mathcal{X}_B$ this can be rewritten as a distribution over the intersection of $A$ and $B$ and the respective complements. The variables in the intersection have to coincide:*

$$p_{A,B}(x_A, y_B) := p_{A \setminus B, A \cap B, B \setminus A, A \cap B}(x_{A \setminus B}, x_{A \cap B}, y_{B \setminus A}, y_{A \cap B}) \tag{1}$$

$$= \delta_{x_{A \cap B}}(y_{A \cap B}) \, p_{A \setminus B, A \cap B, B \setminus A}(x_{A \setminus B}, x_{A \cap B}, y_{B \setminus A}). \tag{2}$$

*Here δ is the Kronecker delta (see Appendix A). If $A \cap B = \emptyset$ and $C = A \cup B$ we also write $p_C(x_A, y_B)$ to keep expressions shorter.*

6. $p_{B|A} : \mathcal{X}_A \times \mathcal{X}_B \to [0,1]$ *with* $(x_A, x_B) \mapsto p_{B|A}(x_B|x_A)$ *as the conditional probability distribution over* $X_B$ *given* $X_A$:

$$p_{B|A}(y_B|x_A) := \frac{p_{A,B}(x_A, y_B)}{p_A(x_A)}. \tag{3}$$

*We also just write* $p_B(x_B|x_A)$ *if it is clear from context what variables we are conditioning on.*

If we are given $p_V$ we can obtain every $p_A$ through marginalisation. In the notation of Definition 1 this is formally written:

$$p_A(x_A) = \sum_{\bar{x}_{V\backslash A} \in \mathcal{X}_{V\backslash A}} p_{A,V\backslash A}(x_A, \bar{x}_{V\backslash A}) \tag{4}$$

$$= \sum_{\bar{x}_{V\backslash A} \in \mathcal{X}_{V\backslash A}} p_V(x_A, \bar{x}_{V\backslash A}). \tag{5}$$

Next we define the partition lattice of a set of random variables. Partition lattices occur as a structure of the set of possible ways to split an object/pattern into parts. Subsets of the partition lattices play an important role in the disintegration theorem.

**Definition 2** (Partition lattice of a set of random variables). *Let* $\{X_i\}_{i \in V}$ *be a set of random variables.*

1. *Then its partition lattice* $\mathfrak{L}(V)$ *is the set of partitions of V partially ordered by refinement (see also Appendix B).*
2. *For two partitions* $\pi, \rho \in \mathfrak{L}(V)$ *we write* $\pi \lhd \rho$ *if* $\pi$ *refines* $\rho$ *and* $\pi \lhd: \rho$ *if* $\pi$ *covers* $\rho$. *The latter means that* $\pi \neq \rho$, $\pi \lhd \rho$, *and there is no* $\xi \in \mathfrak{L}(V)$ *with* $\pi \neq \xi \neq \rho$ *such that* $\pi \lhd \xi \lhd \rho$.
3. *We write* **0** *for the zero element of a partially ordered set (including lattices) and* **1** *for the unit element.*
4. *Given a partition* $\pi \in \mathfrak{L}(V)$ *and a subset* $A \subseteq V$ *we define the restricted partition* $\pi|_A$ *of* $\pi$ *to* $A$ *via:*

$$\pi|_A := \{b \cap A : b \in \pi\}. \tag{6}$$

For some examples of partition lattices see Appendix B and for more background see e.g., Grätzer [28]. For our purpose it is important to note that the partitions of sets of random variables or Bayesian networks we are investigating are partitions of the index set $V$ of these and not partitions of their state spaces $\mathcal{X}_V$.

## 3. Patterns, Entities, Specific, and Complete Local Integration

This section contains the formal part of this contribution.

First we introduce *patterns*. Patterns are the main structures of interest in this publication. Entities are seen as special kinds of patterns. The measures of specific local integration and complete local integration, which we use in our criterion for $\iota$-entities, quantify notions of "oneness" of patterns. We give a brief motivation and show that while each pattern defines a set of "trajectories" of a set of random variables not every such set is defined by a pattern. This justifies studying patterns for their own sake.

Then we motivate briefly the use of specific and complete local integration (SLI and CLI) for an entity criterion on patterns. We then turn to more formal aspects of SLI and CLI. We first prove an upper bound for SLI and construct a candidate for a lower bound. We then go on to define the disintegration hierarchy and its refinement-free version. These structures are used to prove the main result, the *disintegration theorem*. This relates the SLI of whole trajectories of a Bayesian network to the CLI of parts of these trajectories and vice versa.

## 3.1. Patterns

This section introduces the notion of patterns. These form the basic candidate structures for entities.

The structures we are trying to capture by entities should be analogous to spatially and temporally extended objects we encounter in everyday life (e.g., soap bubbles, living organisms). These objects seem to occur in the single history of the universe that also contains us. The purpose of patterns is then to capture arbitrary structures that occur within single trajectories or histories of a multivariate discrete dynamical system (see Figure 2 for an example of a Bayesian network of such a system, any cellular automaton is also such a system).

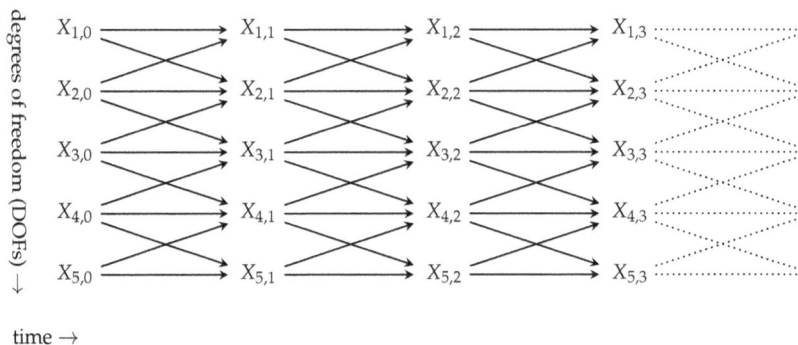

**Figure 2.** First time steps of a Bayesian network representing a multivariate dynamical system (or multivariate Markov chain) $\{X_i\}_{i \in V}$. Here we used $V = J \times T$ with $J$ indicating spatial degrees of freedom and $T$ the temporal extension. Then each node is indexed by a tuple $(j, t)$ as shown. The shown edges are just an example, edges are allowed to point from any node to another one within the same or in the subsequent column.

We emphasise the single trajectory since many structures of interest (e.g., gliders) occur in some trajectories in some "places", in other trajectories in other "places" (compare e.g., Figures 1a and 3a), and in some trajectories not at all. We explicitly want to be able to capture such trajectory dependent structures and therefore choose patterns. Examples of formal structures for which it makes no sense to say that they occur within a trajectory are for example the random variables in a Bayesian network and, as we will see, general sets of trajectories of the Bayesian network.

Unlike entities, which we conceive of as special patterns that fulfil further criteria, patterns are formed by *any* combination of events at arbitrary times and positions. As an example, we might think of cellular automaton again. The time evolutions over multiple steps of the cells attributed to a glider see [14] for a principled way to attribute cells to theseas in Figure 1b,e should be patterns but also arbitrary choices of events in a trajectory as in Figure 3b.

In the more general context of (finite) Bayesian networks there may be no interpretation of time or space. Nonetheless, we can define that a trajectory in this case fixes every random variable to a particular value. We then define patterns formally in the following way.

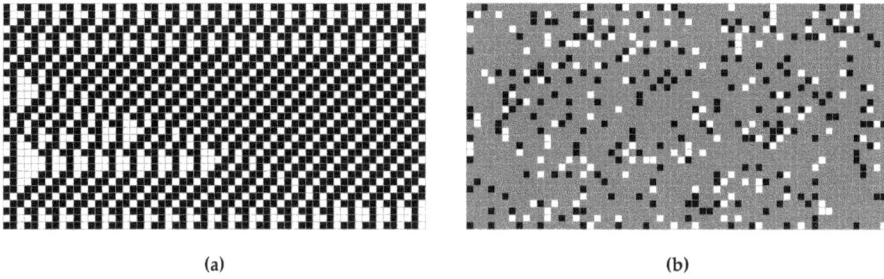

(a)                                                    (b)

**Figure 3.** In (a) we show a trajectory of the same cellular automaton as in Figure 1 with a randomly chosen initial condition. The set of gliders and their paths occurring in this trajectory is clearly different from those in Figure 1a. In (b) we show an example of a random pattern that occurs in the trajectory of (a) and is probably not an entity in any sense.

**Definition 3** (Patterns and trajectories). *Let $\{X_i\}_{i\in V}$ be set of random variables with index set $V$ and state spaces $\{\mathcal{X}_i\}_{i\in V}$ respectively.*

1.  *A pattern at $A \subseteq V$ is an assignment*

$$X_A = x_A \tag{7}$$

   *where $x_A \in \mathcal{X}_A$. If there is no danger of confusion we also just write $x_A$ for the pattern $X_A = x_A$ at $A$.*
2.  *The elements $x_V$ of the joint state space $\mathcal{X}_V$ are isomorphic to the patterns $X_V = x_V$ at $V$ which fix the complete set $\{X_i\}_{i\in V}$ of random variables. Since they will be used repeatedly we refer to them as the trajectories of $\{X_i\}_{i\in V}$.*
3.  *A pattern $x_A$ is said to occur in trajectory $\bar{x}_V \in \mathcal{X}_V$ if $\bar{x}_A = x_A$.*
4.  *Each pattern $x_A$ uniquely defines (or captures) a set of trajectories $\mathcal{T}(x_A)$ via*

$$\mathcal{T}(x_A) = \{\bar{x}_V \in \mathcal{X}_V : \bar{x}_A = x_A\}, \tag{8}$$

   *i.e., the set of trajectories that $x_A$ occurs in.*
5.  *It is convenient to allow the empty pattern $x_\varnothing$ for which we define $\mathcal{T}(x_\varnothing) = \mathcal{X}_V$.*

Remarks:

- Note that for every $x_A \in \mathcal{X}_A$ we can form a pattern $\mathcal{X}_A = x_A$ so the set of all patterns is $\bigcup_{A\subseteq V} \mathcal{X}_A$.
- Our notion of patterns is similar to "patterns" as defined in [29] and to "cylinders" as defined in [30]. More precisely, these other definitions concern (probabilistic) cellular automata where all random variables have identical state spaces $\mathcal{X}_i = \mathcal{X}_j$ for all $i, j \in V$. They also restrict the extent of the patterns or cylinders to a single time-step. Under these conditions our patterns are isomorphic to these other definitions. However, we drop both the identical state space assumption and the restriction to single time-steps.

Our definition is inspired by the usage of the term "spatiotemporal pattern" in [14,31,32]. There is no formal definition of this notion given in these publications but we believe that our definition is a straightforward formalisation. Note that these publications only treat the Game of Life cellular automaton. The assumption of identical state space is therefore implicitly made. At the same time the restriction to single time-steps is explicitly dropped.

Since every pattern defines a subset of $\mathcal{X}_V$, one could think that every subset of $\mathcal{X}_V$ is also a pattern. In that case studying patterns in a set of random variables $\{X_i\}_{i\in V}$ would be the same as studying subsets of its set of trajectories $\mathcal{X}_V$. However, the set of subsets of $\mathcal{X}_V$ defined by patterns and the set of all subsets $2^{\mathcal{X}_V}$ (i.e., the power set) of $\mathcal{X}_V$ of a set of random variables $\{X_i\}_{i \in V}$ are not identical. Formally:

$$\bigcup_{B \subseteq V} \{\mathcal{T}(x_B) \subseteq \mathcal{X}_V : x_B \in \mathcal{X}_B\} \subseteq 2^{\mathcal{X}_V}. \tag{9}$$

While patterns define subsets of $\mathcal{X}_V$, not every subset of $\mathcal{X}_V$ is captured by a pattern. The difference of the two sets is characterised in Theorem 1 below. We first present a simple example of a subset $\mathcal{D} \in 2^{\mathcal{X}_V}$ that cannot be captured by a pattern.

Let $V = \{1, 2\}$ and $\{X_i\}_{i \in V} = \{X_1, X_2\}$ the set of random variables. Let $\mathcal{X}_1 = \mathcal{X}_2 = \{0, 1\}$. Then $\mathcal{X}_V = \{(0, 0), (0, 1), (1, 0), (1, 1)\}$. Now let $A = V = \{1, 2\}$, choose pattern $x_A = (0, 0)$ and pattern $\tilde{x}_A = (1, 1)$. Then let

$$\mathcal{D} := \{x_A \cup \tilde{x}_A\} = \{(0, 0), (1, 1)\}. \tag{10}$$

In this case we can easily list the set of all patterns $\bigcup_{C \subseteq V} \mathcal{X}_C$:

| $C \subseteq V$ | $x_C$ | $\mathcal{T}(x_C)$ |
|---|---|---|
| $\emptyset$ | $x_\emptyset$ | $\mathcal{X}_V$ |
| $\{1\}$ | $(0)$ | $\{(0, 0), (0, 1)\}$ |
| | $(1)$ | $\{(1, 0), (1, 1)\}$ |
| $\{2\}$ | $(0)$ | $\{(0, 0), (1, 0)\}$ |
| | $(1)$ | $\{(0, 1), (1, 1)\}$ |
| $\{1,2\}$ | $(0, 0)$ | $\{(0, 0)\}$ |
| | $(0, 1)$ | $\{(0, 1)\}$ |
| | $(1, 0)$ | $\{(1, 0)\}$ |
| | $(1, 1)$ | $\{(1, 1)\}$ |

$(11)$

and verify that $\mathcal{D}$ is not among them. Before we formally characterise the difference, we define some extra terminology.

**Definition 4.** *Let* $\{X_i\}_{i \in V}$ *be set of random variables with index set $V$ and state spaces $\{\mathcal{X}_i\}_{i \in V}$ respectively. For a subset $\mathcal{D} \subseteq \mathcal{X}_V$ the set $\mathcal{D}_A$ of all patterns at $A$ that occur in one of the trajectories in $\mathcal{D}$ is defined as*

$$\mathcal{D}_A := \{x_A \in \mathcal{X}_A : \exists \tilde{x}_V \in \mathcal{D}, \tilde{x}_A = x_A\}. \tag{12}$$

So in the previous example $\mathcal{D}_{\{1\}} = \{0, 1\}$, $\mathcal{D}_{\{2\}} = \{0, 1\}$, $\mathcal{D}_{\{1,2\}} = \{(0, 0), (1, 1)\}$. In then get the following theorem which establishes the difference between the subsets of $\mathcal{X}_V$ captured by patterns and general subsets.

**Theorem 1.** *Given a set of random variables $\{X_i\}_{i \in V}$, a subset $\mathcal{D} \subseteq \mathcal{X}_V$ cannot be represented by a pattern of $\{X_i\}_{i \in V}$ if and only if there exists $A \subseteq V$ with $\mathcal{D}_A \subset \mathcal{X}_A$ (proper subset) and $|\mathcal{D}_A| > 1$, i.e., if neither all patterns at $A$ are possible nor a unique pattern at $A$ is specified by $\mathcal{D}$.*

**Proof.** See Appendix D. □

We saw that in the previous example the subset $\mathcal{D}$ cannot be captured by a pattern. For $A = \{1\}$ we have $\mathcal{D}_{\{1\}} = \{0, 1\} = \mathcal{X}_{\{1\}}$ and for $A = \{2\}$ we have $\mathcal{D}_{\{2\}} = \{0, 1\} = \mathcal{X}_{\{2\}}$ so these do not fulfil the conditions of Theorem 1. However, for $A = \{1, 2\}$ we have $\mathcal{D}_{\{1,2\}} = \{(0, 0), (1, 1)\} \subset \mathcal{X}_{\{1,2\}}$ and $|\mathcal{D}_{\{1,2\}}| > 1$ so the conditions of Theorem 1 are fulfilled and as expected $\mathcal{D}$ cannot be captured by a pattern.

The proof of the following corollary shows how to construct a subset that cannot be represented by a pattern for all sets of random variables $\{X_i\}_{i \in V}$ with $|\mathcal{X}_V| > 2$.

**Corollary 1.** *Given a set of random variables* $\{X_i\}_{i \in V}$, *if* $|\mathcal{X}_V| > 2$ *then*

$$\bigcup_{B \in V} \{\mathcal{T}(x_B) \subseteq \mathcal{X}_V : x_B \in \mathcal{X}_B\} \subset 2^{\mathcal{X}_V} \tag{13}$$

*(proper subset).*

**Proof.** Choose $\mathcal{D} = \{x_V, y_V\} \in 2^{\mathcal{X}_V}$ with $y_V \in \{\bar{x}_V \in \mathcal{X}_V : \forall i \in V, \bar{x}_i \neq x_i\}$. Then for all $A \subseteq V$ we have $|\mathcal{D}_A| = 2$ and $\mathcal{D}_A \subset \mathcal{X}_A$. So $\mathcal{D}$ cannot be represented by a pattern according to Theorem 1 and so $\mathcal{D} \notin \bigcup_{B \in V} \{\mathcal{T}(x_B) \subseteq \mathcal{X}_V : x_B \in \mathcal{X}_B\}$. □

This means that in every set of random variables that not only consists of a single binary random variable there are subsets of $\mathcal{X}_V$ that cannot be captured by a pattern. We can interpret this result in the following way. Patterns were constructed to be structures that occur within trajectories. It then turned out that each pattern also defines a subset of all trajectories of a system. So for sets of trajectories captured by patterns it could make sense to say they "occur" within one trajectory. However, there are sets of trajectories that are not captured by patterns. For these sets of trajectories it would then not be well-defined to say that they occur within a trajectory. This is the reason we choose to investigate patterns specifically and not sets of trajectories.

### 3.2. Motivation of Complete Local Integration as an Entity Criterion

We proposed to use patterns as the candidate structures for entities since patterns comprise arbitrary structures that occur within single trajectories of multivariate systems. Here we heuristically motivate our choice of using positive complete local integration as a criterion to select entities among patterns. In general such a criterion would give us, for any Bayesian network $\{X_i\}_{i \in V}$ a subset $\mathfrak{E}(\{X_i\}_{i \in V}) \subseteq \bigcup_{A \subseteq V} \mathcal{X}_A$ of the patterns.

So what is an entity? We can rephrase the problem of finding an entity criterion by saying an entity is composed of parts that share the same identity. So if we can define when parts share the same identity we also define entities by finding all parts that share identity with some given part. For the moment, let us decompose (as is often done [33]) the problem of identity into two parts:

1. spatial identity and
2. temporal identity.

Our solution will make no distinction between these two aspects in the end. We note here that conceiving of entities (or objects) as composite of spatial and temporal parts as we do in this paper is referred to as four-dimensionalism or perdurantism in philosophical discussions (see e.g., [34]). The opposing view holds that entities are spatial only and endure over time. This view is called endurantism. Here we will not go into the details of this discussion.

The main intuition behind complete local integration is that every part of an entity should make every other part more probable.

This seems to hold for example for the spatial identity of living organisms. Parts of living organisms rarely exist without the rest of the living organisms also existing. For example, it is rare that an arm exists without a corresponding rest of a human body existing compared to an arm and the rest of a human body existing. The body (without arm) seems to make the existence of the arm more probable and vice versa. Similar relations between parts seem to hold for all living organisms but also for some non-living structures. The best example of a non-living structure we know of for which this is obvious are soap bubbles. Half soap bubbles (or thirds, quarters,...) only ever exist for split seconds whereas entire soap bubbles can persist for up to minutes. Any part of a soap bubble seems to make the existence of the rest more probable. Similarly, parts of hurricanes or tornadoes are rare. So what about spatial parts of structures that are not so entity-like? Does the existence of an arm make things more probable that are not parts of the corresponding body? For example, does the arm make the existence of some piece of rock more probable? Maybe to a small degree as without the existence of any rocks in the universe humans are probably impossible. However, this effect is much smaller than the increase of probability of the existence of the rest of the body due to the arm.

These arguments concerned the spatial identity problem. However, for temporal identity similar arguments hold. The existence of a living organism at one point in time makes it more probable that there is a living organism (in the vicinity) at a subsequent (and preceding) point in time. If we look at structures that are not entity-like with respect to the temporal dimension we find a different situation. An arm at some instance of time does not make the existence of a rock at a subsequent instance much more probable. It does make the existence of a human body at a subsequent instance much more probable. So the human body at the second instance seems to be more like a future part of the arm than the rock. Switching now to patterns in sets of random variables we can easily formalise such intuitions. We required that for an entity every part of the structure, which is now a pattern $x_O$, makes every other part more probable. A part of a pattern is a pattern $x_b$ with $b \subset O$. If we require that every part of a pattern makes every other part more probable then we can write that $x_O$ is an entity if:

$$\min_{b \subset O} \frac{p_{O \backslash b}(x_{O \backslash b} | x_b)}{p_{O \backslash b}(x_{O \backslash b})} > 1. \tag{14}$$

This is equivalent to

$$\min_{b \subset O} \frac{p_O(x_O)}{p_{O \backslash b}(x_{O \backslash b}) p_b(x_b)} > 1. \tag{15}$$

If we write $\mathfrak{L}_2(O)$ for the set of all bipartitions of $O$ we can rewrite this further as

$$\min_{\pi \in \mathfrak{L}_2(O)} \frac{p_O(x_O)}{\prod_{b \in \pi} p_b(x_b)} > 1. \tag{16}$$

We can interpret this form as requiring that for every possible partition $\pi \in \mathfrak{L}_2(O)$ into two parts $x_{b_1}, x_{b_2}$ the probability of the whole pattern $x_O = (x_{b_1}, x_{b_2})$ is bigger than its probability would be if the two parts were independent. To see this, note that if the two parts $x_{b_1}, x_{b_2}$ were independent we would have

$$p_O(x_O) =: p_{b_1, b_2}(x_{b_1}, x_{b_2}) = p_{b_1}(x_{b_1}) p_{b_2}(x_{b_2}). \tag{17}$$

Which would give us

$$\frac{p_O(x_O)}{\prod_{b \in \pi} p_b(x_b)} = 1 \tag{18}$$

for this partition.

From this point of view the choice of bipartitions only seems arbitrary. For example, the existence a partition $\zeta$ into three parts such that

$$p_O(x_O) = \prod_{c \in \zeta} p_c(x_c) \tag{19}$$

seems to suggest that the pattern $x_O$ is not an entity but instead composite of three parts. We can therefore generalise Equation (16) to include all partitions $\mathfrak{L}(O)$ (see Definition 2) of $O$ except the unit partition $\mathbf{1}_O$. Then we would say that $x_O$ is an entity if

$$\min_{\pi \in \mathfrak{L}(O) \backslash \mathbf{1}_O} \frac{p_O(x_O)}{\prod_{b \in \pi} p_b(x_b)} > 1. \tag{20}$$

This measure already results in the same entities as the measure we propose.

However, in order to connect with information theory, log-likelihoods, and related literature we formally introduce the logarithm into this equation. We then arrive at the following entity-criterion

$$\min_{\pi \in \mathfrak{L}(O) \backslash \mathbf{1}_O} \log \frac{p_O(x_O)}{\prod_{b \in \pi} p_b(x_b)} > 0. \tag{21}$$

where the left hand side is the complete local integration (CLI), the function minimised is the specific local integration (SLI), and the inequality provides the criterion for $\iota$-entities. For reference, we define these notions formally. We begin with SLI which quantifies for a given partition $\pi$ of a pattern in how far the probability of the whole pattern is bigger than its probability would be if the blocks of the partition would be independent.

**Definition 5** (Specific local integration (SLI)). *Given a Bayesian network* $\{X\}_{i \in V}$ *and a pattern* $x_O$ *the specific local integration* $\text{mi}_\pi(x_O)$ *of* $x_O$ *with respect to a partition* $\pi$ *of* $O \subseteq V$ *is defined as*

$$\text{mi}_\pi(x_O) := \log \frac{p_O(x_O)}{\prod_{b \in \pi} p_b(x_b)}. \tag{22}$$

*In this paper we use the convention that* $\log \frac{0}{0} := 0$.

**Definition 6** ((Complete) local integration). *Given a Bayesian network* $\{X_i\}_{i \in V}$ *and a pattern* $x_O$ *of this network the complete local integration* $\iota(x_O)$ *of* $x_O$ *is the minimum SLI over the non-unit partitions* $\pi \in \mathfrak{L}(O) \backslash \mathbf{1}_O$:

$$\iota(x_O) := \min_{\pi \in \mathfrak{L}(O) \backslash \mathbf{1}_O} \text{mi}_\pi(x_O). \tag{23}$$

*We call a pattern* $x_O$ *completely locally integrated if* $\iota(x_O) > 0$.

Remarks:

- The reason for excluding the unit partition $\mathbf{1}_O$ of $\mathfrak{L}(O)$ (where $\mathbf{1}_O = \{O\}$ see Definition 2) is that with respect to it every pattern has $\text{mi}_{\mathbf{1}_O}(x_O) = 0$.
- Looking for a partition that minimises a measure of integration is known as the *weakest link approach* [35] to dealing with multiple partitions. We note here that this is not the only approach that is being discussed. Another approach is to look at weighted averages of all integrations. For a further discussion of this point in the case of the expected value of SLI see Ay [35] and references therein. For our interpretation taking the average seems less well suited since requiring a positive average will allow SLI to be negative with respect to some partitions.

**Definition 7** ($\iota$-entity). *Given a multivariate Markov chain* $\{X_i\}_{i \in V}$ *a pattern* $x_O$ *is a* $\iota$-entity *if*

$$\iota(x_O) > 0. \tag{24}$$

The entire set of $\iota$-entities $\mathfrak{E}_\iota(\{X_i\}_{i \in V})$ is then defined as follows.

**Definition 8** (*ci*-entity-set). *Given a multivariate Markov chain* $\{X_i\}_{i \in V}$ *the* $\iota$-entity-set *is the entity-set*

$$\mathfrak{E}_\iota(\{X_i\}_{i \in V}) := \{x_O \in \bigcup_{A \subseteq V} \mathcal{X}_A : \iota(x_O) > 0\}. \tag{25}$$

Next, we look at some interpretations that the introduction of the logarithm allows.

- A first consequence of introducing the logarithm is that we can now formulate the condition of Equation (24) analogously to an old phrase attributed to Aristotle that "the whole is more than the sum of its parts". In our case this would need to be changed to "the log-probability of the (spatiotemporal) whole is greater than the sum of the log-probabilities of its (spatiotemporal) parts". This can easily be seen by rewriting Equation (22) as:

$$\text{mi}_\pi(x_O) = \log p_O(x_O) - \sum_{b \in \pi} \log p_b(x_b). \tag{26}$$

- Another side effect of using the logarithm is that we can interpret Equation (24) in terms of the surprise value (also called information content) $-\log p_O(x_O)$ [36] of the pattern $x_O$ and the surprise value of its parts with respect to any partition $\pi$. Rewriting Equation (22) using properties of the logarithm we get:

$$\text{mi}_\pi(x_O) = \sum_{b \in \pi} (-\log p_b(x_b)) - (-\log p_O(x_O)).$$

Interpreting Equation (24) from this perspective we can then say that a pattern is an entity if the sum of the surprise values of its parts is larger than the surprise value of the whole.
- In coding theory, the Kraft-McMillan theorem [37] tells us that the optimal length (in a uniquely decodable binary code) of a codeword for an event $x$ is $l(x) = -\log p(x)$ if $p(x)$ is the *true* probability of $x$. If the encoding is not based on the true probability of $x$ but instead on a different probability $q(x)$ then the difference between the optimal codeword length and the chosen codeword length is

$$-\log q(x) - (-\log p(x)) = \log \frac{p(x)}{q(x)}. \tag{27}$$

Then we can interpret the specific local integration as a difference in codeword lengths. Say we want to encode what occurs at the nodes/random variables indexed by $O$, i.e., we encode the random variable $X_V$. We can encode every event (now a pattern) $x_O$ based on $p_O(x_O)$. Let's call this the *joint code*. Given a partition $\pi \in \mathfrak{L}(O)$ we can also encode every event $x_O$ based on its product probability $\prod_{b \in \pi_O} p_b(x_b)$. Let's call this the *product code with respect to* $\pi$. For a particular event $x_O$ the difference of the codeword lengths between the joint code and the product code with respect to $\pi$ is then just the specific local integration with respect to $\pi$.

Complete local integration then requires that the joint code codeword is shorter than all possible product code codewords. This means there is no partition with respect to which the product code for the pattern $x_O$ has a shorter codeword than the joint code. So $\iota$-entities are patterns that are shorter to encode with the joint code than a product code. Patterns that have a shorter codeword in a product code associated to a partition $\pi$ have negative SLI with respect to this $\pi$ and are therefore not $\iota$-entities.
- We can relate our measure of identity to other measures in information theory. For this we note that the expectation value of specific local integration with respect to a partition $\pi$ is the multi-information $\text{MI}_\pi(X_O)$ [9,10] with respect to $\pi$, i.e.,

$$\text{MI}_\pi(X_O) := \sum_{x_O \in \mathcal{X}_O} p_O(x_O) \log \frac{p_O(x_O)}{\prod_{b \in \pi} p_b(x_b)} \tag{28}$$

$$= \sum_{x_O \in \mathcal{X}_O} p_O(x_O) \, \text{mi}_\pi(x_O). \tag{29}$$

The multi-information plays a role in measures of complexity and information integration [35]. The generalisation from bipartitions to arbitrary partitions is applied to expectation values similar to the multi-information above in Tononi [38]. The relations of our localised measure (in the sense of [11]) to multi-information and information integration measures also motivates the name specific *local integration*. Relations to these measures will be studied further in the future. Here we note that these are not suited for measuring identity of patterns since they are properties of the random variables $X_O$ and not of patterns $x_O$. We also show in Corollary 2 that if $x_O$ is an $\iota$-entity that $X_O$ (the joint random variable) has a positive $\mathrm{MI}_\pi(X_O)$ for all partitions $\pi$ and is therefore a set of "integrated" random variables.

### 3.3. Properties of Specific Local Integration

This section investigates the specific local integration (SLI) (see Definition 5). After giving its expression for deterministic systems it proves upper bounds constructively and constructs an example of negative SLI.

#### 3.3.1. Deterministic Case

**Theorem 2** (Deterministic specific local integration). *Given a deterministic Bayesian network (Definition A10), a uniform initial distribution over $X_{V_0}$ ($V_0$ is the set of nodes without parents), and a pattern $x_O$ with $O \subseteq V$ the SLI of $x_O$ with respect to partition $\pi$ can be expressed more specifically: Let $N(x_O)$ refer to the number of trajectories in which $x_O$ occurs. Then*

$$\mathrm{mi}_\pi(x_O) = (|\pi| - 1) \log |\mathcal{X}_{V_0}| + \log N(x_O) - \sum_{b \in \pi} \log N(x_b). \tag{30}$$

**Proof.** See Appendix C.2. □

The first term in Equation (30) is always positive if the partition and the set of random variables are not trivial (i.e., have cardinality larger than one) and is a constant for partitions of a given cardinality. The second term is also always non-negative for patterns $x_O$ that actually occur in the system and rises with the number of trajectories that lead to it. The third term is always non-positive and becomes more and more negative the higher the number of trajectories that lead to the parts of the pattern occurring.

This shows that to maximise SLI for fixed partition cardinality we need to find patterns that have a high number of trajectories leading to them and a low number of occurrences for all their parts. Since the number of occurrences of the parts cannot be lower than the number of occurrences of the whole, we should get a maximum SLI for patterns whose parts occur only if the whole occurs. This turns out to be true also for the non-deterministic systems as we prove in Theorem 4.

Conversely, if we can increase the number of occurrences of the parts of the pattern without increasing the occurrences of the whole pattern occurring we minimise the SLI. This leads to the intuition that as often as possible as many parts as possible (i.e., all but one) should co-occur without the whole pattern occurring. This consistently leads to negative SLI as we will show for the non-deterministic case in Theorem 5.

#### 3.3.2. Upper Bounds

In this section we present the upper bounds of SLI. These are of general interest, but the constructive proof also provides an intuition for what kind of patterns have large SLI.

We first show constructively that if we can choose the Bayesian network and the pattern then SLI can be arbitrary large. This construction sets the probabilities of all blocks equal to the probability of the pattern and implies that each of the parts of the pattern occurs only if the entire pattern occurs. The simplest example is one binary random variable determining another to always be in the same state, then the two patterns with both variables equal have this property. In the subsequent theorem we show that this property in general gives the upper bound of SLI if the cardinality of the partition is fixed. A simple extension of this example is used in the proof of the least upper bound. First we prove that there are Bayesian networks that achieve a particular SLI value. This will be used in the proofs that follow. For this we first define the anti-patterns which are patterns that differ to a given pattern at every random variable that is specified.

**Definition 9** (Anti-pattern). *Given a pattern $x_O$ define its set of anti-patterns $\neg(x_O)$ that have values different from those of $x_O$ on all variables in $O$:*

$$\neg(x_O) := \{\tilde{x}_O \in \mathcal{X}_O : \forall i \in O, \tilde{x}_i \neq x_i\}. \tag{31}$$

Remark:

- It is important to note that for an element of $\neg(x_O)$ to occur it is not sufficient that $x_O$ does not occur. Only if *every* random variable $X_i$ with $i \in O$ differs from the value $x_i$ specified by $x_O$ does an element of $\neg(x_O)$ necessarily occur. This is why we call $\neg(x_O)$ the anti-pattern of $x_O$.

**Theorem 3** (Construction of a pattern with maximum SLI). *Given a probability $q \in (0,1)$ and a positive natural number $n$ there is a Bayesian network $\{X_i\}_{i \in V}$ with $|V| \geq n$ and a pattern $x_O$ such that*

$$\mathrm{mi}_\pi(x_O) = -(n-1)\log q. \tag{32}$$

**Proof.** We construct a Bayesian network which realises two conditions on the probability $p_O$. From these two conditions (which can also be realised by other Bayesian networks) we can then derive the theorem.

Choose a Bayesian network $\{X_i\}_{i \in V}$ with binary random variables $\mathcal{X}_i = \{0,1\}$ for all $i \in V$. Choose all nodes in $O$ dependent only on node $j \in O$, the dependence of the nodes in $V \setminus O$ is arbitrary:

- for all $i \in O \subset V$ let $\mathrm{pa}(i) \cap (V \setminus O) = \emptyset$, i.e., nodes in $O$ have no parents in the complement of $O$,
- for a specific $j \in O$ and all other $i \in O \setminus \{j\}$ let $\mathrm{pa}(i) = \{j\}$, i.e., all nodes in $O$ apart from $j$ have $j \in O$ as a parent,
- for all $i \in O \setminus \{j\}$ let $p_i(\tilde{x}_i|b\tilde{x}_j) = \delta_{\tilde{x}_j}(\tilde{x}_i)$, i.e., the state of all nodes in $O$ is always the same as the state of node $j$,
- also choose $p_j(x_j) = q$ and $\sum_{\tilde{x}_j \neq x_j} p_j(x_j) = 1 - q$.

Then it is straightforward to see that:

1. $p_O(x_O) = q$,
2. $\sum_{\tilde{x}_O \in \neg(x_O)} p_O(\tilde{x}_O) = 1 - q$.

Note that there are many Bayesian networks that realise the latter two conditions for some $x_O$. These latter two conditions are the only requirements for the following calculation.

Next note that the two conditions imply that $p_O(\tilde{x}_O) = 0$ if neither $\tilde{x}_O = x_O$ nor $\tilde{x}_O \in \neg(x_O)$. Then for every partition $\pi$ of $O$ with $|\pi| = n$ and $n > 1$ we have

$$\mathrm{mi}_\pi(x_O) = \log \frac{p_O(x_O)}{\prod_{b \in \pi} p_b(x_b)} \tag{33}$$

$$= \log \frac{p_O(x_O)}{\prod_{b \in \pi} \sum_{\tilde{x}_{O \setminus b}} p_O(x_b, \tilde{x}_{O \setminus b})} \tag{34}$$

$$= \log \frac{p_O(x_O)}{\prod_{b \in \pi} \left( p_O(x_O) + \sum_{\tilde{x}_{O \setminus b} \neq x_{O \setminus b}} p_O(x_b, \tilde{x}_{O \setminus b}) \right)} \tag{35}$$

$$= \log \frac{p_O(x_O)}{\prod_{b \in \pi} p_O(x_O)} \tag{36}$$

$$= \log \frac{p_O(x_O)}{p_O(x_O)^n} \tag{37}$$

$$= -(n-1)\log q. \tag{38}$$

$\square$

**Theorem 4** (Upper bound of SLI). *For any Bayesian network* $\{X\}_{i \in V}$ *and pattern* $x_O$ *with fixed* $p_O(x_O) = q$

1.  *The tight upper bound of the SLI with respect to any partition* $\pi$ *with* $|\pi| = n$ *fixed is*

$$\max_{\{\{X_i\}_{i \in V}: \exists x_O, p_O(x_O) = q\}} \max_{\{\pi: |\pi| = n\}} mi_\pi(x_O) \leq -(n-1) \log q. \tag{39}$$

2.  *The upper bound is achieved if and only if for all* $b \in \pi$ *we have*

$$p_b(x_b) = p_O(x_O) = q. \tag{40}$$

3.  *The upper bound is achieved if and only if for all* $b \in \pi$ *we have that* $x_b$ *occurs if and only if* $x_O$ *occurs.*

**Proof. ad 1** By Definition 5 we have

$$mi_\pi(x_O) = \log \frac{p_O(x_O)}{\prod_{b \in \pi} p_b(x_b)}. \tag{41}$$

Now note that for any $x_O$ and $b \subseteq O$

$$p_b(x_b) = \sum_{\tilde{x}_{O \backslash b}} p_O(x_b, \tilde{x}_{O \backslash b}) \tag{42}$$

$$= p_O(x_O) + \sum_{\tilde{x}_{O \backslash b} \neq x_{O \backslash b}} p_O(x_b, \tilde{x}_{O \backslash b}) \tag{43}$$

$$\geq p_O(x_O). \tag{44}$$

Plugging this into Equation (41) for every $p_b(x_b)$ we get

$$mi_\pi(x_O) = \log \frac{p_O(x_O)}{\prod_{b \in \pi} p_b(x_b)} \tag{45}$$

$$\leq \log \frac{p_O(x_O)}{p_O(x_O)^{|\pi|}} \tag{46}$$

$$= -(|\pi| - 1) \log p_O(x_O). \tag{47}$$

This shows that $-(|\pi| - 1) \log p_O(x_O)$ is indeed an upper bound. To show that it is tight we have to show that for a given $p_O(x_O)$ and $|\pi|$ there are Bayesian networks with patterns $x_O$ such that this upper bound is achieved. The construction of such a Bayesian network and a pattern $x_O$ was presented in Theorem 3.

**ad 2** If for all $b \in \pi$ we have $p_b(x_b) = p_O(x_O)$ then clearly $mi_\pi(x_O) = -(|\pi| - 1) \log p_O(x_O)$ and the least upper bound is achieved. If on the other hand $mi_\pi(x_O) = -(|\pi| - 1) \log p_O(x_O)$ then

$$\log \frac{p_O(x_O)}{\prod_{b \in \pi} p_b(x_b)} = -(|\pi| - 1) \log p_O(x_O) \tag{48}$$

$$\Leftrightarrow \quad \log \frac{p_O(x_O)}{\prod_{b \in \pi} p_b(x_b)} = \log \frac{p_O(x_O)}{p_O(x_O)^{|\pi|}} \tag{49}$$

$$\Leftrightarrow \quad \prod_{b \in \pi} p_b(x_b) = p_O(x_O)^{|\pi|}, \tag{50}$$

and because $p_b(x_b) \geq p_O(x_O)$ (Equation (44)) any deviation of any of the $p_b(x_b)$ from $p_O(x_O)$ leads to $\prod_{b \in \pi} p_b(x_b) > p_O(x_O)^{|\pi|}$ such that for all $b \in \pi$ we must have $p_b(x_b) = p_O(x_O)$.

**ad 3** By definition for any $b \in \pi$ we have $b \subseteq O$ such that $x_b$ always occurs if $x_O$ occurs. Now assume $x_b$ occurs and $x_O$ does not occur. In that case there is a positive probability for a pattern $(x_b, \tilde{x}_{O \setminus b})$ with $\tilde{x}_{O \setminus b} \neq x_{O \setminus b}$ i.e., $p_O(x_b, \tilde{x}_{O \setminus b}) > 0$. Recalling Equation (43) we then see that

$$p_b(x_b) = p_O(x_O) + \sum_{\tilde{x}_{O \setminus b} \neq x_{O \setminus b}} p_O(x_b, \tilde{x}_{O \setminus b}) \tag{51}$$

$$> p_O(x_O). \tag{52}$$

which contradicts the fact that $p_b(x_b) = p_O(x_O)$ so $x_b$ cannot occur without $x_O$ occurring as well.
$\square$

Remarks:

- Note that this is the least upper bound for Bayesian networks in general. For a specific Bayesian network there might be no pattern that achieves this bound.
- The least upper bound of SLI increases with the improbability of the pattern and the number of parts that it is split into. If $p_O(x_O) \to 0$ then we can have $\min_\pi(x_O) \to \infty$.
- Using this least upper bound it is easy to see the least upper bound for the SLI of a pattern $x_O$ across all partitions $|\pi|$. We just have to note that $|\pi| \le |O|$.
- Since it is the minimum value of SLI with respect to arbitrary partitions the least upper bound of SLI is also an upper bound for CLI. It may not be the least upper bound however.

### 3.3.3. Negative SLI

This section shows that SLI of a pattern $x_O$ with respect to partition $\pi$ can be negative *independently* of the probability of $x_O$ (as long as it is not 1) and the cardinality of the partition (as long as that is not 1). The construction which achieves this also serves as an example of patterns with low SLI. We conjecture that this construction might provide the greatest lower bound but have not been able to prove this yet. An intuitive description of the construction is that patterns which either occur as a whole or missing exactly one part always have negative SLI.

**Theorem 5.** *For any given probability $q < 1$ and cardinality $|\pi| = n > 1$ of a partition $\pi$ there exists a Bayesian network $\{X_i\}_{i \in V}$ with a pattern $x_O$ such that $q = p_O(x_O)$ and*

$$\min_\pi(x_O) = \log \frac{q}{\left(1 - \frac{1-q}{n}\right)^n} < 0. \tag{53}$$

**Proof.** We construct the probability distribution $p_O : \mathcal{X}_O \to [0, 1]$ and ignore the behaviour of the Bayesian network $\{X_i\}_{i \in V}$ outside of $O \subseteq V$. In any case $\{X_i\}_{i \in O}$ is also by itself a Bayesian network. We define (see remarks below for some intuitions behind these definitions and Definition 9 for $\neg(x_A)$):

1. for all $i \in O$ let $|\mathcal{X}_i| = n$
2. for every block $b \in \pi$ let $|b| = \frac{|O|}{|\pi|}$,
3. for $\tilde{x}_O \in \mathcal{X}_O$ let:

$$p_O(\tilde{x}_O) := \begin{cases} q & \text{if } \tilde{x}_O = x_O, \\ \frac{1-q-d}{\sum_{b \in \pi} |\neg(x_b)|} & \text{if } \exists c \in \pi \text{ s.t. } \tilde{x}_{O \setminus c} = x_{O \setminus c} \land \tilde{x}_c \neq x_c, \\ \frac{d}{|\neg(x_O)|} & \text{if } \tilde{x}_O \in \neg(x_O), \\ 0 & \text{else.} \end{cases} \tag{54}$$

Here $d$ parameterises the probability of any pattern in $\neg(x_O)$ occurring. We will carry it through the calculation but then end up setting it to zero.

Next we calculate the SLI. First note that, according to 1. and 2., we have $|\mathcal{X}_b| = |\mathcal{X}_c|$ for all $b, c \in \pi$ and therefore also $|\neg(x_b)| = |\neg(x_c)|$ for all $b, c \in \pi$. So let $m := |\neg(x_b)|$. Then note that, according to 3, for all $b \in \pi$

$$\sum_{\tilde{x}_{O\backslash b} \neq x_{O\backslash b}} p_O(x_b, \tilde{x}_{O\backslash b}) = \sum_{c \in \pi \backslash b} \sum_{\tilde{x}_c \neq x_c} p_O(x_b, x_{O\backslash(b \cup c)}, \tilde{x}_c) \tag{55}$$

$$= \sum_{c \in \pi \backslash b} \sum_{\tilde{x}_c \neq x_c} \frac{1 - q - d}{\sum_{b \in \pi} |\neg(x_b)|} \tag{56}$$

$$= \sum_{c \in \pi \backslash b} \sum_{\tilde{x}_c \neq x_c} \frac{1 - q - d}{m|\pi|} \tag{57}$$

$$= \sum_{c \in \pi \backslash b} \frac{1 - q - d}{m|\pi|} |\neg(x_c)| \tag{58}$$

$$= \frac{|\pi| - 1}{|\pi|} (1 - q - d) \tag{59}$$

Plug this into the SLI definition:

$$mi_\pi(x_O) = \log \frac{p_O(x_O)}{\prod_{b \in \pi} p_b(x_b)} \tag{60}$$

$$= \log \frac{q}{\prod_{b \in \pi} q + \sum_{\tilde{x}_{O\backslash b} \neq x_{O\backslash b}} p_O(x_b, \tilde{x}_{O\backslash b})} \tag{61}$$

$$= \log \frac{q}{\prod_{b \in \pi} q + \frac{|\pi| - 1}{|\pi|} (1 - q - d)} \tag{62}$$

$$= \log \frac{q}{\left( q + \frac{|\pi| - 1}{|\pi|} (1 - q - d) \right)^{|\pi|}} \tag{63}$$

If we now set $d = 0$ we get:

$$mi_\pi(x_O) = \log \frac{q}{\left( 1 - \frac{1 - q}{|\pi|} \right)^{|\pi|}}. \tag{64}$$

Then we can use Bernoulli's inequality (The authors thank von Eitzen [39] for pointing this out. An example reference for Bernoulli's inequality is Bullen [40]). to prove that this is negative for $0 < q < 1$ and $|\pi| \geq 2$. Bernoulli's inequality is

$$(1 + x)^n \geq 1 + nx \tag{65}$$

for $x \geq -1$ and $n$ a natural number. Replacing $x$ by $-(1 - q)/|\pi|$ we see that

$$\left( 1 - \frac{1 - q}{|\pi|} \right)^{|\pi|} > q \tag{66}$$

such that the argument of the logarithm is smaller than one which gives us negative SLI.  □

Remarks:

- The achieved value in Equation (53) is also our best candidate for a greatest lower bound of SLI for given $p_O(x_O)$ and $|\pi|$. However, we have not been able to prove this yet.
- The construction equidistributes the probability $1 - q$ (left to be distributed after the probability $q$ of the whole pattern occurring is chosen) to the patterns $\tilde{x}_O$ that are *almost* the same as the pattern

$x_O$. These are almost the same in a precise sense: They differ in exactly one of the blocks of $\pi$, i.e., they differ by as little as can possibly be resolved/revealed by the partition $\pi$.

- In order to achieve the negative SLI of Equation (64) the requirement is only that Equation (59) is satisfied. Our construction shows one way how this can be achieved.
- For a pattern and partition such that $|O|/|\pi|$ is not a natural number, the same bound might still be achieved however a little extra effort has to go into the construction 3. of the proof such that Equation (59) still holds. This is not necessary for our purpose here as we only want to show the existence of patterns achieving the negative value.
- Since it is the minimum value of SLI with respect to arbitrary partitions the candidate for the greatest lower bound of SLI is also a candidate for the greatest lower bound of CLI.

### 3.4. Disintegration

In this section we define the disintegration hierarchy and its refinement-free version. We then prove the disintegration theorem which is the main formal result of this paper. It exposes a connection between partitions minimising the SLI of a trajectory and the CLI of the blocks of such partitions. More precisely for a given trajectory the blocks of the *finest* partitions among those leading to a particular value of SLI consist only of completely locally integrated blocks. Conversely, *each* completely locally integrated pattern is a block in such a finest partition leading to a particular value of SLI. The theorem therefore reveals that $\iota$-entities can not only be motivated heuristically as we tried to do in Section 3.2 but in fact play a special role within the trajectories they occur in. Furthermore, this theorem allows additional interpretations of the $\iota$-entities which will be discussed in Section 3.5.

The main tool we use for the proof, the disintegration hierarchy and especially its refinement free version are also interesting structure in their own right since they define a hierarchy among the partitions of trajectories that we did not anticipate. In the case of the refinement free version the disintegration theorem tells us that this hierarchy among partitions of trajectories turns out to be a hierarchy of splits of the trajectory into *ci*-entities.

**Definition 10** (Disintegration hierarchy). *Given a Bayesian network* $\{X_i\}_{i \in V}$ *and a trajectory* $x_V \in \mathcal{X}_V$, *the disintegration hierarchy of* $x_V$ *is the set* $\mathfrak{D}(x_V) = \{\mathfrak{D}_1, \mathfrak{D}_2, \mathfrak{D}_3, ...\}$ *of sets of partitions of* $x_V$ *with:*

1.
$$\mathfrak{D}_1(x_V) := \underset{\pi \in \mathcal{L}(V)}{\arg\min} \operatorname{mi}_\pi(x_V) \tag{67}$$

2. *and for* $i > 1$:
$$\mathfrak{D}_i(x_V) := \underset{\pi \in \mathcal{L}(V) \setminus \mathfrak{D}_{\prec i}(x_V)}{\arg\min} \operatorname{mi}_\pi(x_V). \tag{68}$$

*where* $\mathfrak{D}_{\prec i}(x_V) := \bigcup_{j < i} \mathfrak{D}_j(x_V)$. *We call* $\mathfrak{D}_i(x_V)$ *the i-th disintegration level.*

Remark:

- Note that arg min returns all partitions that achieve the minimum SLI if there is more than one.
- Since the Bayesian networks we use are finite, the partition lattice $\mathcal{L}(V)$ is finite, the set of attained SLI values is finite, and the number $|\mathfrak{D}|$ of disintegration levels is finite.
- In most cases the Bayesian network contains some symmetries among their mechanisms which cause multiple partitions to attain the same SLI value.
- For each trajectory $x_V$ the disintegration hierarchy $\mathfrak{D}$ then partitions the elements of $\mathcal{L}(V)$ into subsets $\mathfrak{D}_i(x_V)$ of equal SLI. The levels of the hierarchy have increasing SLI.

**Definition 11.** *Let* $\mathcal{L}(V)$ *be the lattice of partitions of set* $V$ *and let* $\mathfrak{E}$ *be a subset of* $\mathcal{L}(V)$. *Then for every element* $\pi \in \mathcal{L}(V)$ *we can define the set*

$$\mathfrak{E}_{\lhd \pi} := \{\zeta \in \mathfrak{E} : \zeta \lhd \pi\}. \tag{69}$$

*That is $\mathfrak{E}_{\lhd\pi}$ is the set of partitions in $\mathfrak{E}$ that are refinements of $\pi$.*

**Definition 12** (Refinement-free disintegration hierarchy). *Given a Bayesian network $\{X_i\}_{i\in V}$, a trajectory $x_V \in \mathcal{X}_V$, and its disintegration hierarchy $\mathfrak{D}(x_V)$ the refinement-free disintegration hierarchy of $x_V$ is the set $\mathfrak{D}^{\blacktriangleleft}(x_V) = \{\mathfrak{D}_1^{\blacktriangleleft}, \mathfrak{D}_2^{\blacktriangleleft}, \mathfrak{D}_3^{\blacktriangleleft}, ...\}$ of sets of partitions of $x_V$ with:*

1.

$$\mathfrak{D}_1^{\blacktriangleleft}(x_V) := \{\pi \in \mathfrak{D}_1(x_V) : \mathfrak{D}_1(x_V)_{\lhd\pi} = \varnothing\}, \tag{70}$$

2. *and for $i > 1$:*

$$\mathfrak{D}_i^{\blacktriangleleft}(x_V) := \{\pi \in \mathfrak{D}_i(x_V) : \mathfrak{D}_{\prec i}(x_V)_{\lhd\pi} = \varnothing\} \tag{71}$$

Remark:

- Each level $\mathfrak{D}_i^{\blacktriangleleft}(x_V)$ in the refinement-free disintegration hierarchy $\mathfrak{D}^{\blacktriangleleft}(x_V)$ consists only of those partitions that neither have refinements at their own nor at any of the preceding levels. So each partition that occurs in the refinement-free disintegration hierarchy at the $i$-th level is a finest partition that achieves such a low level of SLI or such a high level of disintegration.
- As we will see below, the blocks of the partitions in the refinement-free disintegration hierarchy are the main reason for defining the refinement-free disintegration hierarchy.

**Theorem 6** (Disintegration theorem). *Let $\{X_i\}_{i\in V}$ be a Bayesian network, $x_V \in \mathcal{X}_V$ one of its trajectories, and $\mathfrak{D}^{\blacktriangleleft}(x_V)$ the associated refinement-free disintegration hierarchy.*

1. *Then for every $\mathfrak{D}_i^{\blacktriangleleft}(x_V) \in \mathfrak{D}^{\blacktriangleleft}(x_V)$ we find for every $b \in \pi$ with $\pi \in \mathfrak{D}_i^{\blacktriangleleft}(x_V)$ that there are only the following possibilities:*

   (a) *$b$ is a singleton, i.e., $b = \{i\}$ for some $i \in V$, or*
   (b) *$x_b$ is completely locally integrated, i.e., $\iota(x_b) > 0$.*

2. *Conversely, for any completely locally integrated pattern $x_A$, there is a partition $\pi^A \in \mathcal{L}(V)$ and a level $\mathfrak{D}_{iA}^{\blacktriangleleft}(x_V) \in \mathfrak{D}^{\blacktriangleleft}(x_V)$ such that $A \in \pi^A$ and $\pi^A \in \mathfrak{D}_{iA}^{\blacktriangleleft}(x_V)$.*

**Proof. ad 1** We prove the theorem by contradiction. For this assume that there is block $b$ in a partition $\pi \in \mathfrak{D}_i^{\blacktriangleleft}(x_V)$ which is neither a singleton nor completely integrated. Let $\pi \in \mathfrak{D}_i^{\blacktriangleleft}(x_V)$ and $b \in \pi$. Assume $b$ is not a singleton i.e., there exist $i \neq j \in V$ such that $i \in b$ and $j \in b$. Also assume that $b$ is not completely integrated i.e., there exists a partition $\xi$ of $b$ with $\xi \neq 1_b$ such that $\text{mi}_\xi(x_b) \leq 0$. Note that a singleton cannot be completely locally integrated as it does not allow for a non-unit partition. So together the two assumptions imply $p_b(x_b) \leq \prod_{d\in\xi} p_d(x_d)$ with $|\xi| > 1$. However, then

$$\text{mi}_\pi(x_V) = \log \frac{p_V(x_V)}{p_b(x_b) \prod_{c\in\pi\setminus b} p_c(x_c)} \tag{72}$$

$$\geq \log \frac{p_V(x_V)}{\prod_{d\in\xi} p_d(x_d) \prod_{c\in\pi\setminus b} p_c(x_c)} \tag{73}$$

We treat the cases of ">" and "=" separately. First, let

$$\text{mi}_\pi(x_V) = \log \frac{p_V(x_V)}{\prod_{d\in\xi} p_d(x_d) \prod_{c\in\pi\setminus b} p_c(x_c)}. \tag{74}$$

Then we can define $\rho := (\pi \setminus b) \cup \xi$ such that

1. $\text{mi}_\rho(x_V) = \text{mi}_\pi(x_V)$ which implies that $\rho \in \mathfrak{D}_i(x_V)$ because $\pi \in \mathfrak{D}_i(x_V)$, and
2. $\rho \lhd \pi$ which contradicts $\pi \in \mathfrak{D}_i^{\blacktriangleleft}(x_V)$.

Second, let

$$\text{mi}_\pi(x_V) > \log \frac{p_V(x_V)}{\prod_{d \in \zeta} p_d(x_d) \prod_{c \in \pi \setminus b} p_c(x_c)}. \tag{75}$$

Then we can define $\rho := (\pi \setminus b) \cup \zeta$ such that

$$\text{mi}_\rho(x_V) < \text{mi}_\pi(x_V), \tag{76}$$

which contradicts $\pi \in \mathfrak{D}_i^\blacktriangleleft(x_V)$.

**ad 2** By assumption $x_A$ is completely locally integrated. Then let $\pi^A := \{A\} \cup \{\{j\}\}_{j \in V \setminus A}$. Since $\pi^A$ is a partition of $V$ it is an element of some disintegration level $\mathfrak{D}_{i^A}$. Then partition $\pi^A$ is also an element of the refinement-free disintegration level $\mathfrak{D}_{i^A}^\blacktriangleleft(x_V)$ as we will see in the following. This is because any refinements must (by construction of $\pi^A$ break up $A$ into further blocks which means that the local specific integration of all such partitions is higher. Then they must be at lower disintegration level $\mathfrak{D}_k(x_V)$ with $k \geq i^A$. Therefore, $\pi^A$ has no refinement at its own or a higher disintegration level. More formally, let $\zeta \in \mathfrak{L}(V), \zeta \neq \pi^A$ and $\zeta \lhd \pi^A$ since $\pi^A$ only contains singletons apart from $A$ the partition $\zeta$ must split the block $A$ into multiple blocks $c \in \zeta|_A$. Since $\iota(x_A) > 0$ we know that

$$\text{mi}_{\zeta|_A}(x_A) = \log \frac{p_A(x_A)}{\prod_{c \in \zeta|_A} p_c(x_c)} > 0 \tag{77}$$

so that $\prod_{c \in \zeta|_A} p_c(x_c) < p_A(x_A)$ and

$$\text{mi}_\zeta(x_V) = \log \frac{p_V(x_V)}{\prod_{c \in \zeta|_A} p_c(x_c) \prod_{i \in V \setminus A} p_i(x_i)} \tag{78}$$

$$> \log \frac{p_V(x_V)}{p_A(x_A) \prod_{i \in V \setminus A} p_i(x_i)} \tag{79}$$

$$= \text{mi}_{\pi^A}(x_V). \tag{80}$$

Therefore $\zeta$ is on a disintegration level $\mathfrak{D}_k(x_V)$ with $k > i^A$, but this is true for any refinement of $\pi^A$ so $\mathfrak{D}_{\lhd i^A}(x_V)_{\lhd \pi^A} = \emptyset$ and $\pi^A \in \mathfrak{D}_{i^A}^\blacktriangleleft(x_V)$.  □

We mentioned in Section 3.2 that the expectation value of SLI $\text{mi}_\pi(x_A)$ is the (specific) multi-information $\text{MI}_\pi(X_A)$. A positive SLI value of $x_A$ implies a positive expectation value $\text{MI}_\pi(X_A)$. Therefore every $\iota$-entity $x_A$ implies positive specific multi-informations $\text{MI}_\pi(X_A)$ with respect to any partition $\pi$. We put this into the following corollary.

**Corollary 2.** *Under the conditions of Theorem 6 and for every $\mathfrak{D}_i^\blacktriangleleft(x_V) \in \mathfrak{D}^\blacktriangleleft(x_V)$ we find for every $b \in \pi$ with $\pi \in \mathfrak{D}_i^\blacktriangleleft(x_V)$ that there are only the following possibilities:*

1. *$b$ is a singleton, i.e., $b = \{i\}$ for some $i \in V$, or*
2. *$X_b$ is completely (not only locally) integrated, i.e., $\text{I}(X_b) > 0$.*

*here*

$$\text{I}(X_A) := \min_{\pi \in \mathfrak{L}(A) \setminus 0} \text{MI}_\pi(X_A). \tag{81}$$

**Proof.** Since $\text{MI}_\pi(X_A)$ is a Kullback–Leibler divergence we know from Gibbs' inequality that $\text{MI}_\pi(X_A) \geq 0$ and $\text{MI}_\pi(X_A) = 0$ if and only if for all $x_A \in \mathcal{X}_A$ we have $p_A(x_A) = \prod_{b \in \pi} p_b(x_b)$. To see that $\text{MI}_\pi(X_A)$ is a Kullback–Leibler divergence note:

$$\mathrm{MI}_\pi(X_A) := \sum_{x_A \in \mathcal{X}_A} p_A(x_A) \, \mathrm{mi}_\pi(x_A) \tag{82}$$

$$= \sum_{x_A \in \mathcal{X}_A} p_A(x_A) \log \frac{p_A(x_A)}{\prod_b p_b(x_b)} \tag{83}$$

$$= \mathrm{KL}[p_A || \prod_{b \in \pi} p_b]. \tag{84}$$

Now let a specific $x_A \in \mathcal{X}_A$ be a $\iota$-entity. Then for all $\pi \in \mathfrak{L}(A) \setminus \mathbf{0}$ we have

$$\log \frac{p_A(x_A)}{\prod_b p_b(x_b)} > 0, \tag{85}$$

which implies that

$$p_A(x_A) \neq \prod_b p_b(x_b) \tag{86}$$

and therefore

$$\mathrm{KL}[p_A || \prod_{b \in \pi} p_b] > 0 \tag{87}$$

which implies $\mathrm{I}(X_A) > 0$. $\square$

### 3.5. Disintegration Interpretation

In Section 3.2 we motivated our choice of positive complete local integration as a criterion for entities. This motivation is purely heuristic and starts from the intuition that an entity is a structure for which every part makes every other part more probable. While this heuristic argument seems sufficiently intuitive to be of a certain value we would much rather have a formal reason why an entity criterion is a "good" entity criterion. In other words we would ideally have a formal problem that is best solved by the entities satisfying the criterion. An example of a measure that has such an associated interpretation is the mutual information whose maximum over the input distributions is the channel capacity. Without a formal problem associated to $\iota$-entities there remains a risk that they (and maybe the whole concept of entities and identity over time) are artefacts of an ill-conceived conceptual approach.

Currently, we are not aware of an analogous formal problem that is solved by $\iota$-entities. However, the different viewpoint provided by the disintegration theorem may be a first step towards finding such a problem. We will now discuss some alternative interpretations of SLI and see how CLI can be seen from a different perspective due to the disintegration theorem. These interpretations also exhibit why we chose to include the logarithm into the definition of SLI.

Using the disintegration theorem (Theorem 6) allows us to take another point of view on $\iota$-entities. The theorem states that for each trajectory $x_V \in \mathcal{X}_V$ of a multivariate Markov chain the refinement-free disintegration hierarchy only contains partitions whose blocks are completely integrated patterns i.e., they only contain $\iota$-entities. At the same time the blocks of all those partitions together are *all* $\iota$-entities that occur in that trajectory.

A partition in the refinement-free disintegration hierarchy is always a minimal/finest partition reaching such a low specific local integration.

Each $\iota$-entity is then a block $x_c$ with $c \in \pi$ of a partition $\pi \in \mathfrak{D}^{\blacktriangleleft}(x_V)$ for some trajectory $x_V \in \mathcal{X}_V$ of the multivariate Markov chain.

Let us recruit the interpretation from coding theory above. If we want to find the optimal encoding for the entire multivariate Markov chain $\{X_i\}_{i \in V}$ this means finding the optimal encoding for the random variable $X_V$ whose values are the trajectories $x_V \in \mathcal{X}_V$. The optimal code has the codeword lengths $-\log p_V(x_V)$ for each trajectory $x_V$. The partitions in the lowest level $\mathfrak{D}_1^{\blacktriangleleft}(x_V)$ in the refinement-free disintegration hierarchy for $x_V$ have minimal specific local integration i.e.,

$$\mathrm{mi}_\pi(x_V) = \log \frac{p_V(x_V)}{\prod_{c\in\pi} p_c(x_c)} \tag{88}$$

is minimal among all partitions. At the same time these partitions are the finest partitions that achieve this low specific local integration. This implies on the one hand that the codeword lengths of the product codes associated to these partitions are the shortest possible for $x_V$ among all partitions. On the other hand these partitions split up the trajectory in as many parts as possible while generating these shortest codewords. In this combined sense the partitions in $\mathfrak{D}_1^{\blacktriangleleft}(x_V)$ generate the "best" product codes for the particular trajectory $x_V$.

Note that the *expected codeword length* of the product code:

$$\sum_{x_V\in\mathcal{X}_V} p_V(x_V)(-\log \prod_{c\in\pi} p_c(x_c)) \tag{89}$$

which is the more important measure for encoding in general, might not be short at all, i.e., it might not be an efficient code for arbitrary trajectories. The product codes based on partitions in $\mathfrak{D}_1^{\blacktriangleleft}(x_V)$ are specifically adapted to assign a short codeword to $x_V$, i.e., to a single trajectory or story of this system. As product codes they are constructed/forced to describe $x_V$ as a composition of stochastically independent parts. More precisely they are constructed in the way that would be optimal for stochastically independent parts.

Nonetheless, the product codes exist (they can be generated using Huffman coding or arithmetic coding [37] based on the product probability) and are uniquely decodable. The parts/blocks of them are the $\iota$-entities. We mentioned before that we would like to find a problem that is solved by $\iota$-entities. This is then equivalent to finding a problem that is solved by the according product codes. Can we construct such a problem? This question is still open. A possible direction for finding such a problem may be the following line of reasoning. Say for some reason the trajectory $x_V$ is more important than any other and that we want to "tell its story" as a story of as many as possible (stochastically) independent parts (that are maybe not really stochastically independent) i.e., say we wanted to encode the trajectory *as if it were* a combination of as many as possible stochastically independent parts/events. And because $x_V$ is more important than all other trajectories we wanted the codeword for $x_V$ to be the shortest possible. Then we would use the product codes of partitions in the refinement-free disintegration hierarchy because those combine exactly these two conditions. The pseudo-stochastically-independent parts would then be the blocks of these partitions which according to the disintegration theorem are exactly the $\iota$-entities occurring in $x_V$.

Speculating about where the two conditions may arise in an actual problem, we mention that the trajectory/history that we (real living humans) live in is more important to us than all other possible trajectories of our universe (if there are any). What happens/happened in this trajectory needs to be communicated more often than what happens/happened in counterfactual trajectories. Furthermore, a good reason to think of a system as composite of as many parts as possible is that this reduces the number of parameters that need to be learned which in turn improves the learning speed (see e.g., [41]). So the entities that mankind has partitioned its history into might be related to the $\iota$-entities of the universe's history. These would compose the shortest product codes for what actually happened. The disintegration level might be chosen to optimise rates of model learning.

Recall that this kind of product code is not the optimal code in general (which would be the one with shortest expected codeword length). It is possibly more of a naive code that does not require deep understanding of the dynamical system but instead can be learned fast and works. The language of physics for example might be more optimal in the sense of shortest expected codeword lengths reflecting a desire to communicate efficiently about all counterfactual possibilities as well.

*3.6. Related Approaches*

We now discuss in some more detail than in Section 1.3 the approaches of Beer [14] and Balduzzi [26].

In Beer [14] the construction of the entities proceeds roughly as follows. First the maps from the Moore neighbourhood to the next state of a cell are classified into five classes of *local processes*. Then these are used to reveal the dynamical structure in the transitions from one time-slice (or temporal part) of a pattern to the next. The used example patterns are the famous block, blinker, and glider and they are considered including their temporal extension. Using both the processes and the spatial patterns/values/components (the black and white values of cells are called components) networks characterising the organisation of the spatiotemporally extended patterns are constructed. These can then be investigated for their *organisational closure*. Organisational closure occurs if the same process-component relations reoccur at a later time. Boundaries of the spatiotemporal patterns are identified by determining the cells around the pattern that have to be fixed to get re-occurrence of the organisation.

Beer [14] mentions that the current version of this method of identifying entities has its limitations. If the closure is perturbed or delayed and then recovered the entity still looses its identity according to this definition. Two possible alternatives are also suggested. The first is to define the *potential for closure* as enough for the ascription of identity. This is questioned as well since a sequence of perturbations can take the entity further and further away from its "defining" organisation and make it hard to still speak of a defining organisation at all. The second alternative is to define that the persistence of any organisational closure indicates identity. It is suggested that this would allow blinkers to transform to gliders.

We note that using the entity criterion we propose does not need similar choices to be made since it is not based on the re-occurrence of any organisation. Later time-slices of $\iota$-entities need no organisational (or any other) similarity to earlier ones. Another, possibly only small, advantage is that our criterion is formalised and reasonably simply to state. Whether this is possible for the organisational closure based entities remains to be seen.

This is related to the philosophical discussion about identity across possible worlds [33].

Some further parallels can be drawn between the present work and Balduzzi [26] especially if we take into account the disintegration theorem. Given a trajectory (entire time-evolution) of the system in both cases a partition is sought which fulfills a particular trajectory-wide optimality criterion. Also in both cases, each block of the trajectory-wide partition fulfills a condition with respect to its own partitions. For our conditions the disintegration theorem exposes the direct connection between the trajectory-wide and the block-specific conditions. Such a connection is not known for other approaches. The main reason for this might be the simpler formal expression of CLI and SLI compared to the IIT approaches.

In how far our approach and the IIT approaches lead to coinciding or contradicting results is beyond the scope of this paper and constitutes future work. One avenue to pursue here are differences with respect to entities occurring in multiple trajectories as well as the possibility of overlapping entities within single trajectories.

## 4. Examples

In this section we investigate the structure of integrated and completely locally integrated spatiotemporal patterns as it is revealed by the disintegration hierarchy. First we take a quick look at the trivial case of a set of independent random variables. Then we look at two very simple multivariate Markov chains. We use the disintegration theorem (Theorem 6) to extract the completely locally integrated spatiotemporal patterns.

*4.1. Set of Independent Random Variables*

Let us first look at a set $\{X_i\}_{i \in V}$ of independently and identically distributed random variables. For each trajectory $x_V \in \mathcal{X}_V$ we can then calculate SLI with respect to a partition $\pi \in \mathfrak{L}(V)$. For every $A \subseteq V$ and every $x_A \in \mathcal{X}_A$ we have $p_A(x_A) = \prod_{i \in A} p_i(x_i)$. Then we find for every $\pi \in \mathfrak{L}(V)$:

$$\mathrm{mi}_\pi(x_V) = 0. \tag{90}$$

This shows that the disintegration hierarchy for each $x_V \in \mathcal{X}_V$ contains only a single disintegration level $\mathfrak{D}(x_V) = \{\mathfrak{D}_1\}$ with $\mathfrak{D}_1 = \mathfrak{L}(V)$. The finest partition of $\mathfrak{L}(V)$ is its zero element $\mathbf{0}$ which then constitutes the only element of the refinement-free disintegration level $\mathfrak{D}_1^{\blacktriangleleft} = \{\mathbf{0}\}$. Recall that the zero element of a partition lattice only consists of singleton sets as blocks. The set of completely locally integrated patterns i.e., the set of $\iota$-entities in a given trajectory $x_V$ is then the set $\{x_i : i \in V\}$.

Next we will look at more structured systems.

*4.2. Two Constant and Independent Binary Random Variables: $MC^=$*

4.2.1. Definition

Define the time- and space-homogeneous multivariate Markov chain $MC^=$ with Bayesian network $\{X_{j,t}\}_{j \in \{1,2\}, t \in \{0,1,2\}}$ and

$$\mathrm{pa}(j,t) = \begin{cases} \varnothing & \text{if } t = 0, \\ \{(j,t-1)\} & \text{else,} \end{cases} \tag{91}$$

$$p_{j,t}(x_{j,t}|x_{j,t-1}) = \delta_{x_{j,t-1}}(x_{j,t}) = \begin{cases} 1 & \text{if } x_{j,t} = x_{j,t-1}, \\ 0 & \text{else,} \end{cases} \tag{92}$$

$$p_{j,0}(x_{j,0}) = 1/4. \tag{93}$$

The Bayesian network can be seen in Figure 4.

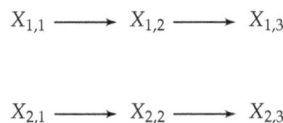

$$X_{1,1} \longrightarrow X_{1,2} \longrightarrow X_{1,3}$$

$$X_{2,1} \longrightarrow X_{2,2} \longrightarrow X_{2,3}$$

**Figure 4.** Bayesian network of $MC^=$. There is no interaction between the two processes.

4.2.2. Trajectories

In order to get the disintegration hierarchy $\mathfrak{D}(x_V)$ we have to choose a trajectory $x_V$ and calculate the SLI of each partition $\pi \in \mathfrak{L}(V)$. There are only four different trajectories possible in $MC^=$ and they are:

$$x_V = (x_{1,0}, x_{2,0}, x_{1,1}, x_{2,1}, x_{1,2}, x_{2,2}) = \begin{cases} (0,0,0,0,0,0) & \text{if } x_{1,0} = 0, x_{2,0} = 0; \\ (0,1,0,1,0,1) & \text{if } x_{1,0} = 0, x_{2,0} = 1; \\ (1,0,1,0,1,0) & \text{if } x_{1,0} = 1, x_{2,0} = 0; \\ (1,1,1,1,1,1) & \text{if } x_{1,0} = 1, x_{2,0} = 1. \end{cases} \tag{94}$$

Each of these trajectories has probability $p_V(x_V) = 1/4$ and all other trajectories have $p_V(x_V) = 0$. We call the four trajectories the *possible trajectories*. We visualise the possible trajectories as a grid with each cell corresponding to one variable. The spatial indices are constant across rows and time-slices $V_t$

correspond to the columns. A white cell indicates a 0 and a black cell indicates a 1. This results in the grids of Figure 5.

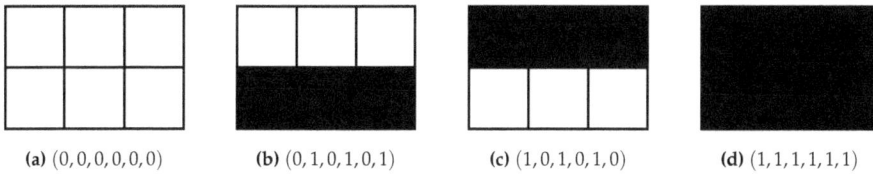

(a) $(0,0,0,0,0,0)$    (b) $(0,1,0,1,0,1)$    (c) $(1,0,1,0,1,0)$    (d) $(1,1,1,1,1,1)$

**Figure 5.** Visualisation of the four possible trajectories of $MC^=$. In each trajectory the time index increases from left to right. There are two rows corresponding to the two random variables at each time step and three columns corresponding to the three time-steps we are considering here.

### 4.2.3. Partitions of Trajectories

The disintegration hierarchy is composed out of all partitions in the lattice of partitions $\mathfrak{L}(V)$. Recall that we are partitioning the entire spatially and temporally extended index set $V$ of the Bayesian network and not only the time-slices. Blocks in the partitions of $\mathfrak{L}(V)$ are then, in general, spatiotemporally and not only spatially extended patterns.

The number of partitions $|\mathfrak{L}(V)|$ of a set of $|V| = 6$ elements is $\mathcal{B}_6 = 203$ ($\mathcal{B}_n$ is the Bell number of $n$). These partitions $\pi$ can be classified according to their cardinality $|\pi|$ (number of blocks in the partition). The number of partitions of a set of cardinality $|V|$ into $|\pi|$ blocks is the Stirling number $\mathcal{S}(|V|, |\pi|)$. For $|V| = 6$ we find the Stirling numbers:

$$
\begin{array}{c|cccccc}
|\pi| & 1 & 2 & 3 & 4 & 5 & 6 \\
\hline
\mathcal{S}(|V|, |\pi|) & 1 & 31 & 90 & 65 & 15 & 1
\end{array}
\tag{95}
$$

It is important to note that the partition lattice $\mathfrak{L}(V)$ is the same for all trajectories as it is composed out of partitions of $V$. On the other hand the values of SLI $mi_\pi(x_V)$ with respect to the partitions in $\mathfrak{L}(V)$ generally depend on the trajectory $x_V$.

### 4.2.4. SLI Values of the Partitions

We can calculate the SLI $mi_\pi(x_V)$ of every trajectory $x_V$ with respect to each partition $\pi \in \mathfrak{L}(V)$ according to Definition 5:

$$
mi_\pi(x_V) := \log \frac{p_V(x_V)}{\prod_{b \in \pi} p_b(x_b)}.
\tag{96}
$$

In the case of $MC^=$ the SLI values with respect to each partition do not depend on the trajectories. For an overview we plotted the values of SLI with respect to each partition $\pi \in \mathfrak{L}(V)$ for any trajectory of $MC^=$ in Figure 6.

We can see in Figure 6 that the cardinality does not determine the value of SLI. At the same time there seems to be a trend to higher values of SLI with increasing cardinality of the partition. We can also observe that only five different values of SLI are attained by partitions on this trajectory. We will collect these classes of partitions with equal SLI values in the disintegration hierarchy next.

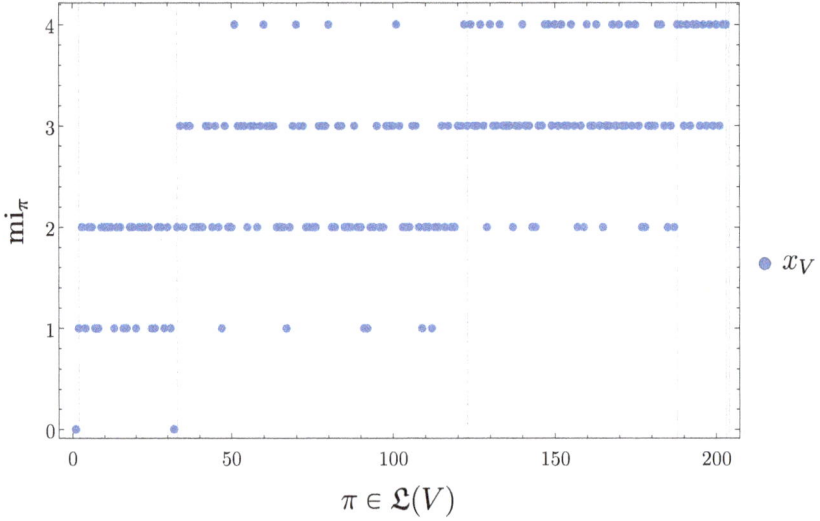

**Figure 6.** Specific local integrations $\mathrm{mi}_\pi(x_V)$ of any of the four trajectories $x_V$ seen in Figure 5 with respect to all $\pi \in \mathfrak{L}(V)$. The partitions are ordered according to an enumeration with increasing cardinality $|\pi|$ ((see Pemmaraju and Skiena [42], Chapter 4.3.3) for the method). We indicate with vertical lines at what partitions the cardinality $|\pi|$ increases by one.

### 4.2.5. Disintegration Hierarchy

In order to get insight into the internal structure of the partitions of a trajectory $x_V$ we obtain the disintegration hierarchy $\mathfrak{D}(x_V)$ (see Definition 10) and look at the Hasse diagrams of each of the disintegration levels $\mathfrak{D}_i(x_V)$ partially ordered by refinement. If we sort the partitions of any trajectory of $MC^=$ according to increasing SLI value we obtain Figure 7. There we see groups of partitions attaining the SLI values $\{0, 1, 2, 3, 4\}$ (precisely) these groups are the disintegration levels $\{\mathfrak{D}_1(x_V), \mathfrak{D}_2(x_V), \mathfrak{D}_3(x_V), \mathfrak{D}_4(x_V), \mathfrak{D}_5(x_V)\}$. The exact numbers of partitions in each of the levels are:

$$
\begin{array}{c|ccccc}
i & 1 & 2 & 3 & 4 & 5 \\
\hline
\mathrm{mi}_\pi & 0 & 1 & 2 & 3 & 4 \\
|\mathfrak{D}_i| & 2 & 18 & 71 & 78 & 34
\end{array}
\tag{97}
$$

Next we look at the Hasse diagram of each of those disintegration levels. Since the disintegration levels are subsets of the partition lattice $\mathfrak{L}(V)$, they are in general not lattices by themselves. The Hasse diagrams (see Appendix B for the definition) visualise the set of partitions in each disintegration level partially ordered by refinement $\lhd$. The Hasse diagrams are shown in Figure 8. We see immediately that within each disintegration level apart from the first and the last the Hasse diagrams contain multiple connected components.

Furthermore, within a disintegration level the connected components often have the same Hasse diagrams. For example, in $\mathfrak{D}_2$ ( Figure 8b) we find six connected components with three partitions each. The identical refinement structure of the connected components is related to the symmetries of the probability distribution over the trajectories. As it requires further notational overhead and is straightforward we do not describe these symmetry properties formally. In order to see the symmetries, however, we visualise the partitions themselves in the Hasse diagrams in Figure 9. We also visualise examples of the different connected components in each disintegration level in Figure 10.

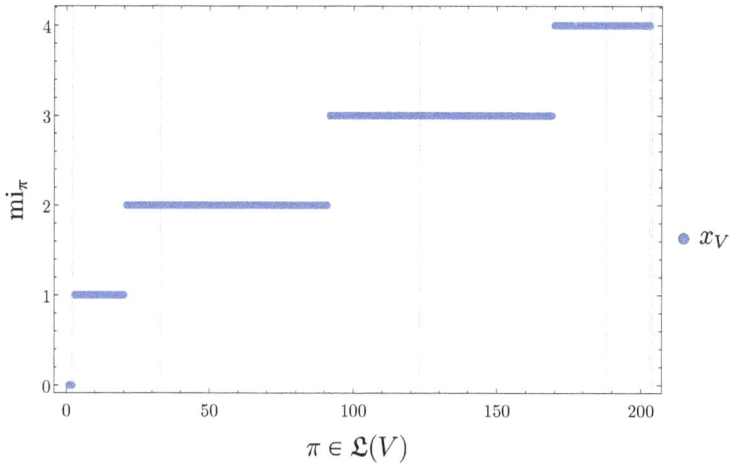

**Figure 7.** Same as Figure 6 but with the partitions sorted according to increasing SLI.

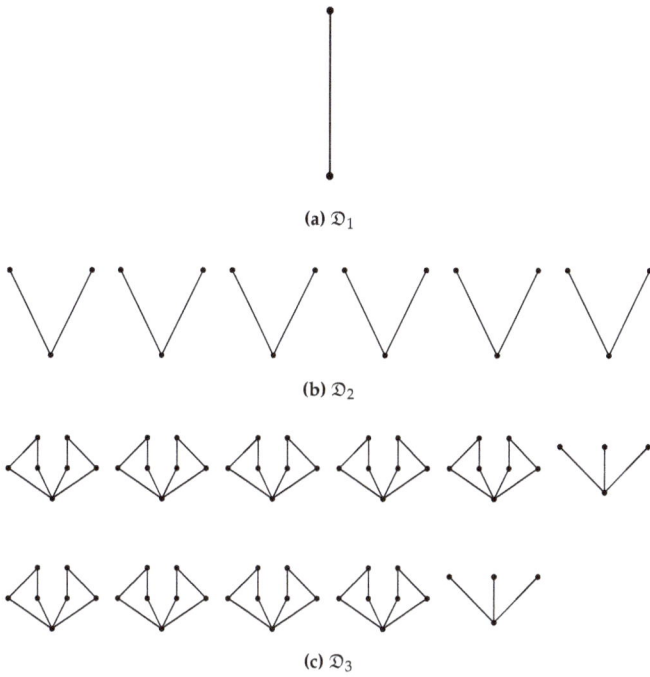

(a) $\mathfrak{D}_1$

(b) $\mathfrak{D}_2$

(c) $\mathfrak{D}_3$

**Figure 8.** *Cont.*

**(d)** $\mathfrak{D}_4$

**(e)** $\mathfrak{D}_5$

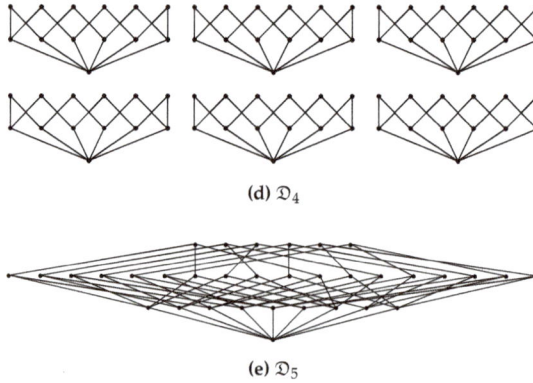

**Figure 8.** Hasse diagrams of the five disintegration levels of the trajectories of $MC^=$. Every vertex corresponds to a partition and edges indicate that the lower partition refines the higher one.

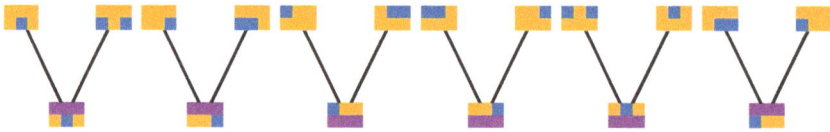

**Figure 9.** Hasse diagram of $\mathfrak{D}_2$ of $MC^=$ trajectories. Here we visualise the partitions at each vertex. The blocks of a partition are the cells of equal colour. Note that we can obtain all six disconnected components from one by symmetry operations that are respected by the joint probability distribution $p_V$. For example, we can shift each row individually to the left or right since every value is constant in each row. We can also switch top and bottom row since they have the same probability distributions even if 1 and 0 are exchanged.

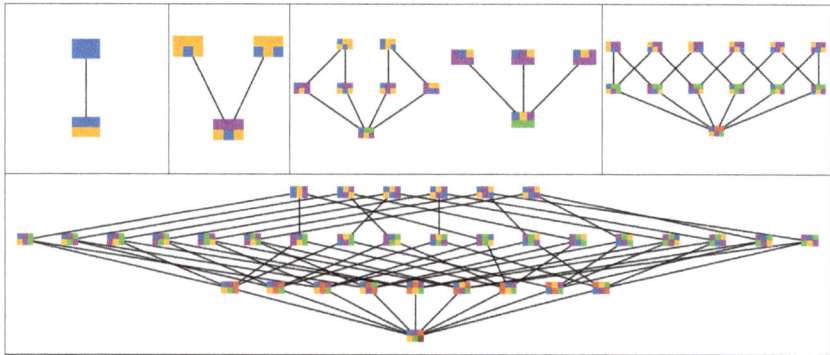

**Figure 10.** For each disintegration level of the trajectories of $MC^=$ we here show example connected components of Hasse diagrams with the partitions at each vertex visualised. The disintegration level increases clockwise from the top left. The blocks of a partition are the cells of equal colour.

Recall that due to the disintegration theorem (Theorem 6) we are interested especially in partitions that do not have refinements at their own or any preceding (i.e., lower indexed) disintegration level. These partitions consist of blocks that are completely integrated. i.e., all possible partitions of each of the blocks results in a positive SLI value or is a single node of the Bayesian network. The refinement-free disintegration hierarchy $\mathfrak{D}^{\blacktriangleleft}(x_V)$ contains only these partitions and is shown in a Hasse diagram in Figure 11.

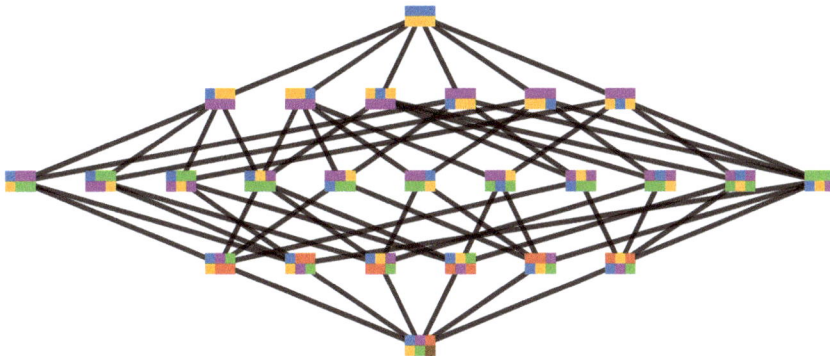

**Figure 11.** Hasse diagrams of the refinement-free disintegration hierarchy $\mathfrak{D}^{\blacktriangleleft}$ of $MC^=$ trajectories. Here we visualise the partitions at each vertex. The blocks of a partition are the cells of equal colour. It turns out that partitions that are on the same *horizontal* level in this diagram correspond exactly to a level in the refinement-free disintegration hierarchy $\mathfrak{D}^{\blacktriangleleft}$. The *i*-th horizontal level starting from the top corresponds to $\mathfrak{D}_i^{\blacktriangleleft}$. Take for example the second horizontal level from the top. The partitions on this level are just the minimal elements of the poset $\mathfrak{D}_2$ which was visualised in Figure 9. To connect this to Figure 8 note that for each disintegration level $\mathcal{D}_i$ shown there as a Hasse diagram, the partitions on the *i*-th horizontal level (counting from the top) in the present figure are the minimal elements of that disintegration level.

### 4.2.6. Completely Integrated Patterns

Having looked at the disintegration hierarchy we now make use of it by extracting the completely (When it is clear from context that we are talking about complete local integration we drop "local" for the sake of readability.) integrated patterns (*ι*-entities) of the four trajectories of $MC^=$. Recall that due to the disintegration theorem (Theorem 6) we know that all blocks in partitions that occur in the refinement-free disintegration hierarchy are either singletons or correspond to *ι*-entities. If we look at the refinement-free disintegration hierarchy in Figure 11 we see that many blocks occur in multiple partitions and across disintegration levels. We also see that there are multiple blocks that are singletons. If we ignore singletons, which are trivially integrated as they cannot be partitioned, we end up with eight different blocks. Since the disintegration hierarchy is the same for all four possible trajectories these blocks are also the same for each of them (note that this is the case for $MC^=$ but not in general as we will see in Section 4.3). However, the patterns that result are different due to the different values within the blocks. We show the eight *ι*-entities and their complete local integration (Definition 6) on the first trajectory in Figure 12 and on the second trajectory in Figure 13.

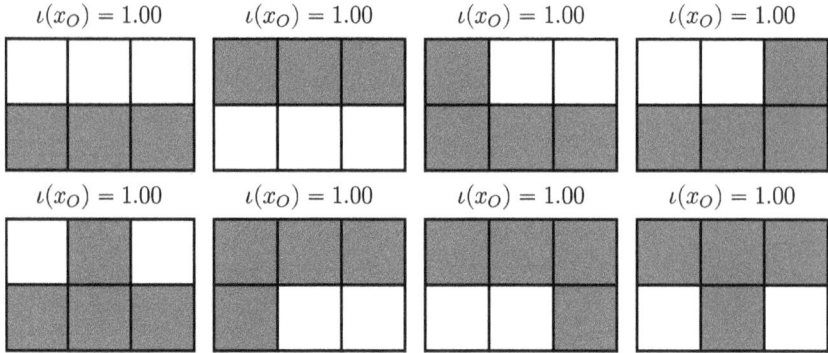

**Figure 12.** All distinct completely integrated composite patterns (singletons are not shown) on the first possible trajectory of $MC^=$. The value of complete local integration is indicated above each pattern. We display patterns by colouring the cells corresponding to random variables that are not fixed to any value by the pattern in grey. Cells corresponding to random variables that are fixed by the pattern are coloured according to the value i.e., white for 0 and black for 1.

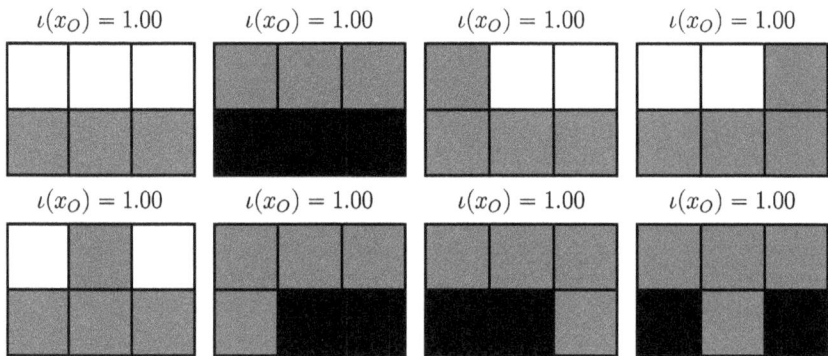

**Figure 13.** All distinct completely integrated composite patterns on the second possible trajectory of $MC^=$. The value of complete local integration is indicated above each pattern.

Since the disintegration hierarchies are the same for the four possible trajectories of $MC^=$ we get the same refinement-free partitions and therefore the same blocks containing the $\iota$-entities. This is apparent when comparing Figures 12 and 13 and noting that each pattern occurring on the first trajectory has a corresponding pattern on the second trajectory that differs (if at all) only in the values of the cells it fixes and not in what values it fixes. More visually speaking, for each pattern in Figure 12 there is a corresponding pattern in Figure 13 leaving the same cells grey.

If we are not interested in a particular trajectory, we can also look at all different $\iota$-entities on any trajectory. For $MC^=$ these are shown in Figure 14.

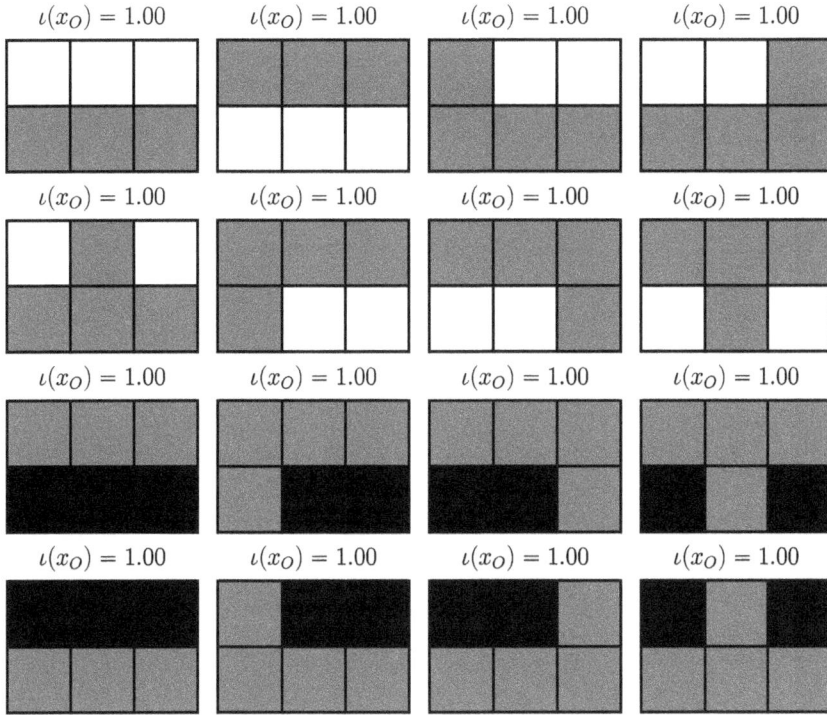

**Figure 14.** All distinct completely integrated composite patterns on all four possible trajectories of $MC^=$. The value of complete local integration is indicated above each pattern.

We see that all $\iota$-entities $x_O$ have the same value of complete local integration $\iota(x_O) = 1$. This can be explained using the deterministic expression for the SLI of Equation (30) and noting that for $MC^=$ if any of the values $x_{j,t}$ is fixed by a pattern then $(x_{j,s})_{s \in T} = x_{j,T}$ are determined since they must be the same value. This means that the number of trajectories $N(x_{j,S})$ in which any pattern $x_{j,S}$ with $S \subseteq T$ occurs is either $N(x_{j,S}) = 0$, if the pattern is impossible, or $N(x_{j,S}) = 2$ since there are two trajectories compatible with it. Note that all blocks $x_b$ in any of the $\iota$-entities and all $\iota$-entities $x_O$ themselves are of the form $x_{j,S}$ with $S \subseteq T$. Let $N(x_{j,S}) =: N$ and plug this into Equation (30) for an arbitrary partition $\pi$:

$$\mathrm{mi}_\pi(x_O) = (|\pi| - 1) \log |\mathcal{X}_{V_0}| - \log \frac{\prod_{b \in \pi} N(x_b)}{N(x_O)} \tag{98}$$

$$= (|\pi| - 1) \log |\mathcal{X}_{V_0}| - \log \frac{N^{|\pi|}}{N} \tag{99}$$

$$= (|\pi| - 1) \log \frac{|\mathcal{X}_{V_0}|}{N}. \tag{100}$$

To get the complete local integration value we have to minimise this with respect to $\pi$ where $|\pi| \geq 2$. So for $|\mathcal{X}_{V_0}| = 4$ and $N = 2$ we get $\iota(x_O) = 1$.

Another observation is that the $\iota$-entities are all limited to one of the two rows. This shows on a simple example that, as we would expect, $\iota$-entities cannot extend from one independent process to another.

### 4.3. Two Random Variables with Small Interactions

In this section we look at a system almost identical to that of Section 4.2 but with a kind of noise introduced. This allows all trajectories to occur and is designed to test whether the spatiotemporal patterns maintain integration in the face of noise.

#### 4.3.1. Definition

We define the time- and space-homogeneous multivariate Markov chain $MC^\epsilon$ via the Markov matrix $P$ with entries

$$P_{f(x_{1,t+1},x_{2,t+1}),f(x_{1,t},x_{2,t})} = p_{J,t+1}(x_{1,t+1}, x_{2,t+1} | x_{1,t}, x_{2,t}) \tag{101}$$

where we define the function $f : \{0,1\}^2 \to [1:4]$ via

$$f(0,0) = 1, f(0,1) = 2, f(1,0) = 3, f(1,1) = 4. \tag{102}$$

With this convention $P$ is

$$P = \begin{pmatrix} 1-3\epsilon & \epsilon & \epsilon & \epsilon \\ \epsilon & 1-3\epsilon & \epsilon & \epsilon \\ \epsilon & \epsilon & 1-3\epsilon & \epsilon \\ \epsilon & \epsilon & \epsilon & 1-3\epsilon \end{pmatrix} \tag{103}$$

This means that the state of *both* random variables remains the same with probability $1 - 3\epsilon$ and transitions into each of the other possible combinations with probability $\epsilon$. The noise then does not act independently on both random variables but disturbs the joint state. This makes $\iota$-entities possible that extend across the two processes. In the following we set $\epsilon = 1/100$. The initial distribution is again the uniform distribution

$$p_{j,0}(x_{j,0}) = 1/4. \tag{104}$$

Writing this multivariate Markov chain as a Bayesian network is possible but the conversion is tedious. The Bayesian network one obtains can be seen in Figure 15.

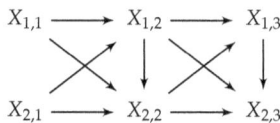

**Figure 15.** Bayesian network of $MC^\epsilon$.

#### 4.3.2. Trajectories

In this system all trajectories are possible trajectories. This means there are $2^6 = 64$ possible trajectories, since every one of the six random variables can be in any of its two states. There are three classes of trajectories with equal probability of occurring. The first class with the highest probability of occurring are the four possible trajectories of $MC^=$. Then there are 24 trajectories that make a single $\epsilon$-transition (i.e., a transition where the next pair is not the same as the current one $(x_{1,t+1}, x_{2,t+1}) \neq (x_{1,t}, x_{2,t})$, these transitions occur with probability $\epsilon$), and 36 trajectories with two $\epsilon$-transitions. We pick only one trajectory from each class. The representative trajectories are shown in Figure 16 and will be denoted $x_V^1, x_V^2,$ and $x_V^3$ respectively. The probabilities are $p_V(x_V^1) = 0.235225, p_V(x_V^2) = 0.0024250, p_V(x_V^3) = 0.000025.$

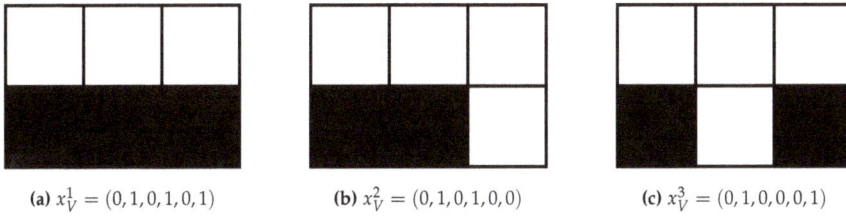

**(a)** $x_V^1 = (0, 1, 0, 1, 0, 1)$      **(b)** $x_V^2 = (0, 1, 0, 1, 0, 0)$      **(c)** $x_V^3 = (0, 1, 0, 0, 0, 1)$

**Figure 16.** Visualisation of three trajectories of $MC^\epsilon$. In each trajectory the time index increases from left to right. There are two rows corresponding to the two random variables at each time step and three columns corresponding to the three time-steps we are considering here. We can see that the first trajectory (in **(a)**) makes no $\epsilon$-transitions, the second (in **(b)**) makes one from $t = 2$ to $t = 3$, and the third (in **(c)**) makes two.

### 4.3.3. SLI Values of the Partitions

Again we calculate the SLI $mi_\pi(x_V)$ of every trajectory $x_V$ with respect to each partition $\pi \in \mathfrak{L}(V)$. In contrast to $MC^=$ the SLI values with respect to each partition of $MC^\epsilon$ do depend on the trajectories. We plot the values of SLI with respect to each partition $\pi \in \mathfrak{L}(V)$ for the three representative trajectories in Figure 17.

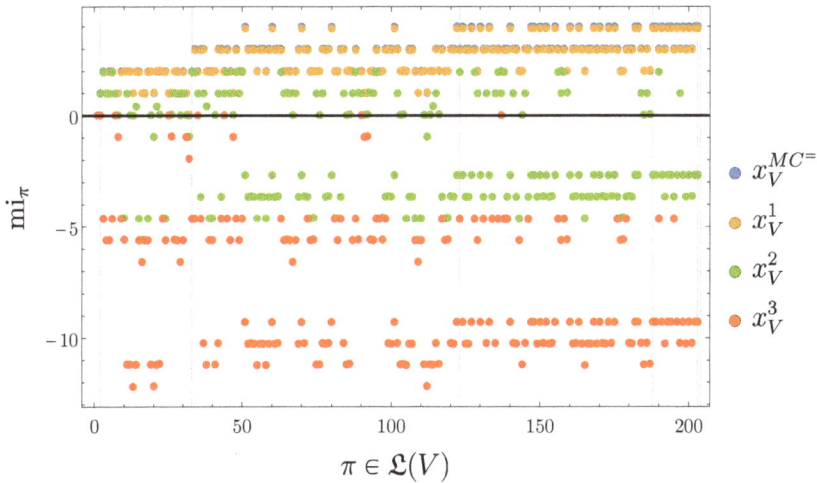

**Figure 17.** Specific local integrations $mi_\pi(x_V)$ of one of the four trajectories of $MC^=$ (measured w.r.t. the probability distribution of $MC^=$), here denoted $x_V^{MC^=}$ (this is the same data as in Figure 6), and the three representative trajectories $x_V^k, x \in \{1, 2, 3\}$ of $MC^\epsilon$ (measured w.r.t. the probability distribution of $MC^\epsilon$) seen in Figure 16 with respect to all $\pi \in \mathfrak{L}(V)$. The partitions are ordered as in Figure 6 with increasing cardinality $|\pi|$. Vertical lines indicate partitions where the cardinality $|\pi|$ increases by one. Note that the values of $x_V^{MC^=}$ are almost completely hidden from view by those of $x_V^1$.

It turns out that the SLI values of $x_V^1$ are almost the same as those of $MC^=$ in Figure 6 with small deviations due to the noise. This should be expected as $x_V^1$ is also a possible trajectory of $MC^=$. Also note that trajectories $x_V^2, x_V^3$ exhibit negative SLI with respect to some partitions. In particular, $x_V^2$ has non-positive SLI values with respect to any partition. This is due to the low probability of this trajectory compared to its parts. The blocks of any partition have so much higher probability than the entire trajectory that the product of their probabilities is still greater or equal to the trajectory probability.

### 4.3.4. Completely Integrated Patterns

In this section we look at the $\iota$-entities for each of the three representative trajectories $x_V^k, k \in \{1, 2, 3\}$. They are visualised together with their complete local integration values in Figures 18–20. In contrast to the situation of $MC^=$ we now have $\iota$-entities with varying values of complete local integration.

On the first trajectory $x_V^1$ we find all the eight patterns that are completely locally integrated in $MC^=$ (see Figure 13). These are also more than an order of magnitude more integrated than the rest of the $\iota$-entities. This is also true for the other two trajectories.

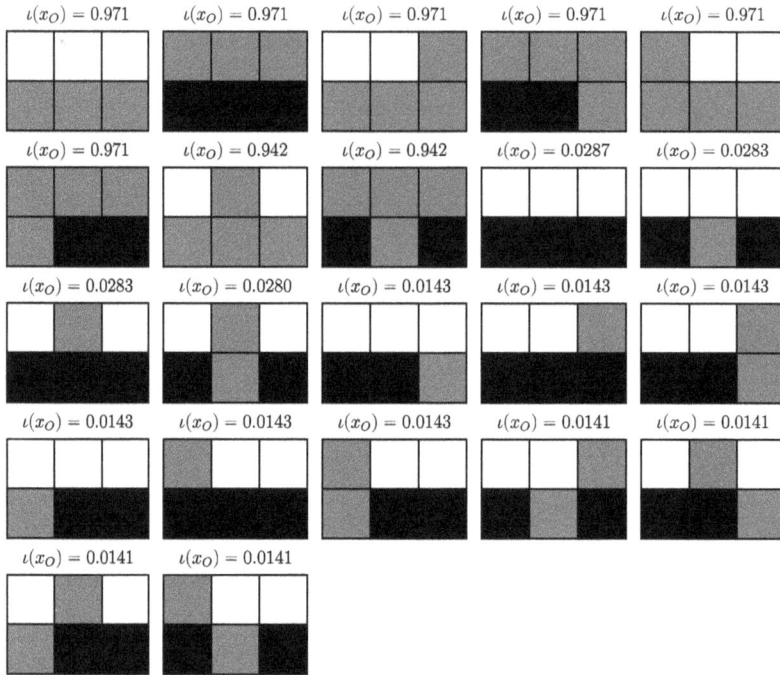

**Figure 18.** All distinct completely integrated composite patterns on the first trajectory $x_V^1$ of $MC^\epsilon$. The value of complete local integration is indicated above each pattern. See Figure 12 for colouring conventions.

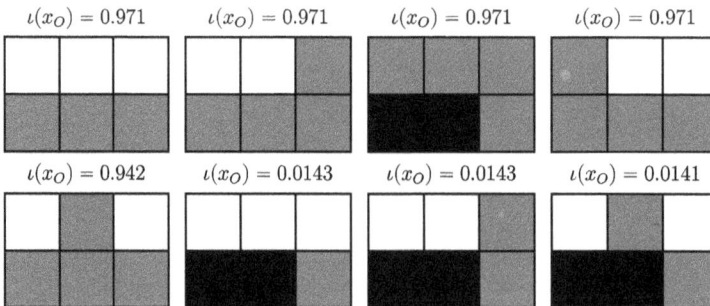

**Figure 19.** All distinct completely integrated composite patterns on the second trajectory $x_V^2$ of $MC^\epsilon$. The value of complete local integration is indicated above each pattern.

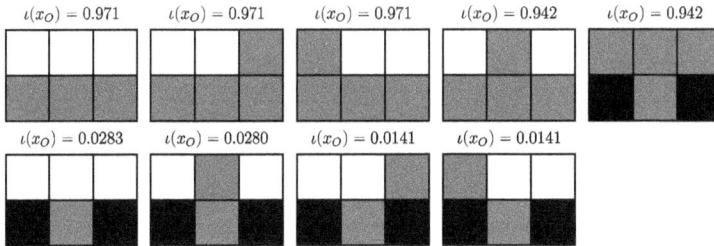

**Figure 20.** All distinct completely integrated composite patterns on the third trajectory $x_V^3$ of $MC^\epsilon$. The value of complete local integration is indicated above each pattern.

## 5. Discussion

In Section 3.1 we have argued for the use of patterns as candidates for entities. Patterns can be composed of arbitrary spatially and temporally extended parts of trajectories. We have seen in Theorem 1 that they are distinct from arbitrary subsets of trajectories. The important insight here is that patterns are structures that occur within trajectories but this cannot be said of sets of trajectories.

One of the main target applications of patterns is in time-unrolled Bayesian networks of cellular automata like those in Figure 1. Patterns in such Bayesian networks become spatiotemporal patterns like those used to describe the glider, block, and blinker in the Game of Life cellular automaton by Beer [14]. We would also like to investigate whether the latter spatiotemporal patterns are $\iota$-entities. However, at the present state of the computational models and, without approximations, this was out of reach computationally. We will discuss this further below.

In Section 3.3 we defined SLI and in Section 3.3 gave its expression for deterministic Bayesian networks (including cellular automata) as well. We also established the least upper bound of SLI with respect to a partition $\pi$ of cardinality $n$ for a pattern $x_A$ with probability $q$. This upper bound is achieved if each of the blocks $x_b$ in the partition $\pi$ occur if and only if the whole pattern $x_O$ occurs. This is compatible with our interpretation of entities since in this case clearly the occurrence of any part of the pattern leads necessarily to the occurrence of the entire pattern (and not only vice versa).

We also presented a candidate for a greatest lower bound of SLI with respect to a partition of cardinality $n$ for a pattern with probability $q$. Whether this is the greatest lower bound or not it shows a case for which SLI is always negative. This happens if either the whole pattern $x_A$ occurs (with probability $q$) or one of the "almost equal" patterns occurs, each with identical probability. A pattern $y_A$ is almost equal to $x_A$ with respect to $\pi$ in this sense if it only differs at one of the blocks $b \in \pi$ i.e., if $y_A = (x_{A \setminus b}, z_b)$ where $z_b \neq x_b$. This construction makes as many parts as possible (i.e., all but one) occur as many times as possible without the whole pattern occurring. This creates large marginalised probabilities $p_b(x_b)$ for each part/block which means that their product probability also becomes large.

Beyond these quantitative interpretations an interpretation of the greatest lower bound candidate seems difficult. A more intuitive candidate for the opposite of an integrated pattern seem to be patterns with independent parts. i.e., zero SLI but quantitatively these are not on the opposite end of the SLI spectrum. A more satisfying interpretation of the presented candidate is still to be found.

We also proved the disintegration theorem which relates states that the refinement-free partitions of a trajectory among those partitions achieving a particular SLI value consist of $\iota$-entities only, where an $\iota$-entity is a pattern with positive CLI. This theorem allows us to interpret the $\iota$-entities in new ways and may lead to a more formal or quantitative justification of $\iota$-entities. It is already a first step in this direction since it establishes a special role of the $\iota$-entities within trajectories of Bayesian networks. A further justification would tell us what in turn the refinement-free partitions can be used for. We have discussed a possible direction for further investigation in detail in Section 3.5. This tried to connect the $\iota$-entities with a coding problem.

In Section 4 we investigated SLI and CLI in three simple example sets of random variables. We found that if the random variables are all independently distributed the according entities are just all the possible $x_j \in \mathcal{X}_j$ of each of the random variables $X_j \in \{X_i\}_{i \in V}$. This is what we would expect from an entity criterion. There are no entities with any further extent than a single random variable and each value corresponds to a different entity.

For the simple Markov chain $MC^=$ composed out of two independent and constant processes we presented the entire disintegration hierarchy and the Hasse diagrams of each disintegration level ordered by refinement. The Hasse diagrams reflected the highly symmetric dynamics of the Markov chain via multiple identical components. For the refinement-free disintegration hierarchy we then get multiple partitions at the same disintegration level as well. Different partitions of the trajectory imply overlapping blocks which in the case of the refinement-free partition are $\iota$-entities. So in general the $\iota$-entities at a particular disintegration level are not necessarily unique and can overlap. We also saw in Figure 11 that the same $\iota$-entities can occur on multiple disintegration levels.

The $\iota$-entities of $MC^=$ included the expected three timestep constant patterns within each of the two independent processes. It also included the two timestep parts of these constant patterns. This may be less expected. It shows that parts of $\iota$-entities can be $\iota$-entities themselves. We note that these "sub-entities (those that are parts of larger entities) are always on a different disintegration level than their" super-entities (the larger entities). We can speculate that the existence of such sub- and super-entities on different disintegration levels may find an interpretation through multicellular organisms or similar structures. However, the overly simplistic examples here only serve as basic models for the potential phenomena, but are still far too simplistic to warrant any concrete interpretation in this direction.

We also looked at a version of $MC^=$ perturbed by noise, denoted $MC^\epsilon$. We found that the entities of $MC^=$ remain the most strongly integrated entities in $MC^\epsilon$. At the same time new entities occur. So we observe that in $MC^\epsilon$ the entities vary from one trajectory to another (Figures 18–20). We also observe spatially extended entities i.e., entities that extend across both (formerly independent) processes. We also observe entities that switch from one process to the other (from top row to bottom row or vice versa). The capacity of entities to exhibit this behaviour may be necessary to capture the movement or metabolism of entities in more realistic scenarios. In Biehl et al. [8] we argued that these properties are important and showed that they hold for a crude approximation of CLI (namely for SLI with respect to $\pi = 0$) but not for the full CLI measure.

We established that the $\iota$-entities:

- correspond to fixed single random variables for a set of independent random variables,
- can vary from one trajectory to another,
- and can change the degrees of freedom that they occupy over time,
- can be ambiguous at a fixed level of disintegration due to symmetries of the system,
- can overlap at the same level of disintegration due to this ambiguity,
- can overlap across multiple levels of disintegration i.e., parts of $\iota$-entities can be $\iota$-entities again.

In general the examples we investigated concretely are too small to sufficiently support the concept of positive CLI as an entity criterion. Due to the extreme computational burden, this may remain the case for a while. For a straightforward calculation of the minimum SLI of a trajectory of a Bayesian network $\{X_i\}_{i \in V}$ with $|V| = k$ nodes we have to calculate the SLI with respect to $\mathcal{B}_k$ partitions. According to (Bruijn [43], p. 108) the Bell numbers $\mathcal{B}_n$ grow super-exponentially. Furthermore, to evaluate the SLI we need the joint probability distribution of the Bayesian network $\{X_i\}_{i \in V}$. Naively, this means we need the probability (a real number between 0 and 1) of each trajectory. If we only have binary random variables, the number of trajectories is $2^{|V|}$ which make the straightforward computation of disintegration hierarchies unrealistic even for quite small systems. If we take a seven by seven grid of the game of life cellular automaton and want to look at three time-steps we have $|V| = 147$. If we use 32 bit floating numbers this gives us around $10^{30}$ petabytes of storage needed for this probability distribution. We are sceptical that the exact evaluation of reasonably large systems can be achieved even with non-naive methods. This suggests that formal proofs may be the more promising way to investigate SLI and CLI further.

**Acknowledgments:** Part of the research was performed during Martin Biehl's time as a JSPS International Research Fellow and as an ELSI Origins Network (EON) long term visitor. Daniel Polani was supported in part by the EC Horizon 2020 H2020-641321 socSMCs FET Proactive project.

**Author Contributions:** Martin Biehl, Takashi Ikegami, and Daniel Polani conceived the problem and the measures, and wrote the paper; Martin Biehl proved the theorems and conceived and calculated the examples. All authors have read and approved the final manuscript.

**Conflicts of Interest:** The authors declare no conflict of interest.

## Abbreviations

The following abbreviations are used in this manuscript:

SLI     Specific local integration
CLI     Complete local integration

## Appendix A. Kronecker Delta

The Kronecker–delta is used in this paper to represent deterministic conditional distributions.

**Definition A1** (Delta). *Let $X$ be a random variable with state space $\mathcal{X}$ then for $x \in \mathcal{X}$ and a subset $C \subset \mathcal{X}$ define*

$$\delta_x(C) := \begin{cases} 1 & \text{if } x \in C, \\ 0 & \text{else.} \end{cases} \tag{A1}$$

*We will abuse this notation if $C$ is a singleton set $C = \{\bar{x}\}$ by writing*

$$\delta_x(\bar{x}) := \begin{cases} 1 & \text{if } x \in \{\bar{x}\}, \\ 0 & \text{else.} \end{cases} \tag{A2}$$

$$= \begin{cases} 1 & \text{if } x = \bar{x}, \\ 0 & \text{else.} \end{cases} \tag{A3}$$

*The second line is a more common definition of the Kronecker–delta.*

Remark:

- Let $X, Y$ be two random variables with state spaces $\mathcal{X}, \mathcal{Y}$ and $f : \mathcal{X} \to \mathcal{Y}$ a function such that

$$p(y|x) = \delta_{f(x)}(y), \tag{A4}$$

then

$$p(y) = \sum_x p_Y(y|x) p_X(x) \tag{A5}$$

$$= \sum_x \delta_{f(x)}(y) p_X(x) \tag{A6}$$

$$= \sum_x \delta_x(f^{-1}(y)) p_X(x) \tag{A7}$$

$$= \sum_{x \in f^{-1}(y)} p_X(x) \tag{A8}$$

$$= p_X(f^{-1}(y)). \tag{A9}$$

## Appendix B. Refinement and Partition Lattice Examples

This appendix recalls the definitions of set partitions, refinement and coarsening of set partitions, and Hasse diagrams. It also shows the Hasse diagram of an example partition lattices. The definitions are due to Grätzer [28].

**Definition A2.** *A (set) partition $\pi$ of a set $\mathcal{X}$ is a set of non-empty subsets (called* blocks*) of $\mathcal{X}$ satisfying*

1. *for all $x_1, x_2 \in \pi$, if $x_1 \neq x_2$, then $x_1 \cap x_2 = \varnothing$,*
2. *$\bigcup_{x \in \pi} = \mathcal{X}$.*

*We write $\mathfrak{L}(\mathcal{X})$ for the set of all partitions of $\mathcal{X}$.*

Remark:

- In words, a partition of a set is a set of disjoint non-empty subsets whose union is the whole set.

**Definition A3.** *If two elements $x_1, x_2 \in \mathcal{X}$ belong to the same block of a partition $\pi$ of $\mathcal{X}$ write $x_1 \equiv_\pi x_2$. Also write $x_1 / \pi$ for the block $\{x_2 \in \mathcal{X} : x_2 \equiv_\pi x_1\}$.*

**Definition A4** (Refinement and coarsening). *We define the binary relation $\unlhd$ between partitions $\pi, \rho \in \mathfrak{L}(\mathcal{X})$ as:*

$$\pi \unlhd \rho \text{ if } x_1 \equiv_\pi x_2 \text{ implies } x_1 \equiv_\rho x_2. \tag{A10}$$

*In this case $\pi$ is called a* refinement *of $\rho$ and $\rho$ is called a* coarsening *of $\pi$.*

Remarks:

- More intuitively, $\pi$ is a refinement of $\rho$ if all blocks of $\pi$ can be obtained by further partitioning the blocks of $\rho$. Conversely, $\rho$ is a coarsening of $\pi$ if all blocks in $\rho$ are unions of blocks in $\pi$.
- Examples are contained in the Hasse diagrams (defined below) shown in Figure A1.

**Definition A5** (Hasse diagram). *A Hasse diagram is a visualisation of a poset (including lattices). Given a poset A ordered by $\lhd$ the Hasse diagram represents the elements of A by dots. The dots representing the elements are arranged in such a way that if $a, b \in A$, $a \neq b$, and $a \lhd b$ then the dot representing a is drawn below the dot representing b. An edge is drawn between two elements $a, b \in A$ if $a \lhd: b$, i.e., if b covers a. If edges cross in the diagram this does not mean that there is an element of A where they cross and edges never pass through a dot representing an element.*

Remarks:

- No edge is drawn between two elements $a, b \in A$ if $a \lhd b$ but not $a \lhd: b$.
- Only drawing edges for the covering relation does not imply a loss of information about the poset since the covering relation determines the partial order completely.
- For an example Hasse diagrams see Figure A1.

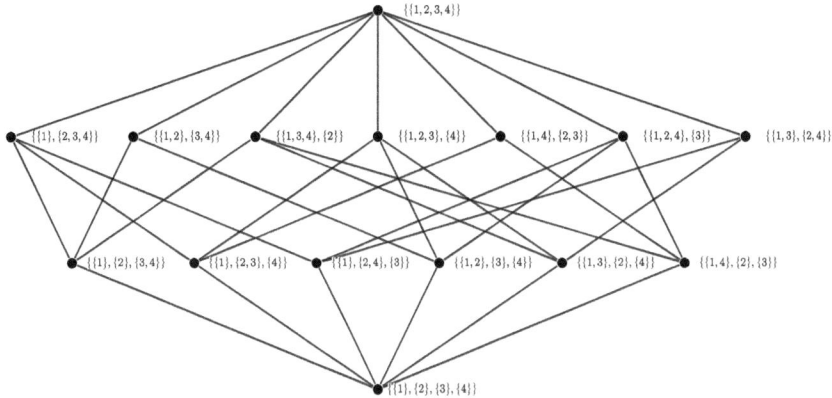

**Figure A1.** Hasse diagrams of the partition lattice of the four element set.

## Appendix C. Bayesian Networks

Intuitively a Bayesian network is a graph representation of the inter-dependencies of a set of random variables. Recall that any joint probability distribution over a set $\{X_i\}_{i \in I}$ with $I = \{1, ..., n\}$ of random variables can always be written as a product of conditional probabilities:

$$p_I(x_1, ..., x_n) = \prod_{i=1}^{n-1} p_i(x_i | x_{i+1}, ..., x_n) p(x_n). \tag{A11}$$

This also holds for any reordering of the indices $i \mapsto f(i)$ with $f : \{1, ...n\} \to \{1, ...n\}$ bijective.

In many cases however this factorisation can be simplified. Often some of the conditional probabilities $p(x_i | x_{i+1}, ..., x_n)$ do not depend on all variables $\{x_{i+1}, ..., x_n\}$ listed in the product of Equation (A11). For example, $X_1$ might only depend on $X_2$ so we would have $p(x_1 | x_2, ..., x_n) = p(x_1 | x_2)$. Note that the latter conditional probability is determined by fixing $|\mathcal{X}_2|(|\mathcal{X}_1| - 1)$ probabilities whereas the former needs $\prod_{i=1}^{n} |\mathcal{X}_{i+1}|(|\mathcal{X}_1| - 1)$ probabilities to be fixed. This means the number of parameters (the probabilities) of the joint distribution $p(x_1, ..., x_n)$ is often much smaller than suggested by Equation (A11). One way to encode this simplification and make sure that we are dealing only with joint probabilities that reflect the dependencies we allow are Bayesian networks. These can be defined as follows. First we define graphs that are factorisation compatible with joint probability distributions over a set of random variables and then define the Bayesian networks as pairs of joint probability distributions and a factorisation compatible graph.

**Definition A6.** *A directed acyclic graph $G = (V, E)$ with nodes $V$ and edges $E$ is* factorisation compatible *with the joint probabilities the probabilities of a probability distribution $p_V : \mathcal{X}_V \to [0, 1]$ iff the latter can be factorised in the way suggested by $G$ which means:*

$$p_V(x_V) = \prod_{i \in V} p(x_i | x_{\mathrm{pa}(i)}). \tag{A12}$$

*where $\mathrm{pa}(i)$ denotes the parents of node $i$ according to $G$.*

Remark:

- In general there are multiple directed acyclic graphs that are factorisation compatible with the same probability distribution. For example, if we choose any total order for the nodes in $V$ and define a graph by $\mathrm{pa}(i) = \{j \in V : j < i\}$ then Equation (A12) becomes Equation (A11) which always holds.

**Definition A7** (Bayesian network). *A Bayesian network is a (here assumed finite) set of random variables $\{X_i\}_{i \in V}$ and a directed acyclic graph $G = (V, E)$ with nodes indexed by $V$ such that the joint probability distribution $p_V : \mathcal{X}_V \to [0, 1]$ of $\{X_i\}_{i \in V}$ is factorisation compatible with $G$. We also refer to the graph set of random variables $\{X_i\}_{i \in V}$ as a Bayesian network implying the graph $G$.*

Remarks:

- On top of constituting the vertices of the graph $G$ the set $V$ is also assumed to be totally ordered in an (arbitrarily) fixed way. Whenever we use a subset $A \subset V$ to index a sequence of variables in the Bayesian network (e.g., in $p_A(x_A)$) we order $A$ according to this total order as well.
- Since $\{X_i\}_{i \in V}$ is finite and $G$ is acyclic there is a set $V_0$ of nodes without parents.

**Definition A8** (Mechanism). *Given a Bayesian network $\{X_i\}_{i \in V}$ with index set $V$ for each node with parents i.e., for each node $i \in V \setminus V_0$ (with $V_0$ the set of nodes without parents) the mechanism of node $i$ or also called the mechanism of random variable $X_i$ is the conditional probability (also called a transition kernel) $p_i : \mathcal{X}_{\mathrm{pa}(i)} \times \mathcal{X}_i \to [0, 1]$ mapping $(x_{\mathrm{pa}(i)}, x_i) \mapsto p_i(x_i | x_{\mathrm{pa}(i)})$. For each $x_{\mathrm{pa}(i)}$ the mechanism defines a probability distribution $p_i(.|x_{\mathrm{pa}(i)}) : \mathcal{X}_i \to [0, 1]$ satisfying (like any other probability distribution)*

$$\sum_{x_i \in \mathcal{X}_i} p_i(x_i | x_{\mathrm{pa}(i)}) = 1. \tag{A13}$$

Remarks:

- We could define the set of all mechanisms to formally also include the mechanisms of the nodes without parents $V_0$. However, in practice it makes sense to separate the nodes without parents as those that we choose an initial probability distribution over (similar to a boundary condition) which is then turned into a probability distribution $p_V$ over the entire Bayesian network $\{X_i\}_{i \in V}$ via Equation (A12). Note that in Equation (A12) the nodes in $V_0$ are not explicit as they are just factors $p_i(x_i | x_{\mathrm{pa}(i)})$ with $\mathrm{pa}(i) = \varnothing$.
- To construct a Bayesian network, take graph $G = (V, E)$ and equip each node $i \in (V \setminus V_0)$ with a mechanism $p_i : \mathcal{X}_{\mathrm{pa}(i)} \times \mathcal{X}_i \to [0, 1]$ and for each node $i \in V_0$ choose a probability distribution $p_i : \mathcal{X}_i \to [0, 1]$. The joint probability distribution is then calculated by the according version of Equation (A12):

$$p_V(x_V) = \prod_{i \in V \setminus V_0} p_i(x_i | x_{\mathrm{pa}(i)}) \prod_{j \in V_0} p_j(x_j). \tag{A14}$$

*Appendix C.1. Deterministic Bayesian Networks*

**Definition A9** (Deterministic mechanism). *A mechanism $p_i : \mathcal{X}_{\mathrm{pa}(i)} \times \mathcal{X}_i \to [0, 1]$ is deterministic if there is a function $f_i : \mathcal{X}_{\mathrm{pa}(i)} \to \mathcal{X}_i$ such that*

$$p_i(x_i | x_{\mathrm{pa}(i)}) = \delta_{f_i(x_{\mathrm{pa}(i)})}(x_i) = \begin{cases} 1 & \text{if } x_i = f_i(x_{\mathrm{pa}(i)}), \\ 0 & \text{else.} \end{cases} \tag{A15}$$

**Definition A10** (Deterministic Bayesian network). *A Bayesian network $\{X_i\}_{i \in V}$ is deterministic if all its mechanisms are deterministic.*

**Theorem A1.** *Given a deterministic Bayesian network $\{X_i\}_{i \in V}$ there exists a function $f_{V \setminus V_0} : \mathcal{X}_{V_0} \to \mathcal{X}_{V \setminus V_0}$ which given a value $x_{V_0}$ of the random variables without parents $X_{V_0}$ returns the value $x_{V \setminus V_0}$ fixing the values of all remaining random variables in the network.*

**Proof.** According to Equation (A12), the definition of conditional probabilities, and using the deterministic mechanisms we have:

$$p_{V\setminus V_0}(x_{V\setminus V_0}|x_{V_0}) = \prod_{i\in V\setminus V_0} p_i(x_i|x_{\mathrm{pa}(i)}). \tag{A16}$$

$$= \prod_{i\in V\setminus V_0} \delta_{f_i(x_{\mathrm{pa}(i)})}(x_i). \tag{A17}$$

For every $x_{V_0}$ the product on the right hand side is a probability distribution and therefore is always greater or equal to zero and maximally one. Also for every $x_{V_0}$ the sum of the probabilities over all $x_{V\setminus V_0} \in \mathcal{X}_{V\setminus V_0}$ is equal to one. As a product of zeros and/or ones the right hand side on the second line can only either be zero or one. This means for every $x_{V_0}$ there must be a unique $x_{V\setminus V_0}$ such that the right hand side is equal to one. Define this as the value of the function $f_{V\setminus V_0}(x_{V_0})$. $\square$

**Theorem A2** (Pattern probability in a deterministic Bayesian network). *Given a deterministic Bayesian network (Definition A10) and uniform initial distribution $p_{V_0} : \mathcal{X}_{V_0} \to [0,1]$ the probability of the occurrence of a pattern $x_A$ is:*

$$p_A(x_A) = \frac{N(x_A)}{|\mathcal{X}_{V_0}|} \tag{A18}$$

*where $N(x_A)$ is the number of trajectories $\bar{x}_V$ in which $x_A$ occurs.*

**Proof.** Recall that in a deterministic Bayesian network we have a function $f_{V\setminus V_0} : \mathcal{X}_{V_0} \to \mathcal{X}_{V\setminus V_0}$ (see Theorem A1) which maps a given value of $x_{V_0}$ to the value of the rest of the network $x_{V\setminus V_0}$. We calculate $p_A(x_A)$ for an arbitrary subset $A \subset V$. To make this more readable let $A \cap V_0 = A_0$, $A \setminus V_0 = A_r$, $B := V \setminus A$, $B \cap V_0 = B_0$, and $B \setminus V_0 = B_r$. Then

$$p_A(x_A) = \sum_{\bar{x}_B} p_V(x_A, \bar{x}_B) \tag{A19}$$

$$= \sum_{\bar{x}_{B_0}, \bar{x}_{B_r}} p_V(x_{A_r}, \bar{x}_{B_r}|x_{A_0}, \bar{x}_{B_0}) p_{V_0}(x_{A_0}, \bar{x}_{B_0}) \tag{A20}$$

$$= \sum_{\bar{x}_{B_0}, \bar{x}_{B_r}} \delta_{f_{V\setminus V_0}(x_{A_0}, \bar{x}_{B_0})}(x_{A_r}, \bar{x}_{B_r}) p_{V_0}(x_{A_0}, \bar{x}_{B_0}) \tag{A21}$$

$$= \sum_{\bar{x}_{B_r}} \sum_{\{\bar{x}_{B_0} : (x_{A_0}, \bar{x}_{B_0}) \in f_{V\setminus V_0}^{-1}(x_{A_r}, \bar{x}_{B_r})\}} p_{V_0}(x_{A_0}, \bar{x}_{B_0}) \tag{A22}$$

$$= \frac{1}{|\mathcal{X}_{V_0}|} \sum_{\bar{x}_{B_r}} |\{\bar{x}_{B_0} \in \mathcal{X}_{B_0} : (x_{A_0}, \bar{x}_{B_0}) \in f_{V\setminus V_0}^{-1}(x_{A_r}, \bar{x}_{B_r})\}| \tag{A23}$$

$$= \frac{1}{|\mathcal{X}_{V_0}|} N(x_A) \tag{A24}$$

In the second to last line we used the uniformity of the initial distribution $p_{V_0}$. The second sum in the second to last line counts all initial conditions that are compatible with $x_{A_0}$ and lead to the occurrence of $x_{A_r}$ together with some $\bar{x}_{B_r}$. The first one then sums over all such $\bar{x}_{B_r}$ to get all initial conditions that are compatible with $x_{A_0}$ and lead to the occurrence of $x_{A_r}$. Together these are all initial conditions compatible with $x_A$. In a deterministic system the number of initial conditions that lead to the occurrence of a pattern $x_A$ is equal to the number of trajectories $N(x_A)$ since every different initial condition will produce a single, unique trajectory. $\square$

Remark:

- Due to the finiteness of the network, deterministic mechanisms, and chosen uniform initial distribution the minimum possible non-zero probability for a pattern $x_A$ is $1/|\mathcal{X}_{V_0}|$. This happens for any pattern that only occurs in a single trajectory. Furthermore, the probability of any pattern is a multiple of $1/|\mathcal{X}_{V_0}|$.

*Appendix C.2. Proof of Theorem 2*

**Proof.** Follows by replacing the probabilities $p_O(x_O)$ and $p_b(x_b)$ in Equation (22) with their deterministic expressions from (Theorem A2), i.e., $p_A(x_A) = N(X_A)/|\mathcal{X}_{V_0}|$. Then:

$$\min_\pi(x_O) := \log \frac{p_O(x_O)}{\prod_{b \in \pi} p_b(x_b)} \tag{A25}$$

$$= \log \frac{\frac{N(x_O)}{|\mathcal{X}_{V_0}|}}{\prod_{b \in \pi} \frac{N(x_b)}{|\mathcal{X}_{V_0}|}} \tag{A26}$$

$$= \log \frac{\frac{N(x_O)}{|\mathcal{X}_{V_0}|}}{|\mathcal{X}_{V_0}|^{-|\pi|} \prod_{b \in \pi} N(x_b)} \tag{A27}$$

$$= \log \frac{|\mathcal{X}_{V_0}|^{|\pi|-1} N(x_O)}{\prod_{b \in \pi} N(x_b)} \tag{A28}$$

$$= (|\pi| - 1) \log |\mathcal{X}_{V_0}| - \log \frac{\prod_{b \in \pi} N(x_b)}{N(x_O)}. \tag{A29}$$

$\square$

## Appendix D. Proof of Theorem 1

**Proof.** Given a set of random variables $\{X_i\}_{i \in V}$, a subset $\mathcal{D} \subseteq \mathcal{X}_V$ cannot be represented by a pattern of $\{X_i\}_{i \in V}$ if and only if there exists $A \subseteq V$ with $\mathcal{D}_A \subset \mathcal{X}_A$ (proper subset) and $|\mathcal{D}_A| > 1$, i.e., if neither all patterns at $A$ are possible nor a unique pattern at $A$ is specified by $\mathcal{D}$.

We first show that if there exists $A \subseteq V$ with $\mathcal{D}_A \subset \mathcal{X}_A$ and $|\mathcal{D}_A| > 1$ then there is no pattern $\tilde{x}_B \in \bigcup_{C \subseteq V} \mathcal{X}_C$ with $\mathcal{D} = \mathcal{T}(\tilde{x}_B)$. Then we show that if no such $A$ exists then there is such a pattern $\tilde{x}_B$.

Since $\mathcal{D}_A > 1$ we have $x_A, \check{x}_A \in \mathcal{D}_A \subset \mathcal{X}_A$ with $x_A \neq \check{x}_A$. Next note that we can write any pattern $\tilde{x}_B$ as

$$\tilde{x}_B = (\tilde{x}_{B \setminus A}, \tilde{x}_{B \cap A}). \tag{A30}$$

If $B \cap A \neq \emptyset$ we can see since $\tilde{x}_{B \cap A}$ must take a single value it cannot contain $\mathcal{D}$ since there are trajectories in $\mathcal{D}$ taking value $x_{B \cap A}$ on $B \cap A$ and trajectories in $\mathcal{D}$ taking values $\tilde{x}_{B \cap A}$. More formally, we must have either $\tilde{x}_{B \cap A} = x_A$ or $\tilde{x}_{B \cap A} \neq x_A$. First, let $\tilde{x}_{B \cap A} = x_A$ but then $\mathcal{T}(\check{x}_A) \nsubseteq \mathcal{T}(\tilde{x}_B)$ so $\mathcal{D} \nsubseteq \mathcal{T}(\tilde{x}_B)$. Next choose $\tilde{x}_{B \cap A} \neq x_A$ but then $\mathcal{T}(x_A) \nsubseteq \mathcal{T}(\tilde{x}_B)$ so also $\mathcal{D} \nsubseteq \mathcal{T}(\tilde{x}_B)$. So we must have $B \cap A = \emptyset$.

Now we show that if $B \cap A = \emptyset$ there are trajectories in $\tilde{x}_B$ that are not in $\mathcal{D}$. We construct one explicitly by fixing its value on $A$ to the value in $\mathcal{X}_A$ that is not in $\mathcal{D}_A$ and its value on $B$ to $\tilde{x}_B$. More formally: choose $y_A \in \mathcal{X}_A \setminus \mathcal{D}_A$, then $y_A \neq x_A$ and $y_A \neq \check{x}_A$. This is always possible since $\mathcal{D}_A \subset \mathcal{X}_A$ (proper subset). Then consider a trajectory $\hat{x}_V = (\tilde{x}_B, y_A, \check{x}_D)$ with arbitrary $\check{x}_D \in \mathcal{X}_D$ where $D = V \setminus (B \cup A)$. Then $\hat{x}_V \in \mathcal{T}(\tilde{x}_B)$ but $\hat{x}_V \notin \mathcal{D}$.

Conversely, we show how to construct $\tilde{x}_B$ if no such $A$ exists. the idea is just to fix all random variables where $|\mathcal{D}_A| = 1$ and leave them unspecified where $\mathcal{D}_A = \mathcal{X}_A$. More formally: if there exists no $A \subseteq V$ with $\mathcal{D}_A \subset \mathcal{X}_A$ and $|\mathcal{D}_A| > 1$, then for each $C \subseteq V$ either $\mathcal{D}_C = \mathcal{X}_C$ or $|\mathcal{D}_C| = 1$. Then let $B = \bigcup \{C \subseteq V : |\mathcal{D}_C| = 1\}$ then $|\mathcal{D}_B| = 1$ so that we can define $\tilde{x}_B$ as the unique element in $\mathcal{D}_B$. Then if $y_V \in \mathcal{D}$ we have $y_B = \tilde{x}_B$ so $\mathcal{D} \subseteq \mathcal{T}(\tilde{x}_B)$. If $z_V \in \mathcal{T}(\tilde{x}_B)$ we have $z_B = \tilde{x}_B \in \mathcal{D}_B$ and for $A \subseteq V$ with $A \cap B = \emptyset$ by construction of $B$ we have $\mathcal{D}_A = \mathcal{X}_A$ such that $\mathcal{D}_{V \setminus B} = \mathcal{X}_{V \setminus B}$ which means $z_{V \setminus B} \in \mathcal{D}_{V \setminus B}$ and therefore $z_V \in \mathcal{D}$ and $\mathcal{T}(\tilde{x}_B) \subseteq \mathcal{D}$. So this gives $\mathcal{T}(\tilde{x}_B) = \mathcal{D}$. $\square$

# References

1. Gallois, A. Identity over Time. In *The Stanford Encyclopedia of Philosophy*; Zalta, E.N., Ed.; Metaphysics Research Laboratory, Stanford University: Stanford, CA, USA, 2012.
2. Grand, S. *Creation: Life and How to Make It*; Harvard University Press: Harvard, MA, USA, 2003.
3. Pascal, R.; Pross, A. Stability and its manifestation in the chemical and biological worlds. *Chem. Commun.* **2015**, *51*, 16160–16165.
4. Orseau, L.; Ring, M. Space-Time Embedded Intelligence. In *Artificial General Intelligence*; Number 7716 in Lecture Notes in Computer Science; Bach, J., Goertzel, B., Iklé, M., Eds.; Springer: Berlin/Heidelberg, Germany, 2012; pp. 209–218.
5. Barandiaran, X.E.; Paolo, E.D.; Rohde, M. Defining Agency: Individuality, Normativity, Asymmetry, and Spatio-temporality in Action. *Adapt. Behav.* **2009**, *17*, 367–386.
6. Legg, S.; Hutter, M. Universal Intelligence: A Definition of Machine Intelligence. *arXiv* **2007**, arXiv: 0712.3329.
7. Boccara, N.; Nasser, J.; Roger, M. Particlelike structures and their interactions in spatiotemporal patterns generated by one-dimensional deterministic cellular-automaton rules. *Phys. Rev. A* **1991**, *44*, 866–875.
8. Biehl, M.; Ikegami, T.; Polani, D. Towards information based spatiotemporal patterns as a foundation for agent representation in dynamical systems. In Proceedings of the Artificial Life Conference, Cancun, Mexico, 2016; The MIT Press: Cambridge, MA, USA, 2016; pp. 722–729.
9. McGill, W.J. Multivariate information transmission. *Psychometrika* **1954**, *19*, 97–116.
10. Amari, S.I. Information geometry on hierarchy of probability distributions. *IEEE Trans. Inf. Theory* **2001**, *47*, 1701–1711.
11. Lizier, J.T. *The Local Information Dynamics of Distributed Computation in Complex Systems*; Springer: Berlin/Heidelberg: Germany, 2012.
12. Tononi, G.; Sporns, O. Measuring information integration. *BMC Neurosci.* **2003**, *4*, 31.
13. Balduzzi, D.; Tononi, G. Integrated Information in Discrete Dynamical Systems: Motivation and Theoretical Framework. *PLoS Comput. Biol.* **2008**, *4*, e1000091.
14. Beer, R.D. Characterizing autopoiesis in the game of life. *Artif. Life* **2014**, *21*, 1–19.
15. Fontana, W.; Buss, L.W. "The arrival of the fittest": Toward a theory of biological organization. *Bull. Math. Biol.* **1994**, *56*, 1–64.
16. Krakauer, D.; Bertschinger, N.; Olbrich, E.; Ay, N.; Flack, J.C. The Information Theory of Individuality. *arXiv* **2014**, arXiv:1412.2447.
17. Bertschinger, N.; Olbrich, E.; Ay, N.; Jost, J. Autonomy: An information theoretic perspective. *Biosystems* **2008**, *91*, 331–345.
18. Shalizi, C.R.; Haslinger, R.; Rouquier, J.B.; Klinkner, K.L.; Moore, C. Automatic filters for the detection of coherent structure in spatiotemporal systems. *Phys. Rev. E* **2006**, *73*, 036104.
19. Wolfram, S. Computation theory of cellular automata. *Commun. Math. Phys.* **1984**, *96*, 15–57.
20. Grassberger, P. Chaos and diffusion in deterministic cellular automata. *Phys. D Nonlinear Phenom.* **1984**, *10*, 52–58.
21. Hanson, J.E.; Crutchfield, J.P. The attractor—Basin portrait of a cellular automaton. *J. Stat. Phys.* **1992**, *66*, 1415–1462.
22. Pivato, M. Defect particle kinematics in one-dimensional cellular automata. *Theor. Comput. Sci.* **2007**, *377*, 205–228.
23. Lizier, J.T.; Prokopenko, M.; Zomaya, A.Y. Local information transfer as a spatiotemporal filter for complex systems. *Phys. Rev. E* **2008**, *77*, 026110.
24. Flecker, B.; Alford, W.; Beggs, J.M.; Williams, P.L.; Beer, R.D. Partial information decomposition as a spatiotemporal filter. *Chaos Interdiscip. J. Nonlinear Sci.* **2011**, *21*, 037104.
25. Friston, K. Life as we know it. *J. R. Soc. Interface* **2013**, *10*, 20130475.
26. Balduzzi, D. Detecting emergent processes in cellular automata with excess information. *arXiv* **2011**, arXiv:1105.0158.
27. Hoel, E.P.; Albantakis, L.; Marshall, W.; Tononi, G. Can the macro beat the micro? Integrated information across spatiotemporal scales. *Neurosci. Conscious.* **2016**, *2016*, niw012.
28. Grätzer, G. *Lattice Theory: Foundation*; Springer: New York, NY, USA, 2011.

29. Ceccherini-Silberstein, T.; Coornaert, M. Cellular Automata and Groups. In *Encyclopedia of Complexity and Systems Science*; Meyers, R.A., Ed.; Springer: New York, NY, USA, 2009; pp. 778–791.

30. Busic, A.; Mairesse, J.; Marcovici, I. Probabilistic cellular automata, invariant measures, and perfect sampling. *arXiv* **2010**, arXiv: 1010.3133.

31. Beer, R.D. The cognitive domain of a glider in the game of life. *Artif. Life* **2014**, *20*, 183–206.

32. Beer, R.R. *Autopoiesis and Enaction in the Game of Life*; The MIT Press: Cambridge, MA, USA, 2016; p. 13.

33. Noonan, H.; Curtis, B. Identity. In *The Stanford Encyclopedia of Philosophy*; Zalta, E.N., Ed.; Metaphysics Research Laboratory, Stanford University: Stanford, CA, USA, 2014.

34. Hawley, K. Temporal Parts. In *The Stanford Encyclopedia of Philosophy*; Zalta, E.N., Ed.; Metaphysics Research Laboratory, Stanford University: Stanford, CA, USA, 2015.

35. Ay, N. Information Geometry on Complexity and Stochastic Interaction. *Entropy* **2015**, *17*, 2432–2458.

36. MacKay, D.J. *Information Theory, Inference and Learning Algorithms*; Cambridge University Press: Cambridge, UK, 2003.

37. Cover, T.M.; Thomas, J.A. *Elements of Information Theory*; Wiley: Hoboken, NJ, USA, 2006.

38. Tononi, G. An information integration theory of consciousness. *BMC Neurosc.* **2004**, *5*, 42.

39. Von Eitzen, H. Prove $(1 - (1 - q)/n)^n \geq q$ for $0 < q < 1$ and $n \geq 2$ a Natural Number. Mathematics Stack Exchange, 2016. Available online: http://math.stackexchange.com/q/1974262 (accessed on 18 October 2016).

40. Bullen, P.S. *Handbook of Means and Their Inequalities*; Springer Science+Business Media: Dordrecht, The Netherlands, 2003.

41. Kolchinsky, A.; Rocha, L.M. Prediction and modularity in dynamical systems. In *Advances in Artificial Life, ECAL*; The MIT Press: Cambridge, MA, USA, 2011; pp. 423–430.

42. Pemmaraju, S.; Skiena, S. *Computational Discrete Mathematics: Combinatorics and Graph Theory with Mathematica®*; Cambridge University Press: Cambridge, UK, 2009.

43. De Bruijn, N.G. *Asymptotic Methods in Analysis*; Dover Publications: New York, NY, USA, 2010.

MDPI

*Article*

# Utility, Revealed Preferences Theory, and Strategic Ambiguity in Iterated Games

Michael Harré

Complex Systems Research Group, Faculty of Engineering and Information Technologies, The University of Sydney, Sydney NSW 2006, Australia; michael.harre@sydney.edu.au; Tel.: +61-2-9114-1124

Academic Editors: Mikhail Prokopenko and Kevin Knuth
Received: 28 February 2017; Accepted: 26 April 2017; Published: 29 April 2017

**Abstract:** Iterated games, in which the same economic interaction is repeatedly played between the same agents, are an important framework for understanding the effectiveness of strategic choices over time. To date, very little work has applied information theory to the information sets used by agents in order to decide what action to take next in such strategic situations. This article looks at the mutual information between previous game states and an agent's next action by introducing two new classes of games: "invertible games" and "cyclical games". By explicitly expanding out the mutual information between past states and the next action we show under what circumstances the explicit values of the utility are irrelevant for iterated games and this is then related to revealed preferences theory of classical economics. These information measures are then applied to the Traveler's Dilemma game and the Prisoner's Dilemma game, the Prisoner's Dilemma being invertible, to illustrate their use. In the Prisoner's Dilemma, a novel connection is made between the computational principles of logic gates and both the structure of games and the agents' decision strategies. This approach is applied to the cyclical game Matching Pennies to analyse the foundations of a behavioural ambiguity between two well studied strategies: "Tit-for-Tat" and "Win-Stay, Lose-Switch".

**Keywords:** information theory; transfer entropy; game theory; logic gates; multilayer perceptrons; strategic behaviour; decision theory

---

## 1. Introduction

Game theory as it was originally framed by von Neumann and Morgenstern [1] and Nash [2] is concerned with agents (decision-makers) selecting actions to take when they interact strategically with other agents. Strategic interactions in the economic sense are situations in which the reward, or utility, one agent receives is based upon the action they choose to take as well as the actions taken by other agents. In much of non-cooperative game theory [3], it is assumed that agents are maximising their personal utility in one-off encounters between agents that know nothing of past behaviours, and that they choose their actions independently of one another, i.e., they do not discuss their strategies or collaborate with one another before choosing their actions. Relaxing these assumptions has been very fruitful in understanding the fundamental principles of strategic interactions: repeated games with learning can lead to chaotic dynamics [4] and spatially structured games can lead to cooperation where cooperation would not usually occur [5] or to a lack of cooperation where cooperation would usually occur [6].

An important approach to broadening the interaction model between agents has been to include interactions over time, as opposed to a single one-off game, and this has a significant impact on the possible outcomes. Each game can be thought to have occurred at a discrete point in time, i.e., agents' moves and utilities are said to occur at time $t$, and then consider what choices the agents then make at time $t + 1$ based on the moves and utilities at time $t$, $t - 1$, etc. These are called iterated games and they have been extensively studied by Axelrod [7], Nowak [8] and many others. In order to explore this approach, Axelrod ran a tournament in which contestants submitted algorithms that would compete against each other in playing a game, the Prisoner's Dilemma (see below for details), in which the algorithm that accumulated the highest utility would be the winner. The winning algorithm, submitted by Anatol Rapport [9], played a very simple strategy called Tit-for-Tat in which the algorithm initially cooperates with its opponent and thereafter chooses the strategy its opponent had used in the prior round. These and subsequent results led to a series of fundamental insights into the complexity of strategic interactions in dynamic games [10–13].

The real-valued utilities in game theory play an important role in many artificial intelligent systems such as those that use reinforcement learning, but in economic theory, it is the agent's behaviour that reveals an agent's subjective preferences. In the introduction to *Rational Decisions* ([14], p. 8), Ken Binmore discusses "revealed preference theory", the current economic orthodoxy on subjective utility. The theory states that if a decision-maker consistently acts to select one option over another, then their subjective preferences are revealed by their behaviour and their acts can be interpreted as if they are maximising a real valued utility function (see Savage's *The Foundations of Statistics* [15] for subjective preferences). The alternative point of view, that decision-makers act consistently because they have an internal real-valued utility function that allows them to order their choices, is called the causal utility fallacy (see pp. 19–22 of *Rational Decisions* [14]). For a review of earlier work and an historical discussion of the central role it has played in the foundations of economic theory see Chapter 1 in [16].

Revealed preferences freed economists from needing to consider the psychological or procedural aspects of choice to instead focus on agent behaviour. As a consequence, it should be possible to infer an agent's preferences from their behaviours alone, independently of the values of the utilities. In the non-cooperative game theory introduced in Section 2, we introduce the utility bi-matrices and prove their redundancy for iterated games in Section 3 using elementary results based on information theory. Corollary 1 specifically shows that the utility can be replaced by the previous acts of both agents in determining the next choice for any strategy in a two person iterated game. This appears to be the first time that this fundamental principle has been derived using information theory. The Traveller's Dilemma and the Prisoner's Dilemma illustrate this point in Section 4.

Regardless of the revealed preferences theory, significant work has focused on the neuro-computational processes that result in particular strategic behaviours [16–18] and so the relationship between observed behaviour and the computations that underlie that behaviour is of practical interest. However, it has been noted [19] that there are certain games for which an agent's choices are consistent with multiple different cognitive processes, and we show in Section 4 that these can be of distinctly different levels of complexity. This does not refute revealed preference theory; it only shows that the theory is limited insofar as it cannot distinguish between different internal processes, see [20], particularly footnote 8 and Conclusions, for a neuro-economic view of the internal cognitive states that are considered irrelevant to revealed preferences. In order to analyse the coupled interaction between games and strategies, they both need to be expressed in the same formal language, and this is done by describing games and strategies in terms of logic gates via truth tables and the computational processes that result in this ambiguity are described. In Section 4, we apply two well studied strategies (Tit-for-Tat and Win-Stay, Lose-Shift) to the Matching Pennies game and show that identical behaviours would require qualitatively different artificial neural networks to minimally implement. These two strategies, applied to iterated games as we do in this article, have been pivotal to the modern understanding of strategic evolution in game theory (see [11] and Chapters 4–5 of [8]). This illustrates a key limitation in revealed preferences theory: under certain conditions, it is not possible to know a market's composition of "herders" and "fundamentalists" by observing behaviour alone, an important factor in financial market collapse [21,22]. These points are discussed at the end of Section 5.

## 2. Non-Cooperative Game Theory

A normal form, non-cooperative game is composed of $i = 1, \ldots, N$ agents who are each able to select an act (often called a "pure strategy" in game theory) $a^i \in A^i$ ($A = A^1 \times \ldots \times A^N$) where the joint acts of all agents collectively determines the utility for each agent $i$, $u^i : A \to I\!R$. An act $a^{i*}$ is said to be preferable to an act $a^i$ via the utility function if we have: $u^i(a^1, \ldots, a^{i*}, \ldots, a^N) > u^i(a^1, \ldots, a^i, \ldots, a^N)$. We use $u^i$ to denote agent $i$'s utility *function*, taking joint action $a = a^1 \times \ldots \times a^N$ as an argument, and $u_n$ as the utility *value* (a real number) for agent $i$ in the $n$-th round of an iterated game (see below), if there is any uncertainty as to which agent's utility value we are referring to, we use $u_n^i$. We will represent the actions available to the agents and their subsequent utility values using the conventional bi-matrix notation—for example, the Prisoner's Dilemma [23] is given by the following payoff bi-matrix in Table 1 (by convention, $i$ indexes the agent being considered, $-i$ indexes the other agent):

**Table 1.** The Prisoner's Dilemma payoff table.

|  |  | Agent $-i$ | |
|---|---|---|---|
|  |  | Co-op | Defect |
| agent $i$ | Co-op | (1 year, 1 year) | (5 years, 0 year) |
|  | Defect | (0 year, 5 years) | (3 years, 3 years) |

In this game, there are two agents (prisoners held in two different jail cells) who have been arrested for a crime and they are being questioned (independently) by the police regarding the crime. Each agent chooses from the same set of possible acts: cooperate with their fellow detainee and remain silent about the crime or defect and tell the police about the crime and implicate their fellow detainee. If they both cooperate with each other, they each get one year in jail, if they both defect, they each get three years in jail, if one defects and one cooperates the defector gets no jail time (zero years) while the cooperator gets five years in jail. We also define a game's state space $S^i$ for agent $i$, a specific set of acts and utilities (variables) for agent $i$—for example, $S^i = \{$co-op, defect, five years$\}$ is a specific state space of variables when $i$ cooperates and $-i$ defects. The variables in $S^i$ are deterministically related to one another via the bi-matrix of the game being played. The following definitions form the two sub-classes of games that we will use in the following section:

**Definition 1.** *Invertible Games: An invertible game for agent $i$ is a game for which each of $i$'s real-valued utilities uniquely defines the joint actions of all agents. A game is invertible if it is invertible for all agents.*

For example, the Prisoner's Dilemma and Stag Hunt games are invertible but the Matching Pennies and Rock–Paper–Scissors games are not. Matching Pennies and Prisoner's Dilemma are discussed in detail below, and see, for example, [24] for Rock, Paper, Scissors and [25] for Stag Hunt.

**Definition 2.** *Cyclical Games: Given the cardinality of variables in agent $i$'s state space $\phi = |S^i|$, a cyclical game for agent $i$ is a game for which knowing any combination of $\phi - 1$ variables of $S^i$ is both necessary and sufficient to derive the remaining variable. A game is cyclical if it is cyclical for all agents.*

The Prisoner's Dilemma is not cyclical, Matching Pennies and Rock-Paper-Scissors are cyclical but the Traveler's Dilemma (see Section 4: Example Games) is neither invertible nor cyclical.

## 3. Information Theory

Information theory measures the degree of stochastic variability and dependency in a system based on the probability distributions that describe the relationships between the elements of the system. For a discrete stochastic variable $x \in \{x^1, \ldots, x^j\} = X$, the (Shannon) Entropy is:

$$H(x) = - \sum_{x^i \in X} p(x = x^i) \log_2 p(x = x^i),$$

measured in bits as $\log_2$ assumed throughout, and this is maximised if $x$ is uniformly distributed over all $x^i \in X$ and zero if $p(x = x^i) = 1$ for any $i$. We will write $p(x^i)$ or $p(x)$ if there is no confusion. There are many possible extensions to the notion of Entropy, in this work, we will make use of the Mutual Information and the conditional Mutual Information, respectively defined as [26]:

$$
\begin{aligned}
I(x:y) &= \sum_{x,y} p(x,y) \log \left( \frac{p(x,y)}{p(x)p(y)} \right) \\
&= \sum_{x,y} p(x,y) \log \left( \frac{p(x|y)}{p(x)} \right), \tag{1}
\end{aligned}
$$

$$
\begin{aligned}
I(x:y|z) &= \sum_{x,y,z} p(x,y,z) \log \left( \frac{p(x,y|z)}{p(x|z)p(y|z)} \right) \\
&= \sum_{x,y,z} p(x,y,z) \log \left( \frac{p(x|y,z)}{p(x|z)} \right). \tag{2}
\end{aligned}
$$

If, in Equation (1), there is a (deterministic, linear or nonlinear) mapping $y \to x$, then $p(x|y) \in \{0,1\}, \forall \{x,y\}$ and $I(x:y) = H(x)$. Similarly, for Equation (2), if there is a (deterministic, linear or nonlinear) mapping $y \to x$ such that $p(x|y,z) = p(x|z) \in \{0,1\}$, then the summation reduces to a weighted sum over $\log(\frac{1}{1}) = 0$ and $\log(\frac{0}{0}) = 0$ and so $I(x:y|z) = 0$ bits.

A useful special case of the conditional mutual information is the Transfer Entropy (TE) [27]. For a system with two stochastic variables $X$ and $Y$ that have discrete realisations $x_n$ and $y_n$ at time points $n \in \{1,2,\ldots\}$, then the TE of the joint time series: $\{\{x_n, y_n\}, \{x_{n+1}, y_{n+1}\}, \ldots\}$ is the mutual information between one variable's current state and the second variable's next state conditional on the second variable's current state:

$$
\begin{aligned}
T_{Y \to X} &= I(y_n : x_{n+1}|x_n) \\
&= \sum_{x_n, y_n, x_{n+1}} p(x_n, y_n, x_{n+1}) \log \left( \frac{p(x_{n+1}|y_n, x_n)}{p(x_{n+1}|x_n)} \right). \tag{3}
\end{aligned}
$$

This is interpreted as the degree to which the previous state of $Y$ is informative about the next state of $X$ excluding the past information $X$ carries about its next state. This is one of many different specifications of TE, generalisations based on history length appears in Schreiber's original article [27], and further considerations of delays [28] as well as averaging the TE over whole systems of stochastic variables [29] have also been developed (for a recent review, see [30]).

In iterated games, we wish to know how much information passes from each variable in agent $i$'s state space $S_n^i$ at time $n$ to their next act $a_{n+1}^i$ and what simplifications can be made. It is assumed that each agent $i$ has a strategy $Z^i(\ldots)$ that maps previous system states to actions at time $n+1$, and, in general, this may depend on system states of an arbitrary length of time $l+1$ into the past:

$$Z^i(S_n^i, S_{n-1}^i, \ldots, S_{n-l}^i) = a_{n+1}^i, \tag{4}$$

where $\{S_n^i, S_{n-1}^i, \ldots, S_{n-l}^i\}$ is assumed to be the totality of an agent's information set used to make a decision. For example, one of the simplest 1-step strategies is the Tit-for-Tat (TfT) strategy, whereby agent $i$ simply copies the previous act of the other agent [8,9]:

$$a_{n+1}^i = Z_0^i(a_n^{-i}) = a_n^{-i}. \tag{5}$$

It is possible for $Z^i$ to take no arguments in which case the agent chooses from a distribution over their next act that is independent of any information set. This is the case for the Matching Pennies game considered below: two agents simultaneously toss one coin each and the outcome, either matched coins or mismatched coins, decides the payoff to either agent and so for this "0-step strategy" each action $a_n^i$ is independent of all past states of the game for both agents. Given a maximum history length of $l$, we will say that a strategy $Z^i$ is an $l$-step Markovian strategy and an $l$-step Markovian game is a sequence of the same game played repeatedly for which each agent has an $l$-step Markovian strategy.

*Information Chain Rule and Iterated Games*

We are interested in constructing probability distributions over the variables in $S_n^i$ for iterated games. As $n$ indexes time, for large $n$, a probability of an event is the frequency of occurrence of the event (an element in $S_n^i$) divided by $n$. Other probabilities, such as conditional or joint probabilities, are natural extensions of this approach. For the moment, we make no assumptions about the relationship between elements of $S_n^i$ and $a_{n+1}^i$, except that they are not statistically independent of each other. Given a 2-agent, 1-step Markovian game with states $S_n^i = \{a_n^i, a_n^{-i}, u_n\}$, from the chain rule for information ([26], Theorem 2.5.2), we have the following six identities for the total information between the previous game state and agent $i$'s subsequent act $a_{n+1}^i$:

$$I(S_n^i : a_{n+1}^i) = I(a_n^i, a_n^{-i}, u_n : a_{n+1}^i)$$

$$= I(a_n^i : a_{n+1}^i | u_n, a_n^{-i}) + I(u_n : a_{n+1}^i | a_n^{-i}) + I(a_n^{-i} : a_{n+1}^i) \tag{6a}$$

$$= I(a_n^{-i} : a_{n+1}^i | u_n, a_n^i) + I(u_n : a_{n+1}^i | a_n^i) + I(a_n^i : a_{n+1}^i) \tag{6b}$$

$$= I(a_n^{-i} : a_{n+1}^i | u_n, a_n^i) + I(a_n^i : a_{n+1}^i | u_n) + I(u_n : a_{n+1}^i) \tag{6c}$$

$$= I(a_n^i : a_{n+1}^i | u_n, a_n^{-i}) + I(a_n^{-i} : a_{n+1}^i | u_n) + I(u_n : a_{n+1}^i) \tag{6d}$$

$$= I(u_n : a_{n+1}^i | a_n^{-i}, a_n^i) + I(a_n^{-i} : a_{n+1}^i | a_n^i) + I(a_n^i : a_{n+1}^i) \tag{6e}$$

$$= I(u_n : a_{n+1}^i | a_n^{-i}, a_n^i) + I(a_n^i : a_{n+1}^i | a_n^{-i}) + I(a_n^{-i} : a_{n+1}^i). \tag{6f}$$

We will say that $I(S_n^i : a_{n+1}^i)$ measures the amount of information $S_n^i$ shares with $a_{n+1}^i$. These expressions for the shared information between game states and next actions can be simplified in useful ways as follows:

**Theorem 1.** *For any normal form, non-cooperative 1-step Markovian game (not necessarily cyclical or invertible):* $I(u_n^i : a_{n+1}^i | a_n) = 0$ *for the joint act* $a_n = a_n^1 \times \ldots \times a_n^N$.

**Proof.** From the definition of a non-cooperative game, the utility value $u^i(a_n) = u_n$ is determined by the joint strategies so the log term in Equation (2) can be written: $\log \left( \frac{p(u_n | a_{n+1}^i, a_n)}{p(u_n | a_n)} \right)$. In this case, both conditional probabilities in the log term will be either 0 or 1 and note the discussion following Equation (2). □

**Remark 1.** *From Theorem 1, the first terms in Equations (6e) and (6f) are zero. While the joint actions of the agents unambiguously identifies a utility value, in general the utility value does not unambiguously identify a joint action. See the Traveler's Dilemma example below.*

**Corollary 1.** *For a 2-agent, 1-step Markovian game, the total information from agent $i$'s previous state space to $i$'s next act is encoded in the sum of agent $i$'s Transfer Entropy from agent $-i$'s previous act to agent $i$'s current act and agent $i$'s "active memory" [31] of their own past acts:* $I(S_n^i : a_{n+1}^i) = T_{a^{-i} \to a^i} + I(a_n^i : a_{n+1}^i)$.

**Proof.** By Theorem 1 and Equation (6e). □

**Theorem 2.** *For a 2-agent, invertible 1-step Markovian game* $I(a_n^i : a_{n+1}^{-i} | u_n, \ldots) = I(a_n^i : a_{n+1}^{-i} | a_n) = 0$.

**Proof.** This follows the same approach as Theorem 1 where knowing the joint act $a_n$ implies knowing any single agent's act $a_n^i$ and from the definition of an invertible game for which $[u_i]^{-1}$ is the necessary deterministic map. □

**Remark 2.** *From Theorem 2, the first terms in Equations (6a)–(6f) are zero for invertible 1-step Markovian games.*

**Corollary 2.** *For an invertible 1-step Markovian game, the total information from agent $i$'s previous state space to agent $i$'s next act is encoded in agent $i$'s active memory between the previous utility and their subsequent act:* $I(S_n^i : a_{n+1}^i) = I(u_n : a_{n+1}^i)$.

**Proof.** By Theorem 2 for Equation (6c): $I(a_n^{-i} : a_{n+1}^i | u_n, a_n^i) = I(a_n^i : a_{n+1}^i | u_n) = 0$ and so: $I(S_n^i : a_{n+1}^i) = I(u_n : a_{n+1}^i)$. □

**Theorem 3.** *For a cyclical 1-step Markovian game, $I(x : a_{n+1}^{-i} | S_n^i \setminus x) = 0, \ \forall \, x \in S_n^i$.*

**Proof.** This follows the same approach as Theorem 1 and note that, for a three element $S_n^i$ in a cyclical game, there is a $c \in S_n^i$ and distinct $\{a, b\} = S_n^i \setminus c$, for which there is a deterministic relationship between any pairing of $a$ or $b$ with $c$ and, consequently, the conditional mutual information in Theorem 3 is zero. □

**Remark 3.** *From Theorem 3, the first term in Equations (6a)– (6f) are zero for cyclical 1-step Markovian games.*

**Corollary 3.** *For a cyclical or an invertible 2-agent, 1-step Markovian game:*

$$I(u_n : a_{n+1}^i | a_n^{-i}) = I(a_n^i : a_{n+1}^i | a_n^{-i}), \tag{7}$$

$$I(u_n : a_{n+1}^i | a_n^i) = I(a_n^{-i} : a_{n+1}^i | a_n^i), \tag{8}$$

$$I(a_n^i : a_{n+1}^i | u_n) = I(a_n^{-i} : a_{n+1}^i | u_n). \tag{9}$$

**Proof.** This comes from Theorems 2 and 3 and comparing Equation (6a) with (6f), Equation (6b) with (6e) and Equation (6c) with (6d). □

**Remark 4.** *Equality (8) is notable, and it states that the transfer entropy from the utility of agent $i$ to their next act is the same as the transfer entropy from agent $-i$'s act to agent $i$'s next act.*

## 4. Example Games

In this section, we consider in some detail three games belonging to three distinct classes. The first example, the Traveler's Dilemma, belongs to the class of games that are neither invertible nor cyclical. The second example, the Prisoner's Dilemma, is an invertible game and the final example, the Matching Pennies game, is a cyclical game. The Traveler's Dilemma and the Prisoner's Dilemma illustrate the information theoretical aspects of iterated games and the Matching Pennies game illustrates the computational properties of cyclical games.

### 4.1. The Traveler's Dilemma

The story describing the Traveler's Dilemma (TD) game is often told in the following form [32]:

> "Lucy and Pete, returning from a remote Pacific island, find that the airline has damaged the identical antiques that each had purchased. An airline manager says that he is happy to compensate them but is handicapped by being clueless about the value of these strange objects. Simply asking the travelers for the price is hopeless, he figures, for they will inflate it.
>
> Instead, he devises a more complicated scheme. He asks each of them to write down the price of the antique as any dollar integer between $2 and $100 without conferring together.

If both write the same number, he will take that to be the true price, and he will pay each of them that amount. However, if they write different numbers, he will assume that the lower one is the actual price and that the person writing the higher number is cheating. In that case, he will pay both of them the lower number along with a bonus and a penalty–the person who wrote the lower number will get $2 more as a reward for honesty and the one who wrote the higher number will get $2 less as a punishment. For instance, if Lucy writes $46 and Pete writes $100, Lucy will get $48 and Pete will get $44."

The utility function for the TD game can be defined using the Heaviside step function $\Theta(z)$:

$$
\begin{aligned}
\Theta(z) &= 0, \ \forall \ z < 0, \\
\Theta(z) &= 1/2, \text{ if } z = 0, \\
\Theta(z) &= 1, \ \forall \ z > 0,
\end{aligned}
$$

and agent $i$'s action $a^i$ is the cost $i$ claims for the vase and agent $-i$'s action $a^{-i}$ is the cost $-i$ claims for the vase, the utility for agent $i$ is [33]:

$$
u^i(a^i, a^{-i}) = (a^i + 2)\Theta(a^{-i} - a^i) + (a^{-i} - 2)\Theta(a^i - a^{-i}). \tag{10}
$$

This game is not invertible because, in the case where the two agents disagree with each other, the agent $i$ who offers the lowest value $a^i$ has a utility of $u^i = a^i + 2$, whereas the other agent has a utility of $u^{-i} = a^i - 2$, i.e., agent $-i$'s utility is independent of the precise value offered; therefore, the agent utility is invertible for agent $i$ but not for agent $-i$, so the game is not invertible. For iterated TD games, the simplest calculation of the information shared between the previous state and the next action is: $I(S_n : a^i_{n+1}) = T_{a^{-i} \to a^i} + I(a^i_n : a^i_{n+1})$.

## 4.2. Prisoner's Dilemma

The Prisoner's Dilemma is one of the most well studied economic games. It was originally an experimental test in a repeated game format by researchers at the RAND Corporation (Santa Monica, CA, USA) who were skeptical of Nash's notion of equilibrium on the basis that no real players would play in the fashion Nash had proposed [23]. Here, we use it to illustrate that utility can be used exclusively to learn strategies just as behaviour can be used to exclusively learn better strategies. In the second part of this example, we use Nowak's approach [8] to understand the relationship between information sets, game structure and strategies. The logic gate approach allows us to express games and strategies in the same language, allowing a direct comparison of game structure and strategies. The Prisoner's Dilemma (PD) was introduced earlier, see the payoff matrix in Table 1.

Because it is invertible for both agents (from the uniqueness of each of the four utilities for each agent), the total information from the previous time step to the next act for either agent is encoded in the Mutual Information between the previous utility and the next act: $I(S^i_n : a^i_{n+1}) = I(u_n : a^i_{n+1})$. For a 1-step Markovian strategy, no other past information is needed to select the next act. In this sense, the utility acts as a reward that an agent can use to distinguish between the actions of other players, and the entire joint action space of the game's previous state can be reconstructed by agent $i$ using only the utility value.

To explore this further, we follow Nowak ([8], pp. 78–89) in defining two deterministic strategies in the iterated version of the PD game in terms of previous game states and subsequent acts. In the iterated PD, there are four possible joint acts at time $n$: $Cc, Cd, Dc$ and $Dd$. The action of the agent we are referring to, usually indexed as $i$, has their act in capital letters i.e., $C$ and $D$ for cooperate and defect, whereas the other agent $-i$ has lower case acts $c$ or $d$. A vector of these joint acts is mapped to a vector of acts at time $n + 1$ in the following way: $[Cc, Cd, Dc, Dd] \to [x_1, x_2, x_3, x_4]$, where $x_i \in \{C, D\}$. In the Tit-for-Tat strategy (TfT): $[Cc, Cd, Dc, Dd] \to [C, D, C, D]$, in which agent $i$ copies the other agent's previous act, and in the Win-Stay, Lose-Switch strategy (WSLS): $[Cc, Cd, Dc, Dd] \to [C, D, D, C]$, in which agent $i$ repeats their act if they receive a high payoff (a 'win' of either 1 or 0 years) but changes their act if they receive a low payoff (a 'loss' of three or five years). This interpretation of the WSLS strategy is not an accurate representation of the strategy though: WSLS monitors the success (measured by utility) of the previous act and adjusts the next act accordingly i.e., WSLS uses the information set $\{a_n^i, u_n\}$ to select $a_{n+1}^i$, not $\{a_n^i, a_n^{-i}\}$ as is implied by the representation $[Cc, Cd, Dc, Dd] \rightarrow [C, D, D, C]$. This distinction is not important for Nowak's analysis of PD [8] (p. 88), but it is important in the analysis below.

It can be seen that these two strategies have the same amount of shared information between their previous game states and their next acts: TfT $\simeq$ WSLS. Note that we consider agent $-i$'s actions to be uniformly distributed across their possible acts, whereas agent $i$ follows one of the deterministic strategies described next. We use $\simeq$ and $\equiv$ to distinguish between having the same quantity of information ($\simeq$) and having the same observed behaviour ($\equiv$). For TfT, the shared information between $S_n^i$ and $a_{n+1}^i$ is: $I(S_n^i : a_{n+1}^i) = I(a_n^{-i} : a_{n+1}^i)_{TfT} = 1$ bit. The two expansions of $I(S_n^i, a_{n+1}^i)$ that contain this mutual information term are Equations (6a) and (6f), so the first two terms in these two equations must be zero for PD. The first term is zero by Theorem 2, and we observe that in TfT $a_n^{-i}$ explicitly determines $a_{n+1}^i$, so the second conditional entropy terms in which $a_n^{-i}$ conditions $a_{n+1}^i$ must also be zero. For WSLS, we measure $I(S_n^i : a_{n+1}^i)$ in terms of $\{a_n^i, u_n\}$ and then show that this is equivalent to the TfT result. By Theorem 2 for invertible games, the first term in any expansion of $I(S_n^i : a_{n+1}^i)$ is zero, and the equations that then only have $\{a_n^i, u_n\}$ in their expression are Equations (6b) and (6c). These equations measure the shared information between the information set of WSLS and the agent's next act, i.e., $I(a_n^i, u_n : a_{n+1}^i)$. Equation (6c) is then: $I(a_n^i, u_n : a_n^i) = I(a_n^i : a_{n+1}^i | u_n) + I(u_n : a_{n+1}^i)$ and by Corollary 2 $I(a_n^i, u_n : a_{n+1}^i) = I(u_n : a_{n+1}^i)_{WSLS}$—by definition, Equation (6c) $\simeq$ Equation (6a) and, therefore, $I(u_n : a_{n+1}^i)_{WSLS} \simeq I(a_n^{-i} : a_{n+1}^i)_{TfT}$. However, TfT is not behaviourally equivalent to WSLS for the PD: WSLS $\not\equiv$ TfT as can be seen by direct comparison of the strategies described above.

We now consider the relationship between WSLS and TfT for the PD game by making the following substitutions: $C, c \rightarrow 0$, $D, d \rightarrow 1$, a "win" $\rightarrow 1$, and a "loss" $\rightarrow 0$. Replacing the utilities with a win–loss binary variable is justified as WSLS (the only strategy that uses the utilities) only considers wins and losses, not the numerical values of the utilities. It then be seen that this modified version of the Prisoner's Dilemma is equivalent to a NOT logic gate for agent $i$ that inverts the action of agent $-i$ (see Table 2):

**Table 2.** Prisoner's Dilemma as a NOT logic gate for agent $i$.

| | $S_n^i$ | | $a_{n+1}^i$ | |
| --- | --- | --- | --- | --- |
| $a_n^i$ | $a_n^{-i}$ | $i$'s Win-Loss | WSLS | TfT |
| 0 (C) | 0 (c) | 1 (win) | 0 | 0 |
| 0 (C) | 1 (d) | 0 (loss) | 1 | 1 |
| 1 (D) | 0 (c) | 1 (win) | 1 | 0 |
| 1 (D) | 1 (d) | 0 (loss) | 0 | 1 |

$a_n^{-i}$ ▷○ win-loss for $i$ in round $n$

The first three columns in Table 2 are the actions of the two agents in round $n$ of an iterated game and the outcome of the game based on these actions, i.e., each row is an instance of the 3-element set $S_n^i$. The next two columns of Table 2 are the actions of agent $i$ in round $n+1$ for WSLS and TfT. The diagram shows the relationship between the elements in $S_n^i$ when the game is interpreted as a logic gate relating inputs (agent actions) and outputs (wins and losses). In this case, the $4 \times 3$ matrix made up of the four rows of possible combinations of $S_n^i$ is the "truth table" of the PD game. Note that, while both agents always have an incentive to defect irrespective of what the other agent does (defection is said to strictly dominate cooperation, see ([34], p. 10)), it is only agent $-i$'s action that decides whether or not agent $i$ will win or lose. Whether this is a large or a small win or loss is controlled by agent $i$ though, and this aspect was lost when the utility was made into a binary win–loss outcome in modifying the PD game in the table. Figure 1 represents these relationships for TfT being played in an iterated PD game.

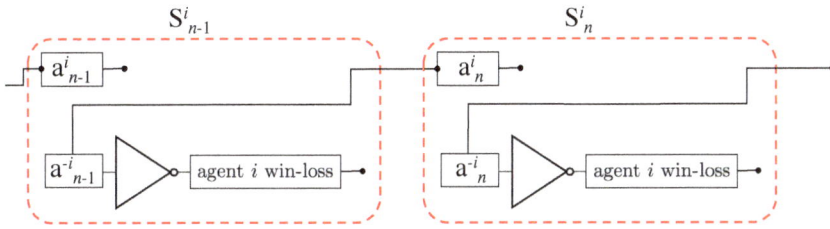

**Figure 1.** The modified Prisoner's Dilemma game for agent $i$ based on wins and losses using the Tit-for-Tat (TfT) strategy. $S_{n-1}^i$ and $S_n^i$ are the game states at time $n-1$ and $n$, the NOT gate is the logical operator that connects the variable $a_n^{-i}$ to the win–loss status of the game, in this sense the NOT operator *is* the logic of the modified PD game. The only connection between successive game states for the TfT strategy is a direct connection between $a_{n-1}^{-i}$ and $a_n^i$, and neither agent $i$'s act nor the win–loss status of the game is in the information set of the TfT strategy. The $a_n^i$ variable influences the win–loss outcome in a similar diagram for agent $-i$ just as $a_n^{-i}$ influences the win–loss outcome for agent $i$ in this diagram.

Next, we show that WSLS is a nonlinear function of its two inputs while TfT is a linear function of its inputs. To see this, note that TfT $\simeq$ WSLS and TfT is linearly (anti-)correlated with win–loss; however, there is zero pairwise linear correlation between WSLS actions $a_{n+1}^i$ and either $a_n^i$ or $u_n$. Consequently, because mutual information measures both linear and nonlinear relationships between variables, all of the shared information between $S_n^i$ and $a_{n+1}^i$ is either linear for TfT or nonlinear for WSLS.

A truth table can also be constructed from the information set of WSLS (columns 1 and 3 in Table 2) and the WSLS action in the next round (column 4 in Table 2) and the truth table is that of an XNOR (exclusive-nor) logic gate in which matching inputs (00 or 11) are mapped to 0 and mismatched inputs (10 or 01) are mapped to 1. Encoding the utilities of the PD as win–loss and representing it as a NOT gate makes clear that learning the relationships between the variables of $S^i$ for the PD game (as encoded in the first three columns of the table above) is a linearly separable task; if agent $i$ wants to learn the PD, it only needs to divide the state space of agent $-i$'s actions up into wins and losses for agent $i$, a problem that can be solved by single layer perceptrons [35]. However, learning the WSLS strategy, equivalent to learning an XNOR operation, is not a linearly separable task and requires a multi-layer perceptron [36]. Figure 2 represents these relationships for WSLS being played in an iterated PD game.

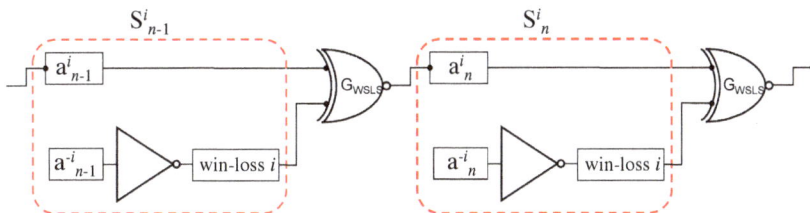

**Figure 2.** The modified Prisoner's Dilemma game for agent $i$ based on wins and losses and using the Win-Stay, Lose-Switch (WSLS) strategy. The WSLS strategy is equivalent to an XNOR (exclusive-nor) gate that has as its information set (inputs) $\{a_{n-1}^i,$ win–loss$\}$ and outputs $a_n^i$.

In the modified PD, the utility values were converted to binary variables, and this works because TfT does not have the utility value in its information set and WSLS has the utility value in its information set but only considers the outcome as a binary win–loss variable. In the next example, these substitutions are unnecessary. We take a similar approach to establish that the Matching Pennies game is not a linearly separable problem and that WSLS is both informationally and behaviourally identical to TfT for one agent but not the other.

### 4.3. Matching Pennies

The Matching Pennies (MP) game is an important example used in laboratory studies [19] of economic choice, learning and strategy. In the MP game, two agents have one coin each and each coin has two states, either Heads (H) or Tails (T). When the agents compare coins one agent wins if the two coins match and the other agent wins if they do not match. In the usual description of the game, the two agents toss their coins before comparing them; this randomising of the coin states can be interpreted as one step of an iterated game using a 0-step Markovian strategy. In what follows, though, we use the name "Matching Pennies" to describe the iterated game where the WSLS and TfT are deterministic strategies that use the action-utility bi-matrix given by (see Table 3):

**Table 3.** Payoff table for the Matching Pennies game.

|          |       | Agent $-i$ |        |
|----------|-------|-----------|--------|
|          |       | Heads     | Tails  |
| agent $i$ | Heads | $(1,0)$  | $(0,1)$ |
|          | Tails | $(0,1)$  | $(1,0)$ |

In the same fashion as for the PD game above, we can map agents' acts to numerical values: $H, h \rightarrow 1$, $T, t \rightarrow 0$, and a win = 1, a loss = 0 (unlike the PD game, we do not need to simplify the utility values and so the game is not modified as the PD was). Then, the MP game can be interpreted as XNOR (exclusive-nor) and XOR (exclusive-or) gates in the following two tables (see Tables 4 and 5) for the two agents:

**Table 4.** XNOR logic gate for agent $i$.

| | $S_n^i$ | | $a_{n+1}^i$ | |
|---|---|---|---|---|
| $a_n^i$ | $a_n^{-i}$ | $u_n^i$ | WSLS | TfT |
| 1 (H) | 1 (h) | 1 (win) | 1 | 1 |
| 1 (H) | 0 (t) | 0 (loss) | 0 | 0 |
| 0 (T) | 1 (h) | 0 (loss) | 1 | 1 |
| 0 (T) | 0 (t) | 1 (win) | 0 | 0 |

agent $i$ ⟩ agent $i$ utility
agent $-i$

**Table 5.** XOR logic gate for agent $-i$.

| | $S_n^{-i}$ | | $a_{n+1}^{-i}$ | |
|---|---|---|---|---|
| $a_n^i$ | $a_n^{-i}$ | $u_n^{-i}$ | WSLS | TfT |
| 1 (H) | 1 (h) | 0 (loss) | 0 | 1 |
| 1 (H) | 0 (t) | 1 (win) | 1 | 0 |
| 0 (T) | 1 (h) | 1 (win) | 0 | 1 |
| 0 (T) | 0 (t) | 0 (loss) | 1 | 0 |

agent $i$ ⟩ agent $-i$ utility
agent $-i$

As for the PD game, the first three columns of each row in Tables 4 and 5 represent an instance of $S_n^i$ (Table 4) and $S_n^{-i}$ (Table 5), and the four possible permutations of the variables of $S_n^i$ and $S_n^{-i}$ forms a 4 × 3 table that can be interpreted as the truth table of a logic gate. However, unlike in the PD game, both agents now have an input into each other's logic gate (see Figures 3 and 4 for schematic representations for agent $i$). Just as in the PD, for MP, the WSLS strategy is not linearly correlated with any element of its information set and TfT is linearly correlated with its information set; recall the WSLS information set for agent $i$'s act $a^i_{n+1}$ is $\{a^i_n, u_n\}$ and the TfT information set is $\{a^{-i}\}$. However, unlike the PD, now WSLS is perfectly (linearly) correlated with TfT for agent $i$ and perfectly anti-correlated for agent $-i$.

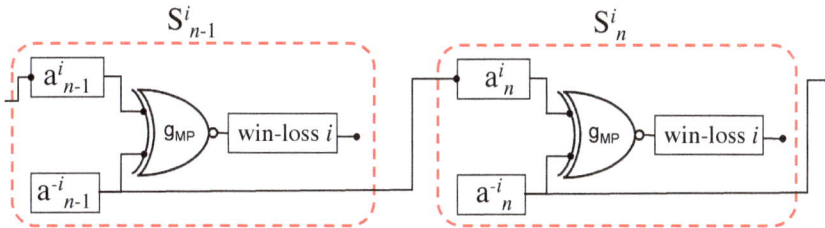

**Figure 3.** Two steps in the Matching Pennies (MP) iterated game, for agent $i$ the elements of $S^i_n$ are related to one another via an XNOR logic gate $g_{MP}$ representation of the MP game. TfT is the copy operation (identity logic gate) from $a^{-i}_{n-1}$ to $a^i_n$

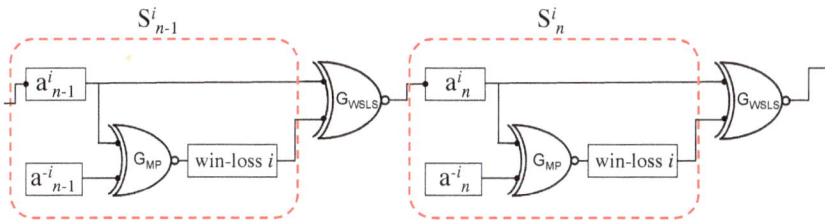

**Figure 4.** Two steps in the MP iterated game. As for Figure 3, the elements of $S^i_n$ are related to one another via an XNOR logic gate $G_{MP}$ representation of the MP game. The WSLS strategy is represented as an XNOR logic gate $G_{WSLS}$ between inputs $\{a^i_{n-1}, \text{win–loss}\}$ and output $a^i_n$. For agent $-i$, the $G_{MP}$ gate would be an XOR gate.

It can also be seen that knowing any two variables in $S^i_n = \{a^i_n, a^{-i}_n, u^i_n\}$ is sufficient to derive the third variable. This is a property of the XOR ($\oplus$) and XNOR ($\overline{\oplus}$) logic operations; given the three variables $a^i_n, a^{-i}_n, u^i_n, \in \{0, 1\}$ related by the XNOR operator, then: $a^i_n \overline{\oplus} a^{-i}_n = u^i_n$, $a^{-i}_n \overline{\oplus} u^i_n = a^i_n$, and $u^i_n \overline{\oplus} a^i_n = a^{-i}_n$, and we will refer to this as the cyclical property of cyclical games. We can now prove the following relationship:

**Theorem 4.** *TfT $\equiv$ WSLS for agent i: for the Matching Pennies iterated game, the Tit-for-Tat strategy is behaviourally indistinguishable from the Win-Stay-Lose-Shift strategy for agent i.*

**Proof.** By the cyclical property for Matching Pennies, if: $G_{MP} : a^i_{n-1} \overline{\oplus} a^{-i}_{n-1} = u^i_{n-1}$, then $G_{MP} :$ $a^i_{n-1} \overline{\oplus} u^i_{n-1} = a^{-i}_{n-1}$. The WSLS strategy $G_{WSLS}$ takes these same inputs $\{a^i_{n-1}, u^i_{n-1}\}$ but outputs $a^i_n$: $G_{WSLS} : a^i_{n-1} \overline{\oplus} u^i_{n-1} = a^i_n$. Because $G_{MP}$ and $G_{WSLS}$ are equivalent to XNOR logic gates, for the same input, they produce the same output: $G_{MP} \equiv G_{WSLS}$ and therefore $a^{-i}_{n-1} \equiv a^i_n$. By definition, the TfT strategy is $a^{-i}_{n-1} \equiv a^i_n$ and so TfT $\equiv$ WSLS. $\square$

**Remark 5.** *TfT is a prototypical herd-like strategy: it simply follows the behaviour of another agent. In contrast, WSLS is a prototypical fundamental strategy: it uses past actions and their payoffs to decide whether to stay with the current strategy or change. The fact that they are indistinguishable from each other is not guaranteed; for the Prisoner's Dilemma, TfT is behaviourally distinguishable from WSLS, and the indistinguishability in the Matching Pennies game comes from identifying the MP game and the WSLS strategy with the same (XNOR) logic gate and the cyclical property of $G_{MP}$.*

Before the next result, we introduce a variation on the WSLS strategy, called "Win-Switch-Lose-Stay": WSLS$^{-1}$ in which an agent changes strategy if they win but will stay with a losing strategy.

**Corollary 4.** *TfT* $\equiv$ *WSLS*$^{-1}$ *for agent* $-i$.

**Proof.** The XOR logic gate for agent $-i$ shows the WSLS strategy is anti-correlated with the TfT strategy; this is the opposite behaviour of the XNOR logic gate for agent $i$, so flipping the behaviours of WSLS: $0 \rightarrow 1$ and $1 \rightarrow 0$ results in the strategy WSLS$^{-1}$ that is behaviourally the same as TfT. In Nowak's notation in the Matching Pennies game for agent $-i$ (who wins if pennies are mismatched), this results in an inverted WSLS: $[Hh, Ht, Th, Tt] \rightarrow [H, T, H, T]$, which is equivalent to $[11, 10, 01, 00] \rightarrow [1, 0, 1, 0]$.  $\square$

The TfT strategy for the MP game is a linearly separable learning task, given $S^i_{n-1}$, a single layer perceptron is sufficient to map $a^{-i}_{n-1}$ to the correct $a^i_n$ output [35]. The WSLS strategy for the MP game is not a linearly separable task because it is equivalent to an XNOR gate [36]: given $S^i_{n-1}$, a multi-layer perceptron is necessary to map $a^i_{n-1}$ and $u^i_{n-1}$ to $a^i_n$. This difference in the complexity of the computational task and the indistinguishable character of the subsequent behaviour of the agents suggests that understanding cognitive decision-making processes is not easily untangled by observing behaviour.

## 5. Discussion

This article uses Transfer Entropy in the analysis of iterated games while also contributing theoretical concepts to the analysis of the computational foundations of games and strategies (work on Matching Pennies can be found in [37] using a different measure of 'information flow'). In some respects, the analysis of games discussed here is very similar to the Elementary Cellular Automata (ECAs) work of Lizier and colleagues [38,39]. In ECAs, the number of agents is much larger than the two agent games considered here, and they form a potentially infinite spatial array of locally connected agents that switch states based on their own states and the states of their neighbours. ECAs are simpler than the agents considered in this article as they do not have a utility function associated with their collective behaviour, an added complexity that has been shown here to be either informative or redundant depending on the game and the strategy being played. The possibility that utility values are redundant is important in reward based learning and it is a poorly studied area. This can be seen in economic theory in which "revealed preference theory" plays a significant role in understanding the connection between behaviour and reward. The analogy between iterated games and elementary cellular automata has been studied earlier by in [40,41]. An important property of these ECAs is that they are massively parallel computational systems and some, such as Wolfram's rule 110, are capable of universal computation [42]. This is very different to conventional approaches to understanding economic foundations. The focus in economics is often on finding equilibrium solutions [43], whereas, in ECAs, the emphasis is on the dynamical properties of the system. From this perspective, ECAs have been studied as dynamical systems made up of parallel logic gates (see [44], p. 81) while logic gates have recently played an important role in information theory [45] and the computational biology of neural networks [46].

The classification of games and strategies as logic gates also has analogies in ECAs where Rule 90 [47] (using Wolfram's numbering system for ECAs) is based on the XOR operation, and it is the simplest non-trivial ECA [48]. If we label an agent in a Rule 90 ECA as $i$ and its two nearest neighbours as $i - 1$ and $i + 1$, and the state of $i$ at time $t$ as $s^i_t$, then Rule 90 updates agent $i$'s state at $t + 1$ to: $s^{1-i}_t \oplus s^{1+i}_t = s^i_{t+1}$. This can be seen as an elementary version of the more complex interactions depicted in the figures illustrated above and emphasises the point that both game theory and strategies in iterated games can be seen as computational processes, and potentially universal Turing processes, just as ECAs are [49]. From this point of view, if an idealised economy is seen as a collection of agents strategically interacting in a game-theoretical fashion, this can be interpreted as a large network of parallel and sequential computational operations that continuously "computes" the output of an economy.

These results are also important for understanding strategic behaviour at two different levels. At an individual's cognitive level, the neural processing of information can be represented as a combination of logic operations implemented by a (biological) neural network [50]. Similarly, earlier neuro-economic work experimentally connected the level of individual neural recordings with adaptive learning and behaviour in the iterated matching pennies game [18,19]. In [51], Lee and collaborators identified the ambiguity between Win-Stay, Lose-Switch and copying the other agent's previous move (Tit-for-Tat) but did not consider the issue any further. Previously, Nowak [8] had examined the strategic effectiveness of Win-Stay, Lose-Switch relative to Tit-for-Tat for the Prisoner's Dilemma but did not note the possibility of an ambiguity in Matching Pennies. The results here show why ambiguities occur and an approach for understanding how different computational processes (decision strategies), in combination with the particular game being played, results in a fundamental indeterminacy in differentiating "internal" strategies by observing "external" behaviour.

At the larger scale of collective behaviour, we would like to understand, and distinguish between, the strategies of those who are following the behaviour of others and those who are processing information in a more complex fashion based on their own past experience. It seems likely that realistic strategies would include a combination of both social influences and fundamental computations because both aspects play a part in the pay-off an agent receives, but splitting these processes out provides an insight into the foundations of collective behaviour. Herding behaviour and financial contagion have been suggested as a possible mechanism that drives financial market collapses [52,53], and so it is important both theoretically and empirically to understand the limits of our ability to detect these two different behaviours. Previously, information theory has been used to measure abrupt transitions in general [29,54] and in financial market collapses specifically [55,56], but little progress has been made in relating information sets and strategic computation in economic theory, particularly as it relates to fundamentalists versus herders [21].

**Acknowledgments:** This work was produced in support of ARC grant DP170102927.

**Conflicts of Interest:** The author declares no conflict of interest.

## References

1. Von Neumann, J.; Morgenstern, O. *Theory of Games and Economic Behavior*; Princeton University Press: Princeton, NJ, USA, 2007.
2. Nash, J. Non-cooperative games. *Ann. Math.* **1951**, *54*, 286–295.
3. Osborne, M.J.; Rubinstein, A. *A Course in Game Theory*; MIT Press: Cambridge, MA, USA, 1994.
4. Sato, Y.; Crutchfield, J.P. Coupled replicator equations for the dynamics of learning in multiagent systems. *Phys. Rev. E* **2003**, *67*, 015206.
5. Nowak, M.A.; May, R.M. Evolutionary games and spatial chaos. *Nature* **1992**, *359*, 826–829.
6. Hauert, C.; Doebeli, M. Spatial structure often inhibits the evolution of cooperation in the snowdrift game. *Nature* **2004**, *428*, 643–646.
7. Axelrod, R.M. *The Evolution of Cooperation*; Basic Books: New York, NY, USA, 2006.
8. Nowak, M.A. *Evolutionary Dynamics*; Harvard University Press: Harvard, MA, USA, 2006.
9. Axelrod, R. Effective choice in the prisoner's dilemma. *J. Confl. Resolut.* **1980**, *24*, 3–25.
10. Axelrod, R.M. *The Complexity of Cooperation: Agent-Based Models of Competition and Collaboration*; Princeton University Press: Princeton, NJ, USA, 1997.
11. Nowak, M.; Sigmund, K. A strategy of win-stay, lose-shift that outperforms tit-for-tat in the Prisoner's Dilemma game. *Nature* **1993**, *364*, 56–58.
12. Arthur, W.B. Inductive reasoning and bounded rationality. *Am. Econ. Rev.* **1994**, *84*, 406–411.
13. Tesfatsion, L. Agent-based computational economics: Modeling economies as complex adaptive systems. *Inf. Sci.* **2003**, *149*, 262–268.

14. Binmore, K. *Rational Decisions*; Princeton University Press: Princeton, NJ, USA, 2008.
15. Savage, L.J. *The Foundations of Statistics*; Courier Corporation: North Chelmsford, MA, USA, 1954.
16. Glimcher, P.W.; Fehr, E. *Neuroeconomics: Decision Making and the Brain*; Academic Press: Cambridge, MA, USA, 2013.
17. Sanfey, A.G. Social decision-making: Insights from game theory and neuroscience. *Science* **2007**, *318*, 598–602.
18. Lee, D. Game theory and neural basis of social decision making. *Nat. Neurosci.* **2008**, *11*, 404–409.
19. Barraclough, D.J.; Conroy, M.L.; Lee, D. Prefrontal cortex and decision making in a mixed-strategy game. *Nat. Neurosci.* **2004**, *7*, 404–410.
20. Camerer, C.; Loewenstein, G.; Prelec, D. Neuroeconomics: How neuroscience can inform economics. *J. Econ. Lit.* **2005**, *43*, 9–64.
21. Lux, T.; Marchesi, M. Scaling and criticality in a stochastic multi-agent model of a financial market. *Nature* **1999**, *397*, 498–500.
22. Tedeschi, G.; Iori, G.; Gallegati, M. Herding effects in order driven markets: The rise and fall of gurus. *J. Econ. Behav. Organ.* **2012**, *81*, 82–96.
23. Goeree, J.K.; Holt, C.A. Ten little treasures of game theory and ten intuitive contradictions. *Am. Econ. Rev.* **2001**, *91*, 1402–1422.
24. Kerr, B.; Riley, M.A.; Feldman, M.W.; Bohannan, B.J. Local dispersal promotes biodiversity in a real-life game of rock–paper–scissors. *Nature* **2002**, *418*, 171–174.
25. Skyrms, B. *The Stag Hunt and the Evolution of Social Structure*; Cambridge University Press: Cambridge, UK, 2004.
26. Cover, T.M.; Thomas, J.A. *Elements of Information Theory*; John Wiley & Sons: Hoboken, NJ, USA, 2012.
27. Schreiber, T. Measuring information transfer. *Phys. Rev. Lett.* **2000**, *85*, 461.
28. Wibral, M.; Pampu, N.; Priesemann, V.; Siebenhühner, F.; Seiwert, H.; Lindner, M.; Lizier, J.T.; Vicente, R. Measuring information-transfer delays. *PLoS ONE* **2013**, *8*, e55809.
29. Barnett, L.; Lizier, J.T.; Harré, M.; Seth, A.K.; Bossomaier, T. Information flow in a kinetic Ising model peaks in the disordered phase. *Phys. Rev. Lett.* **2013**, *111*, 177203.
30. Bossomaier, T.; Barnett, L.; Harré, M.; Lizier, J.T. *An Introduction to Transfer Entropy: Information Flow in Complex Systems*; Springer: Cham, Switzerland, 2016.
31. Lizier, J.T.; Prokopenko, M.; Zomaya, A.Y. Local measures of information storage in complex distributed computation. *Inf. Sci.* **2012**, *208*, 39–54.
32. Basu, K. The traveler's dilemma: Paradoxes of rationality in game theory. *Am. Econ. Rev.* **1994**, *84*, 391–395.
33. Wolpert, D.; Jamison, J.; Newth, D.; Harré, M. Strategic choice of preferences: The persona model. *BE J. Theor. Econ.* **2011**, *11*, doi:10.2202/1935-1704.1593.
34. Weibull, J.W. *Evolutionary Game Theory*; MIT Press: Cambridge, MA, USA, 1997.
35. Minsky, M.; Papert, S. *Perceptrons*; MIT Press: Cambridge, MA, USA, 1969.
36. Rumelhart, D.E.; Hinton, G.E.; Williams, R.J. *Learning Internal Representations by Error Propagation*; Technical Report; DTIC Document; DTIC: Fort Belvoir, VA, USA, 1985.
37. Sato, Y.; Ay, N. Information flow in learning a coin-tossing game. *Nonlinear Theory Its Appl. IEICE* **2016**, *7*, 118–125.
38. Lizier, J.T.; Prokopenko, M.; Zomaya, A.Y. Local information transfer as a spatiotemporal filter for complex systems. *Phys. Rev. E* **2008**, *77*, 026110.
39. Lizier, J.T.; Prokopenko, M.; Zomaya, A.Y. Information modification and particle collisions in distributed computation. *Chaos Interdiscip. J. Nonlinear Sci.* **2010**, *20*, 037109.
40. Albin, P.S.; Foley, D.K. *Barriers and Bounds to Rationality: Essays on Economic Complexity and Dynamics in Interactive Systems*; Princeton University Press: Princeton, NJ, USA, 1998.
41. Schumann, A. Payoff Cellular Automata and Reflexive Games. *J. Cell. Autom.* **2014**, *9*, 287–313.
42. Cook, M. Universality in elementary cellular automata. *Complex Syst.* **2004**, *15*, 1–40.
43. Farmer, J.D.; Geanakoplos, J. The virtues and vices of equilibrium and the future of financial economics. *Complexity* **2009**, *14*, 11–38.
44. Schiff, J.L. *Cellular Automata: A Discrete View of the World*; John Wiley & Sons: Hoboken, NJ, USA, 2011; Volume 45.
45. Griffith, V.; Koch, C. Quantifying synergistic mutual information. In *Guided Self-Organization: Inception*; Springer: Berlin, Germany, 2014; pp. 159–190.

46. Narayanan, N.S.; Kimchi, E.Y.; Laubach, M. Redundancy and synergy of neuronal ensembles in motor cortex. *J. Neurosci.* **2005**, *25*, 4207–4216.

47. Wolfram, S. Statistical mechanics of cellular automata. *Rev. Mod. Phys.* **1983**, *55*, 601.

48. Martin, O.; Odlyzko, A.M.; Wolfram, S. Algebraic properties of cellular automata. *Commun. Math. Phys.* **1984**, *93*, 219–258.

49. Langton, C.G. Self-reproduction in cellular automata. *Phys. D Nonlinear Phenom.* **1984**, *10*, 135–144.

50. Wibral, M.; Priesemann, V.; Kay, J.W.; Lizier, J.T.; Phillips, W.A. Partial information decomposition as a unified approach to the specification of neural goal functions. *Brain Cogn.* **2015**, doi:10.1016/j.bandc.2015.09.004.

51. Lee, D.; Conroy, M.L.; McGreevy, B.P.; Barraclough, D.J. Reinforcement learning and decision making in monkeys during a competitive game. *Cogn. Brain Res.* **2004**, *22*, 45–58.

52. Devenow, A.; Welch, I. Rational herding in financial economics. *Eur. Econ. Rev.* **1996**, *40*, 603–615.

53. Bekaert, G.; Ehrmann, M.; Fratzscher, M.; Mehl, A. The global crisis and equity market contagion. *J. Financ.* **2014**, *69*, 2597–2649.

54. Langton, C.G. Computation at the edge of chaos: Phase transitions and emergent computation. *Phys. D Nonlinear Phenom.* **1990**, *42*, 12–37.

55. Harré, M.; Bossomaier, T. Phase-transition—Like behaviour of information measures in financial markets. *EPL Europhys. Lett.* **2009**, *87*, 18009.

56. Harré, M. Entropy and Transfer Entropy: The Dow Jones and the Build up to the 1997 Asian Crisis. In *Proceedings of the International Conference on Social Modeling and Simulation, plus Econophysics Colloquium 2014*; Springer: Cham, Switzerland, 2015; pp. 15–25.

*entropy*

MDPI

*Article*

# Consensus of Second Order Multi-Agent Systems with Exogenous Disturbance Generated by Unknown Exosystems

Xuxi Zhang [1], Qidan Zhu [2,*] and Xianping Liu [1]

[1]   College of Science, Harbin Engineering University, Harbin 150001, China; zxx@hrbeu.edu.cn (X.Z.); lxp@hrbeu.edu.cn (X.L.)
[2]   College of Automation, Harbin Engineering University, Harbin 150001, China
*    Correspondence: zhuqidan@hrbeu.edu.cn; Tel.: +86-451-8251-8940

Academic Editor: Mikhail Prokopenko
Received: 25 July 2016; Accepted: 23 November 2016; Published: 25 November 2016

**Abstract:** This paper is concerned with consensus problem of a class of second-order multi-agent systems subjecting to external disturbance generated from some unknown exosystems. In comparison with the case where the disturbance is generated from some known exosystems, we need to combine adaptive control and internal model design to deal with the external disturbance generated from the unknown exosystems. With the help of the internal model, an adaptive protocol is proposed for the consensus problem of the multi-agent systems. Finally, one numerical example is provided to demonstrate the effectiveness of the control design.

**Keywords:** consensus; multi-agent systems; internal model; disturbance; unknown exosystems

## 1. Introduction

The consensus problem of multi-agent systems has received increasing attention in recent years due to its broad applications in such areas as cooperative control of unmanned aircrafts and underwater vehicles, flocking of mobile vehicles, communication among wireless sensor networks, rendezvous, formation control, and so on, see [1–15]. In the past years, many researches have been firstly concerned with consensus problems of first order multi-agent systems [16–20]. In [16], the authors proposed a systematic framework to study the consensus problem of first-order multi-agent systems and showed that the consensus can be achieved if the diagraph is strongly connected. In [17], the authors extended the results obtained in [16] and further presented some improved conditions for state agreement under dynamically changing directed topology. In [18], the authors discussed average consensus problem by using a linear matrix inequality method in undirected networks of dynamic agents with fixed and switching topologies as well as multiple time-varying communication delays.

Recently, the consensus problem of second order multi-agent systems has received increasing attention due to the fact that second order dynamics can be used to model more complicated processes in reality [21–26]. In reality, many practical individual systems, especially mechanical systems, can be presented as second-order multi-agent systems; for instance, networks of mass-spring systems[27], coupled pendulum systems[28], harmonic oscillators [29] and frequency control of power systems [30]. In [21], the authors pointed out that the existence of a directed spanning tree is a necessary rather than a sufficient condition to reach the second order consensus. In [22], the authors discussed the consensus problems for undirected networks of point mass dynamic agents with fixed or switching topology. In [23], the authors proposed a Lyapunov-based approach to consider multi-agent systems with switching jointly connected interconnection. In [24], the authors presented some necessary and sufficient conditions for second order consensus in multi-agent dynamical systems. In [25], the authors studied the exponential second order consensus problem of a network of inertial agents using passive decomposition approach with time-varying coupling delays and variable balanced topologies.

However, there are few results that have considered the second order consensus problem for multi-agent systems with exogenous disturbance [31,32]. In [31], by using linear matrix inequality method, the authors studied the consensus problem of second order multi-agent systems with exogenous disturbances generated from linear exogenous system under the assumption that the coefficient matrix of the exogenous system can be used for designing a disturbance observer, and a disturbance observer based protocol was proposed to achieve consensus for the second order multi-agent systems. In [32], by using the input-to-state stability and dynamic gain technique, Zhang et al. further investigated the consensus problem of second order multi-agent systems with exogenous disturbances generated from linear exogenous system and nonlinear exogenous system, respectively.

Nevertheless, the case when consensus problem of multi-agent systems with exogenous disturbance generated from linear unknown exogenous system seems more realistic and has greater practical significance [33–35]. In this paper, we will consider the consensus problem of second order multi-agent systems with exogenous disturbance generated from linear unknown exogenous system. It is worth noting that, unlike [31,32], since the disturbances are generated from some linear unknown exogenous systems and the information of the coefficient matrix of the exogenous system can not be used for designing of disturbance observer and feedback control, we cannot apply the approaches developed in [31,32] to solve the present problem. Meanwhile, the method that used in [34] to solve the problem of asymptotic rejection of unknown sinusoidal disturbances can not be used directly to tackle the consensus problem of multi-agent system, because the multi-agent system is multi-input and multi-output. Therefore, to overcome this difficulty, we need to develop a different technique.

The remainder of this paper is organized as follows. In Section 2, some preliminaries are briefly reviewed and the problem formulation is presented. Some internal models, which are used to deal with the disturbances generated from some linear unknown exosystems, are designed in Section 3. Based on the internal models proposed in Section 3, an adaptive consensus protocol is presented for the second order multi-agent systems in Section 4. In Section 5, an example will be given to illustrate our design. Finally, the conclusions are drawn in Section 6.

## 2. Preliminaries and Problem Formulation

Assuming that each agent can be viewed as a node, and the interaction topology of information exchange between $n$ nodes can be described by a graph $\mathcal{G} = (\mathcal{V}, \mathcal{E}, \mathcal{A})$, where $\mathcal{V} = \{1, \cdots, n\}$ be an index set of $n$ nodes with $i$ representing the $i$th node, $\mathcal{E} \subseteq \mathcal{V} \times \mathcal{V}$ is the set of edges of paired nodes and $\mathcal{A} = [a_{ij}] \in \mathbb{R}^{N \times N}$ with non-negative adjacency elements $a_{ij}$ is the weighted adjacency matrix of the graph $\mathcal{G}$. An edge of $\mathcal{G}$ is denoted by $(i, j)$, representing that node $i$ can get information from node $j$. The adjacency elements associated with the edges are positive, i.e., $(i, j) \in \mathcal{E}$ if and only if $a_{ij} > 0$. Moreover, it is assumed that $a_{ii} = 0$ for all $i \in \mathcal{V}$. A graph is called an undirected graph if the graph has the property that $a_{ij} = a_{ji}$ for any $i, j \in \mathcal{V}$. The neighborhood of node $i$ is denoted by $\mathcal{N}_i = \{j \in \mathcal{V} : (j, i) \in \mathcal{E}\}$. A path on $\mathcal{G}$ from node $i_1$ to node $i_n$ is a sequence of ordered edges of the form $(i_k, i_{k+1}) \in \mathcal{E}, k = 1, \cdots, n - 1$, and $i_k$'s are distinct. A graph $\mathcal{G}$ is said to be connected if there exists a path from node $i$ to node $j$ for any two nodes $i, j \in \mathcal{E}$. A diagonal matrix $\mathcal{D} = diag\{d_1, \cdots, d_n\}$ is a degree matrix of graph $\mathcal{G}$, whose diagonal matrix elements $d_i = \sum_{j \in \mathcal{N}_i} a_{ij}, i \in \mathcal{V}$. Then, the Laplacian matrix of a weighted graph can be defined as $\mathcal{L} = \mathcal{D} - \mathcal{A}$, which is a symmetric positive semi-definite matrix.

Considering a group of agents, the dynamics of the $i$th agent is given by

$$\begin{aligned} \dot{x}_i &= v_i, \\ \dot{v}_i &= -\sum_{j \in \mathcal{N}_i} a_{ij}[(x_i - x_j) + \gamma(v_i - v_j)] + u_i + g_i d_i, \quad i \in \mathcal{V}, \end{aligned} \tag{1}$$

where $x_i \in \mathbb{R}^m$ and $v_i \in \mathbb{R}^m$ are the position and velocity of agent $i$, respectively. $a_{ij}$ is the $(i,j)$th entry of the adjacency matrix, and $\gamma > 0$ denotes a scaling factor. $u_i \in \mathbb{R}^m$ and $g_i \in \mathbb{R}^m$ denote the control input and a coefficient matrix, respectively. $d_i \in \mathbb{R}$ is the external disturbance, which is generated from the following unknown exosystem

$$
\begin{aligned}
\dot{\zeta}_i &= A_i \zeta_i, \\
d_i &= C_i \zeta_i,
\end{aligned}
\tag{2}
$$

where $\zeta_i \in \mathbb{R}^{m_i}$, $A_i \in \mathbb{R}^{m_i \times m_i}$ and $C_i \in \mathbb{R}^{1 \times m_i}$ are the coefficient matrices.

As in [31] and [32], we assume that the desired state is described by

$$
\dot{\bar{x}} = \bar{v},
\tag{3}
$$

where $\bar{x} \in \mathbb{R}^m$ and $\bar{v} \in \mathbb{R}^m$ are the position and velocity of the leader agent, respectively.

**Definition 1.** *The consensus problem of the multi-agent systems* (1) *is formulated as follows: For the multi-agent systems* (1), *design an adaptive consensus protocol such that the states of the close-loop system exist and are bounded, and the states of agents satisfy*

$$
\lim_{t \to \infty} \|x_i - \bar{x}\| = 0, \lim_{t \to \infty} \|v_i - \bar{v}\| = 0,
\tag{4}
$$

*for any initial values* $x_i(0)$ *and* $v_i(0), i \in \mathcal{V}$.

**Remark 1.** *Note that, unlike the cases in [31,32], we allow that the disturbance* $d_i, i \in \mathcal{V}$ *is generated from different unknown exosystems, which makes our problem more challenging and realistic.*

## 3. Designing of Internal Models

In this section, in order to deal with the external disturbances, we will design some internal models. To this end, let

$$
\begin{aligned}
s_i(t) &= [x_i^T(t), v_i^T(t)]^T, \\
s(t) &= [x_1^T(t), \cdots, x_n^T(t), v_1^T(t), \cdots, v_n^T(t)]^T.
\end{aligned}
$$

Then, it follows from system (1) that

$$
\dot{s}_i = L_i s(t) + H_i u_i + G_i d_i
\tag{5}
$$

where $H_i = [0_{m \times m}, I_m]^T \in \mathbb{R}^{2m \times m}$, $G_i = [0_{m \times 1}^T, g_i^T]^T \in \mathbb{R}^{2m \times 1}$, $L_i$ is a matrix with its rows are chosen from rows $(i-1)m+1$ to $im$ and from rows $mn+(i-1)m+1$ to $mn+im$ of the following matrix

$$
\begin{bmatrix}
0_{mn \times mn} & I_n \otimes I_m \\
-\mathcal{L} \otimes I_m & -\gamma \mathcal{L} \otimes I_m
\end{bmatrix}.
$$

Before proceeding further, some standard assumptions are introduced as follows:

**Assumption 1.** *The matrix pair* $(A_i, C_i), i \in \mathcal{V}$ *is observable, and the eigenvalues of* $A_i, i \in \mathcal{V}$ *are with zero real parts and are distinct.*

**Assumption 2.** *There exists a function* $h_i(s_i) : \mathbb{R}^{2m} \to \mathbb{R}^{m_i}, i \in \mathcal{V}$, *such that* $\frac{\partial h_i(s_i)}{\partial s_i} G_i = N_i, i \in \mathcal{V}$, *where* $N_i, i \in \mathcal{V}$ *is a nonzero constant vector in* $\mathbb{R}^{m_i}$.

Under the Assumptions 1 and 2, for any nonzero vector $N_i, i \in \mathscr{V}$, there exists a Hurwitz matrix $M_i, i \in \mathscr{V}$ such that $(M_i, N_i), i \in \mathscr{V}$ is controllable.

Now, we can define a dynamic system of the following form

$$\dot{z}_i = M_i z_i + M_i h_i(s_i) - \frac{\partial h_i(s_i)}{\partial s_i}(L_i s + H_i u_i), i \in \mathscr{V}, \tag{6}$$

which is called an internal model and can be used to handle the disturbance $d_i$ generated from (2).

Furthermore, there exists a nonsingular matrix $T_i, i \in \mathscr{V}$ satisfying the following Sylvester equation

$$T_i A_i - M_i T_i = N_i C_i, \; i \in \mathscr{V}, \tag{7}$$

because the pair $(M_i, N_i), i \in \mathscr{V}$, is controllable with $M_i$ being Hurwitz, and the pair $(M_i, N_i), i \in \mathscr{V}$, is observable.

With the internal model (6) and Sylvester Equation (7) ready, the biased error can be defined by

$$e_i = T_i \xi_i - z_i - h_i(s_i), i \in \mathscr{V}. \tag{8}$$

Then, it can be verified that the internal model (6) and the biased error (8) have a nice property as described in the following lemma.

**Lemma 1.** *There exist some positive constants $d_{e_i}$ and $\lambda_{e_i}$ such that the biased error defined by (8) satisfies the following inequality*

$$\|e_i\| \leq d_{e_i} e^{-\lambda_{e_i} t}, i \in \mathscr{V}, \tag{9}$$

*which implies that $e_i$ is exponentially stable.*

**Proof.** Firstly, by Equations (2), (5) and (6), a straightforward computation shows that

$$\dot{e}_i = T_i \dot{\xi}_i - \dot{z}_i - \frac{\partial h_i(s_i)}{\partial s_i} \dot{s}_i$$

$$= T_i A_i \xi_i - [M_i z_i + M_i h_i(s_i) - \frac{\partial h_i(s_i)}{\partial s_i}(L_i s + H_i u_i)]$$

$$- \frac{\partial h_i(s_i)}{\partial s_i}[L_i s(t) + H_i u_i + G_i d_i].$$

Then, under the Assumption 2, by the Sylvester Equation (7), one has

$$\dot{e}_i = (M_i T_i + N_i C_i)\xi_i - [M_i z_i + M_i h_i(s_i) - \frac{\partial h_i(s_i)}{\partial s_i}(L_i s + H_i u_i)]$$

$$- \frac{\partial h_i(s_i)}{\partial s_i}[L_i s(t) + H_i u_i + G_i d_i]$$

$$= (M_i T_i + N_i C_i)\xi_i - [M_i z_i + M_i h_i(s_i)] - N_i d_i$$

$$= M_i e_i. \tag{10}$$

Next, using the Lyapunov stability theory of [36], it is easy to verify that the solution of $e_i$ system (10) can be given as

$$e_i(t) = e_i(0) e^{M_i t}.$$

Furthermore, owing to $M_i$ is a Hurwitz matrix, it follows that there exist some positive constants $d_{i0}$ and $\lambda_{i0}$ such that

$$\|e^{M_i t}\| \leq d_{i0} e^{-\lambda_{i0} t},$$

which implies that

$$\|e_i(t)\| \leq d_{ei} e^{-\lambda_{ei} t},$$

where $d_{ei} = d_{i0}\|e_i(0)\|$ and $\lambda_{ei} = \lambda_{i0}$. $\square$

**Remark 2.** *Based on Lemma 1, it can be shown that, for the following first-order system*

$$\dot{\bar{e}}_i = -\lambda_{ei}\bar{e}_i, \bar{e}_i(0) = d_{ei}, i \in \mathcal{V},\tag{11}$$

*where $d_{ei}$ and $\lambda_{ei}$ are the same positive constants given in Lemma 1, the following nice property*

$$\|e_i(t)\| \le \bar{e}_i(t), i \in \mathcal{V},\tag{12}$$

*is hold, which is very useful for managing the disturbances caused by the unknown exosystems (2).*

## 4. Main Result

In this section, we will present an adaptive protocol for solving the consensus problem of the multi-agent systems (1). To do this, we further make one more standard assumption and recall one lemma which can be found in [31,32,37].

**Assumption 3.** *The graph $\mathcal{G}$ describing the interaction topology is connected.*

**Lemma 2.** *Under Assumption 3, suppose that $\gamma > 0$ is a positive real number, then the following matrix*

$$\begin{bmatrix} 0 & I_n \\ -(\mathcal{L} + \mathcal{B}) & -\gamma(\mathcal{L} + \mathcal{B}) \end{bmatrix}$$

*is Hurwitz, where $\mathcal{B} = diag\{b_1, \cdots, b_n\}$ with $b_i$ is the control gain of control law (16), and $b_i > 0$ if agent is pinned, otherwise, $b_i = 0$.*

In order to make the problem more tractable, let

$$\eta_i = T_i \xi_i,\tag{13}$$

where $T_i$ is the nonsingular matrix satisfying the Sylvester Equation (7). Then, we have

$$\dot{\eta}_i = T_i A_i \xi_i.\tag{14}$$

By (7) and (13), one can obtain that

$$\begin{aligned} \dot{\eta}_i &= M_i \eta_i + N_i \psi_i \eta_i, \\ d_i &= \psi_i \eta_i, i \in \mathcal{V}, \end{aligned}\tag{15}$$

where $\psi_i = C_i T_i^{-1}$ is unknown vector since $C_i$ and $T_i$ are unknown matrices.

**Remark 3.** *It is worth pointing out that, after the linear transformation (13), the external disturbance $d_i, i \in \mathcal{V}$, can be generated from the system (13), in which $M_i$ is a known Hurwitz matrix and only the matrix $\psi_i, i \in \mathcal{V}$, is unknown. Thus, one can estimate the disturbance $d_i$ through estimating the unknown constant vector $\psi_i, i \in \mathcal{V}$.*

Now, we are ready to state our main result.

**Theorem 1.** *Under Assumptions 1–3, with the help of the internal model presented in (6), the adaptive protocol given by*

$$u_i = -b_i[(x_i - \bar{x}) + \gamma(v_i - \bar{v})] - g_i\hat{\psi}_i(z_i + h_i(s_i)),$$
$$\dot{\hat{\psi}}_i = \varrho_i\omega^T P_{n+i}g_i(z_i + h_i(s_i))^T, i \in \mathscr{V}, \tag{16}$$

*where the control gain $b_i > 0$ if agent is pinned, otherwise, $b_i = 0$, $\hat{\psi}_i$ is the estimation of the unknown vector $\psi_i$, $\varrho_i$ is a positive constant which is used to modify the update rate, $\omega$ and $P_{n+i}$ are defined by (18) and (22), respectively, solves the consensus problem of second order multi-agent systems (1) with external disturbance generated from linear unknown exosystem (2).*

**Proof.** let

$$\tilde{x}_i(t) = x_i(t) - \bar{x},$$
$$\tilde{v}_i(t) = v_i(t) - \bar{v}, i \in \mathscr{V}, \tag{17}$$

and

$$\tilde{x}(t) = [\tilde{x}_1^T(t), \cdots, \tilde{x}_n^T(t)]^T,$$
$$\tilde{v}(t) = [\tilde{v}_1^T(t), \cdots, \tilde{v}_n^T(t)]^T,$$
$$w = [\tilde{x}^T, \tilde{v}^T]^T,$$
$$\eta = [\eta_1^T, \cdots, \eta_n^T]^T,$$
$$z = [z_1^T, \cdots, z_n^T]^T, \tag{18}$$
$$h = [h_1^T, \cdots, h_n^T]^T,$$
$$\psi = diag(\psi_1, \cdots, \psi_n),$$
$$\hat{\psi} = diag(\hat{\psi}_1, \cdots, \hat{\psi}_n).$$

Then, by combining Equations (1), (2), (15) and (16), the following system can be derived

$$\dot{w} = \mathscr{L}w + \Psi[\psi\eta - \hat{\psi}(z + h)], \tag{19}$$

where

$$\mathscr{L} = \begin{bmatrix} 0 & I_n \otimes I_m \\ -(\mathscr{L} + \mathscr{B}) \otimes I_m & -\gamma(\mathscr{L} + \mathscr{B}) \otimes I_m \end{bmatrix},$$
$$\Psi = \begin{bmatrix} I_n \otimes 0_{m \times 1} \\ g \end{bmatrix},$$
$$g = blockdiag(g_1, \cdots, g_n).$$

Furthermore, as noted in [32], according to the Lemma 2 and Theorem 4.2.12 of [38], it follows that the matrix $\mathscr{L}$ is Hurwitz.

Next, consider the following Lyapunov function candidate

$$V = \omega^T P\omega + \frac{2}{\varrho_i}\sum_{i=1}^{n}\tilde{\psi}_i\tilde{\psi}_i^T + \frac{1}{2}\sum_{i=1}^{n}c_i\bar{e}_i^2, \tag{20}$$

where $\tilde{\psi}_i = \psi_i - \hat{\psi}_i$, $c_i, i \in \mathscr{V}$, is a positive real constant number which will be specified later, $\bar{e}_i$ is the state defined by (11), and $P$ is a positive definite matrix satisfying the following Lyapunov equation

$$P\mathscr{L} + \mathscr{L}^T P = -I.$$

The existence of the matrix $P$ is due to the Hurwitzness of $\mathcal{L}$.

Then, taking the derivative of $V$ along the system composed of (11), (16) and (19) gives

$$
\begin{aligned}
\dot{V} &= \dot{\omega}^T P \omega + \omega^T P \dot{\omega} + \frac{2}{\varrho_i} \sum_{i=1}^n \tilde{\psi}_i \dot{\tilde{\psi}}_i^T + \sum_{i=1}^n c_i \bar{e}_i \dot{\bar{e}}_i \\
&= -\|\omega\|^2 + 2\omega^T P \Psi [\psi \eta - \hat{\psi}(z+h)] \\
&\quad - \frac{2}{\varrho_i} \sum_{i=1}^n \tilde{\psi}_i \dot{\hat{\psi}}_i^T - \sum_{i=1}^n c_i \lambda_{e_i} \bar{e}_i^2.
\end{aligned}
\tag{21}
$$

Now, in order to overcome the difficulties caused by the unknown vectors $\psi_i, i \in \mathcal{V}$, let us split the matrix $P$ as

$$
P = [P_1, \cdots, P_n, P_{n+1}, \cdots, P_{2n}],
\tag{22}
$$

where $P_i, i = 1, 2, \cdots, 2n$, are $2mn \times m$ blocks.

Then, in light of (8) and (22), it follows from (21) that

$$
\begin{aligned}
\dot{V} &= -\|\omega\|^2 + 2\sum_{i=1}^n \omega^T P_{n+i} g_i \tilde{\psi}_i (z_i + h_i) \\
&\quad + 2\sum_{i=1}^n \omega^T P_{n+i} g_i \psi_i e_i - \frac{2}{\varrho_i} \sum_{i=1}^n \tilde{\psi}_i \dot{\hat{\psi}}_i^T - \sum_{i=1}^n c_i \lambda_{e_i} \bar{e}_i^2.
\end{aligned}
\tag{23}
$$

Furthermore, substituting the adaptive law $\dot{\hat{\psi}}_i$ proposed by (16) into (23) yields that

$$
\begin{aligned}
\dot{V} &= -\|\omega\|^2 + 2\sum_{i=1}^n \omega^T P_{n+i} g_i \psi_i e_i - \sum_{i=1}^n c_i \lambda_{e_i} \bar{e}_i^2 \\
&\leq -\|\omega\|^2 + \sum_{i=1}^n [\epsilon_i \|\omega\|^2 + \frac{1}{\epsilon_i} \|P_{n+i} g_i \psi_i\|^2 \|e_i\|^2] - \sum_{i=1}^n c_i \lambda_{e_i} \bar{e}_i^2,
\end{aligned}
\tag{24}
$$

where $\epsilon_i, i = 1, 2, \cdots, n$, are any positive real constants.

From the inequality (12), we obtain that

$$
\dot{V} \leq -(1-\epsilon)\|\omega\|^2 + \sum_{i=1}^n \left(\frac{1}{\epsilon_i} \|P_{n+i} g_i \psi_i\|^2 - c_i \lambda_{e_i}\right) \bar{e}_i^2,
\tag{25}
$$

where $\epsilon = \sum_{i=1}^n \epsilon_i$.

Next, choosing $\epsilon \leq \frac{1}{2}$, and $c_i = \frac{2}{\lambda_{e_i} \epsilon_i} \|P_{n+i} g_i \psi_i\|^2$, which will lead to

$$
\dot{V} \leq -\frac{1}{2}\|w\|^2.
\tag{26}
$$

Hence, we can conclude that all the variables are bounded. Finally, by invoking the Barbalat's Lemma, one can obtain that

$$
\lim_{t\to\infty} \tilde{x}(t) = 0,
$$

$$
\lim_{t\to\infty} \tilde{v}(t) = 0,
$$

which complete this proof. □

**Remark 4.** *It is worth pointing out that, distributed proportional-integral control law was also studied in [30] for second-order multi-agent systems with constant disturbances. Unlike the results in [30], the disturbances in this paper are assumed to be generated from some unknown exosystems, which include the constant disturbance as special case. In addition, the results in this paper are proved by combining Lyapunov-based method and adaptive control technique, which are totally different proof techniques from that used in [30].*

## 5. Illustrative Example

In this section, an example will be provided to illustrate our design. The model parameters are taken from [31,32] with some adjustments. We assume that there are ten agents with an undirected communication graph $\mathcal{G}$ shown in Figure 1. The gain $\gamma$ is set to 1 and the coefficient matrix of system (1) is $g_i = 1$, respectively. The desired consensus state is described by $\dot{x} = 0.08$. However, unlike [31,32], we assume the disturbance $d_i$ is generated from

$$\dot{\xi}_i = \begin{bmatrix} 0 & \sigma_i \\ -\sigma_i & 0 \end{bmatrix} \xi_i,$$

$$d_i = \begin{bmatrix} 1 & 0 \end{bmatrix} \xi, \ i \in \mathcal{V}.$$

Then, we have

$$A_i = \begin{bmatrix} 0 & \sigma_i \\ -\sigma_i & 0 \end{bmatrix}, \ C_i = \begin{bmatrix} 1 & 0 \end{bmatrix}, \ G_i = \begin{bmatrix} 0 \\ 1 \end{bmatrix}. \tag{27}$$

Let $h_i(s_i) = s_i$, one has

$$\frac{\partial h_i(s_i)}{s_i} G_i = \begin{bmatrix} 0 \\ 1 \end{bmatrix}.$$

Therefore, Assumptions 1–3 are satisfied.

Furthermore, select

$$M_i = \begin{bmatrix} 0 & 1 \\ -9 & -8 \end{bmatrix}, \ i \in \mathcal{V},$$

such that $(M_i, N_i)$ is controllable with $M_i$ being Hurwitz.

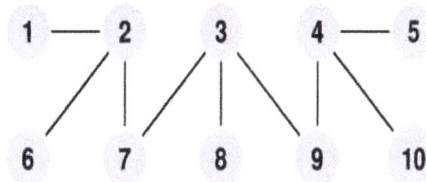

**Figure 1.** Communication graph $\mathcal{G}$.

Then, based on the proposed approach, the internal model (6) and adaptive protocol (16) can be designed. Numerical simulations are conducted to show the performance of the presented control law. Some of the results are depicted in Figures 2 and 3 with initial conditions of states and initial velocities of agents are chosen randomly from $[0,4]$ and $[0,5]$, respectively. The unknown parameters of the exosystems are set as $\sigma_1 = 0.1, \sigma_2 = 0.2, \sigma_3 = 0.3, \sigma_4 = 0.4, \sigma_5 = 0.5, \sigma_6 = 0.6, \sigma_7 = 0.7, \sigma_8 = 0.8,$ $\sigma_9 = 0.9, \sigma_{10} = 1$, and the initial conditions of the exosystem are all set as $\xi_i(0) = [0.5 \sin 1, 0.5 \cos 1]^T$. The pining control gains are selected as $b_2 = b_4 = 1$. All the other initial conditions in the controller are set to zero. From Figures 2 and 3, it can be seen that the consensus protocol proposed in this paper allows the agents to reach consensus, in the presence of external disturbance generated from some unknown exosystems.

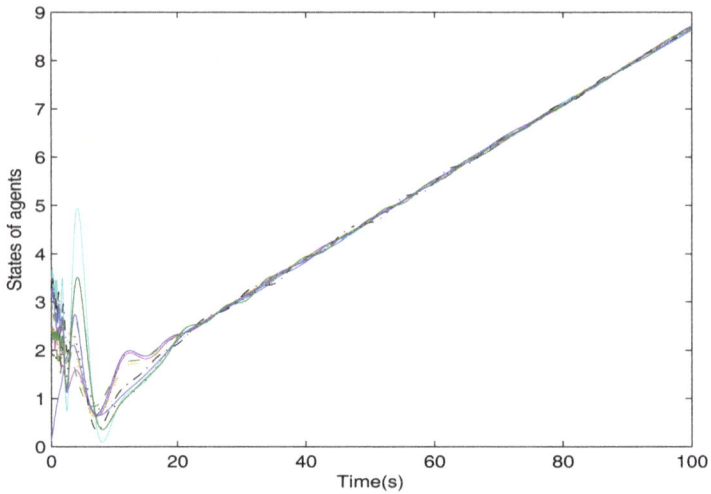

**Figure 2.** States of the agents.

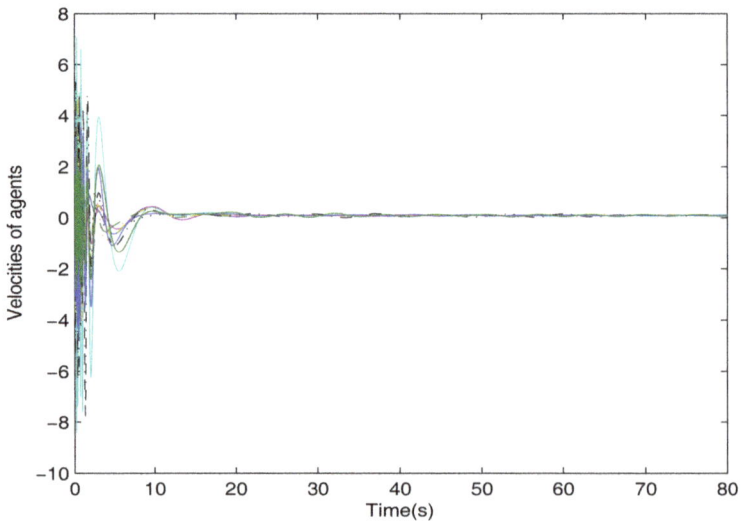

**Figure 3.** Velocities of the agents.

## 6. Conclusions

This paper address a consensus problem of second order multi-agent systems with exogenous disturbance generated by unknown exosystems. A class of internal model was proposed for deal with the disturbance caused by the unknown exosystems. Based on the internal model, an adaptive consensus protocol was presented for the second order multi-agent systems. Finally, the effectiveness of our results is validated by numerical simulations.

**Acknowledgments:** This work is supported partly by the NNSF of China under Grant 61503089, the Natural Science Foundation of Heilongjiang Province under Grant QC2015077, the Fundamental Research for the Central Universities under Grant HEUCFM161101.

**Author Contributions:** Xuxi Zhang and Qidan Zhu conceived and designed the control law; Xianping Liu performed the experiments. All authors have read and approved the final manuscript.

**Conflicts of Interest:** The authors declare no conflict of interest.

## References

1. Chen, J.; Cao, X.; Cheng, P.; Xiao, Y.; Sun, Y. Distributed collaborative control for industrial automation with wireless sensor and actuator networks. *IEEE Trans. Ind. Electron.* **2010**, *57*, 4219–4230.

2. Cruz, D.; McClintock, J.; Perteet, B.; Orqueda, O.A.; Cao, Y.; Fierro, R. Decentralized cooperative control—A multivehicle platform for research in networked embedded systems. *IEEE Control Syst. Mag.* **2007**, *27*, 58–78.

3. Ni, W.; Cheng, D. Leader-following consensus of multi-agent systems under fixed and switching topologies. *Syst. Control Lett.* **2010**, *59*, 209–217.

4. Qin, J.; Yu, C.; Hirche, S. Stationary consensus of asynchronous discrete-time second-order multi-agent systems under switching topology. *IEEE Trans. Ind. Inform.* **2012**, *8*, 986–994.

5. Ren, W.; Beard, R.W.; Atkins, E.M. Information consensus in multivehicle cooperative control. *IEEE Control Syst. Mag.* **2007**, *2*, 71–82.

6. Wang, X.; Yadav, V.; Balakrishnan, S. Cooperative UAV formation flying with obstacle/collision avoidance. *IEEE Trans. Control Syst. Technol.* **2007**, *15*, 672–679.

7. Zhang, H.; Lewis, F.L.; Qu, Z. Lyapunov, adaptive, and optimal design techniques for cooperative systems on directed communication graphs. *IEEE Trans. Ind. Electron.* **2012**, *59*, 3026–3041.

8. Olfati-Saber, R.; Murray, R.M. Consensus protocols for networks of dynamic agents. In Proceedings of the 2003 American Controls Conference, Denver, CO, USA, 4–6 June 2003.

9. De la Prieta, F.; Heras, S.; Palanca, J.; Rodríguez, S.; Bajo, J.; Julián, V. Real-time agreement and fulfilment of SLAs in Cloud Computing environments. *AI Commun.* **2015**, *28*, 403–426.

10. Garcia, A.; Sanchez-Pi, N.; Correia, L.; Molina, J.M. Multi-agent simulations for emergency situations in an airport scenario. *Adv. Distrib. Comput. Artif. Intell. J.* **2012**, *1*, 69–73.

11. Song, J. Observer-Based Consensus Control for Networked Multi-Agent Systems with Delays and Packet-Dropouts. *Int. J. Innov. Comput. Inf. Control* **2016**, *12*, 1287–1302.

12. Liang, H.; Kang, F. Artificial Immune Intelligent Modeling for UUV Underwater Navigation System Based on Immune Multi-Agents Network. *Int. J. Innov. Comput. Inf. Control* **2015**, *11*, 1525–1537.

13. Wu, Y.; Su, H.; Shi, P.; Shu, Z.; Wu, Z.-G. Consensus of multi-agent systems using aperiodic sampled-data control. *IEEE Trans. Cybern.* **2016**, *46*, 2132–2143.

14. Shi, P.; Shen, Q. Cooperative control of multi-agent systems with unknown state-dependent controlling effects. *IEEE Trans. Autom. Sci. Eng.* **2015**, *12*, 827–834.

15. Modares, H.; Nageshrao, S.P.; Delgado Lopesb, G.A.; Babuškab, R.; Lewisa, F.L. Optimal model-free output synchronization of heterogeneous systems using off-policy reinforcement learning. *Automatica* **2016**, *71*, 334–341.

16. Olfati-Saber, R.; Murray, R.M. Consensus problems in networks of agents with switching topology and time-delays. *IEEE Trans. Autom. Control* **2004**, *49*, 1520–1533.

17. Ren, W.; Beard, R.W. Consensus seeking in multiagent systems under dynamically changing interaction topologies. *IEEE Trans. Autom. Control* **2005**, *50*, 655–661.

18. Sun, Y.G.; Wang, L.; Xie, G. Average consensus in networks of dynamic agents with switching topologies and multiple time-varying delays. *Syst. Control Lett.* **2008**, *57*, 175–183.

19. Jadbabaie, A.; Lin, J.; Morse, A.S. Coordination of groups of mobile autonomous agents using nearest neighbor rules. *IEEE Trans. Autom. Control* **2002**, *48*, 988–1001.

20. Bliman, P.-A.; Ferrari-Trecate, G. Average consensus problems in networks of agents with delayed communications. *Automatica* **2008**, *44*, 1985–1995.

21. Ren, W.; Atkins, E. Distributed multi-vehicle coordinated control via local information exchange. *Int. J. Robust Nonlinear Control* **2007**, *17*, 1002–1033.

22. Xie, G.; Wang, L. Consensus control for a class of networks of dynamic agents. *Int. J. Robust Nonlinear Control* **2007**, *17*, 941–959.

23. Hong, Y.; Gao, L.; Cheng, D.; Hu, J. Lyapunov-based approach to multiagent systems with switching jointly connected interconnection. *IEEE Trans. Autom. Control* **2007**, *52*, 943–948.

24. Yu, W.; Chen, G.; Cao, M. Some necessary and sufficient conditions for second-order consensus in multi-agent dynamical systems. *Automatica* **2010**, *46*, 1089–1095.

25. Chen, G.; Lewis, F.L. Robust consensus of multiple inertial agents with coupling delays and variable topologies *Int. J. Robust Nonlinear Control* **2011**, *21*, 666–685.

26. Shen, Q.; Shi, P. Output consensus control of multi-agent systems with unknown nonlinear dead-zone. *IEEE Trans. Syst. Man Cybern. Syst.* **2016**, *46*, 1329–1337.

27. Li, Z.; Duan, Z.; Xie, L.; Liu, X. Distributed robust control of linear multi-agent systems with parameter uncertainties. *Int. J. Control* **2012**, *85*, 1039–1050.

28. Amster, P.; Mariani, M. Some results on the forced pendulum equation. *Nonlinear Anal. Theory Methods Appl.* **2008**, *68*, 1874–1880.

29. Ren, W. Synchronization of coupled harmonic oscillators with local interaction. *Automatica* **2008**, *44*, 3195–3200.

30. Andreasson, M.; Sandberg, H.; Dimarogonas, D.V.; Johansson, K.H. Distributed integral action: Stability analysis and frequency control of power systems. In Proceedings of the IEEE 51st Annual Conference on Decision and Control, Maui, HI, USA, 10–13 December 2012; pp. 2077–2083.

31. Yang, H.; Zhang, Z.; Zhang, S. Consensus of second-order multi-agent systems with exogenous disturbances. *Int. J. Robust Nonlinear Control* **2011**, *21*, 945–956.

32. Zhang, X.; Liu, X. Further results on consensus of second-order multi-agent systems with exogenous disturbance. *IEEE Trans. Circuits Syst. I Regul. Pap.* **2013**, *60*, 3215–3226.

33. Nikiforov, V. Adaptive non-linear tracking with complete compensation of unknown disturbances. *Eur. J. Control* **1998**, *4*, 132–139.

34. Ding, Z. Asymptotic rejection of unknown sinusoidal disturbances in nonlinear systems. *Automatica* **2007**, *43*, 174–177.

35. Liu, L.; Chen, Z.; Huang, J. Parameter convergence and minimal internal model with an adaptive output regulation problem. *Automatica* **2009**, *45*, 1306–1311.

36. Khalil, H.K. *Nonlinear Systems*, 3rd ed.; Prentice Hall: Upper Saddle River, NJ, USA, 2001.

37. Ren, W. On consensus algorithms for double-integrator dynamics. *IEEE Trans. Autom. Control* **2008**, *53*, 1503–1509.

38. Horn, R.A.; Johnson, C.R. *Topics in Matrix Analysis*; Cambridge UP: New York, NY, USA, 1991.

*Article*

# Criticality and Information Dynamics in Epidemiological Models

**E. Yagmur Erten** [1,*,†], **Joseph T. Lizier** [1], **Mahendra Piraveenan** [1] **and Mikhail Prokopenko** [1,2]

[1] Centre for Complex Systems, Faculty of Engineering and IT, University of Sydney,
Sydney, NSW 2006, Australia; joseph.lizier@sydney.edu.au (J.T.L.);
mahendrarajah.piraveenan@sydney.edu.au (Ma.P.); mikhail.prokopenko@sydney.edu.au (Mi.P.)

[2] Marie Bashir Institute for Infectious Diseases and Biosecurity, University of Sydney,
Westmead, NSW 2145, Australia

* Correspondence: yagmur.erten@ieu.uzh.ch; Tel.: +41-44-635-6118

† Current address: Department of Evolutionary Biology and Environmental Studies, University of Zurich,
Winterthurerstrasse 190, 8057 Zurich, Switzerland.

Academic Editor: J. A. Tenreiro Machado
Received: 2 March 2017; Accepted: 25 April 2017; Published: 27 April 2017

**Abstract:** Understanding epidemic dynamics has always been a challenge. As witnessed from the ongoing Zika or the seasonal Influenza epidemics, we still need to improve our analytical methods to better understand and control epidemics. While the emergence of complex sciences in the turn of the millennium have resulted in their implementation in modelling epidemics, there is still a need for improving our understanding of critical dynamics in epidemics. In this study, using agent-based modelling, we simulate a Susceptible-Infected-Susceptible (SIS) epidemic on a homogeneous network. We use transfer entropy and active information storage from information dynamics framework to characterise the critical transition in epidemiological models. Our study shows that both (bias-corrected) transfer entropy and active information storage maximise after the critical threshold ($R_0 = 1$). This is the first step toward an information dynamics approach to epidemics. Understanding the dynamics around the criticality in epidemiological models can provide us insights about emergent diseases and disease control.

**Keywords:** epidemiology; criticality; information dynamics; phase transitions; agent-based simulation

## 1. Introduction

The mathematical modelling of epidemics dates back to mid-18th century [1], while it was Kermack and McKendrick [2] who studied Susceptible-Infected-Recovered (SIR) model, being the first to use the formal models of epidemics, known generally as compartmental mean-field models [3]. In these models, the population is categorised into distinct groups, depending on their infection status: susceptible individuals are the ones who has never had the infection and can have it upon contact with infected individuals, infected individuals have the infection and can transmit it to the susceptible individuals, and recovered individuals are those have recovered from the infection and are immune since then. This baseline model is captured mathematically by the following set of ordinary differential equations (ODEs):

$$\frac{dS}{dt} = -\beta SI \tag{1}$$

$$\frac{dI}{dt} = \beta SI - \gamma I \tag{2}$$

$$\frac{dR}{dt} = \gamma I \tag{3}$$

where $\beta$ is the transmission rate, $\gamma$ is the recovery rate, and $S$, $I$ and $R$ are the proportion of susceptible, infected and recovered individuals respectively. These equations can be further modified to account for various other factors (e.g., pathogen-induced mortality) by the inclusion of more parameters in the ODEs, or they can be adapted to reflect different infectious dynamics (e.g., waning immunity).

Compartmental mean-field models are simplistic: for instance, they do not account for the contact structure in the population, which can influence epidemic dynamics (e.g., [4,5]). However, they were crucial in understanding the epidemic threshold: an epidemic with the dynamics as described in Equations (1)–(3) can invade a population only if the initial fraction of the susceptible population is higher than $\gamma/\beta$ [2,3,6]. The inverse of this value is called 'basic reproductive ratio' ($R_0$) which is defined as "the expected number of secondary cases produced by a typical infected individual during its entire infectious period, in a population consisting of susceptibles only" [7]. An infection will cause an epidemic only if $R_0 > 1$ (*endemic equilibrium* in models with population dynamics) and will die out if $R_0 < 1$ (*disease-free equilibrium* or *extinction*) [8]. This is rather intuitive: if each individual transmits the disease to less than one other individual on average, the infected individuals will be removed from the population (for instance, due to recovery) faster than they transmit the disease to the susceptible individuals [9]. However, recurring contacts between a typical infective individual and other previously infected individuals may cause $R_0$ to overestimate the mean number of secondary infections, so more exact measures may sometimes be required [10].

The epidemic threshold is a well-known result in epidemiology, and can be related to the concepts of phase transitions and critical thresholds in statistical physics [11]. A phase transition is "a sharp change in the properties (state) of a substance (system)" that happens when "there is a singularity in the free energy or one of its derivatives" and occurs through changing a control parameter $\lambda$ [11,12]. It is observed in various systems, such as fluid or magnetic phase transitions [12]. The phases are distinguished by the order parameter $\rho$, which is typically zero in one phase and attains a non-zero value in the other [11]. When the control parameter takes the value $\lambda_c$, known as critical point, the phase transition occurs such that when $\lambda \leq \lambda_c$, $\rho = 0$ while $\lambda > \lambda_c$, $\rho > 0$ [11]. For instance, in the case of liquid-gas transition, the difference between the densities of liquid and gas becomes zero as the temperature increases above the critical temperature [12]. Relating back to the analogy in epidemiology, we see that an epidemic occurs only if $R_0 > 1$, as shown in Figure 1. The control parameter, in this case, is $R_0$, while $\lambda_c = 1$ and the order parameter is the final size of the epidemic (which can alternatively be the density of the infected individuals or prevalence [11]).

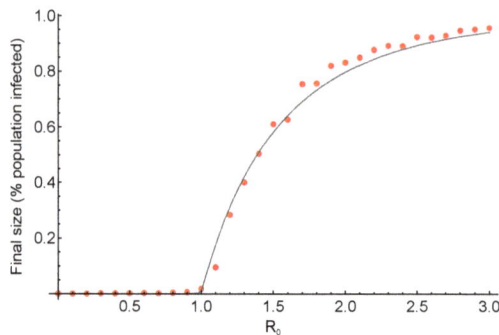

**Figure 1.** Epidemic phase transition. Final size of an epidemic as a function of its basic reproductive ratio $R_0$, for a susceptible-infected-recovered (SIR) model with a homogeneous network structure, with a number of connections ($k$) of 4 for each individual. Transmission rate $\beta$ varies between 0 and 3 with recovery rate $\gamma = 1$, resulting in $R_0$ ranging between 0 and 3. The line depicts the analytical results whereas the red dots show the results from stochastic simulations with a population size of $10^4$. The epidemic does not occur for $R_0 < 1$, whereas the final size increases as a function of $R_0$ for values higher than 1. The analytical results and the simulations are in good agreement.

The basic reproductive ratio is widely used for modelling, predictions and control of epidemics (see [9] for examples). Furthermore, understanding the dynamics when a pathogen is near the epidemic threshold is crucial. A maladapted pathogen with $R_0 < 1$ can cause an epidemic if its $R_0$ exceeds 1 (due to, for instance, mutations or changes in the host population), which is how new pathogens can emerge by crossing the species barrier [13]. From a phase transition point-of-view, Reference [14] have studied the change in epidemiological quantities while approaching the critical point, in order to see if they can be used for anticipating the emergence of criticality and potential elimination of such dynamics. Their results suggest that theoretically, we can predict critical thresholds in epidemics [14], which could be of a substantial value in health care. In practice, one of the challenges lies in pinpointing such critical thresholds in finite-size systems, where a precise identification of phase transitions requires an estimation of the rate of change of the order parameter, often from finite and/or distributed data [15,16].

Information dynamics [17–21] is a recently-developed framework based on information theory [22], which shows promise in characterising phase transitions in dynamical systems. For instance, in studies of the order-chaos phase transitions in random Boolean networks (driven by changes in activity [23] and/or network structure [24]), information storage has been shown to peak just on the ordered side, whereas information transfer was found to maximise on the other (chaotic) side of the critical threshold. Information transfer, from a collection of sources, has also been measured to peak prior to the phase transition in the Ising model, when the critical point is approached from the disordered (high-temperature) side [25]. And similarly, both information storage and transfer were measured to be maximised near the critical state in echo-state networks (a type of recurrent neural network) [26], where such networks are argued to be best placed for general-purpose computation.

Furthermore, one of the features of complex computation is a coherent information structure defined as a pattern appearing in a state-space formed by information-theoretic quantities, such as transfer entropy and active information storage [27]. The "information dynamics" state-space diagrams are known to provide insights which are not immediately visible when the measures are considered in isolation [28]. For example, critical spatiotemporal dynamics of Cellular Automata (CA) were characterised via state-space diagrams formed by transfer entropy and active information storage. These state-space diagrams highlighted how the complex distributed computation carried out by CA interlinks the communication and memory operations [27].

Local information dynamics were also shown to have maxima that relate to the spread of cascading failures in energy networks [29]. We, therefore, expect that this framework will prove to be applicable not only to explaining epidemic dynamics around the critical threshold, but also in developing new predictive methods during emergence of diseases, evolution of pathogens and spillage from zoonotic reservoirs, as well as applications of stochastic epidemiological models to computer virus spreading and other similar scenarios [30].

Here we study the phase transition in a SIS model of epidemics using the information dynamics framework. We model a homogeneous network where a pathogen can spread through contact between the neighbours. Changing the transmissibility of the pathogen between different realisations of the same network while keeping the recovery rate fixed, we use basic reproductive ratio as the control parameter. We track the prevalence of the infection as our order parameter and use the infection status of individuals to calculate the transfer entropy and active information storage.

## 2. Materials and Methods

### 2.1. Model Description

We focus on Susceptible-Infected-Susceptible (SIS) dynamics [31] which were originally defined using the following ODE mean-field model:

$$\frac{\mathrm{d}S}{\mathrm{d}t} = -\beta SI + \gamma I \tag{4}$$

$$\frac{dI}{dt} = \beta SI - \gamma I \tag{5}$$

where $\beta$ is the transmission rate, $\gamma$ is the recovery rate.

As described in Section 1 however, such ODE models do not account for contact structure in the population, and so we use the following network model. We simulate a population of $N = 10^4$ individuals, where each is connected to four neighbours (randomly selected from the network), and assume the network is undirected. Susceptible ($S$) individuals can become infected through contact with their infected ($I$) neighbours with transmission rate $\beta$ and infected individuals can recover with recovery rate $\gamma$. Each individual has the same rate of transmission and recovery. We scale the transmission rate by $c$ to calculate per contact transmission rate in the event-based simulations. It is a closed population, therefore for the total population $S(t) + I(t) = N$. Also, we assume there is no mortality due to infection and $N$ remains constant through the simulations (i.e., we neglect births and deaths in the population). At the beginning of the simulation one random individual gets infected and the epidemiological process is simulated using the Gillespie's Direct Algorithm [32]. For each parameter set (summarised in Table 1), we ran either 500 replicates or 10 complete runs (i.e., simulations that ran for $10^3$ time steps), depending on whichever is attained first. We binned the continuous time to discrete time steps for our information dynamic analysis and recorded the infectious status of each individual and their four neighbours once in every time step.

**Table 1.** Simulation parameters.

| Parameter | Value |
|---|---|
| Time steps ($t$) | $10^3$ |
| Population size ($N$) | $10^4$ |
| Number of contacts | 4 |
| Transmission rate ($\mu$) | 0.7–2.0 (with step size 0.1) |
| Coefficient for per contact transmission rate ($c$) | 0.33 |
| Recovery rate ($\mu$) | 1.0 |

### 2.2. Information Dynamics

Disease spreading could be considered as a computational process by which a population "computes" how far a disease will spread, and what the final states (susceptible or recovered) of the various individuals will be once the disease has run its course. The framework of local information dynamics [17–21] studies how information is intrinsically processed within a system while it is "computing" its new state. Specifically, the information dynamics framework measures how information is stored, transferred and modified within a system during such a computational process. It is a recently-developed framework, based on information theory [22], and involving a well-established quantity for measuring the uncertainty of a random variable $X$, known as *Shannon entropy*, which is written as:

$$H(X) = -\sum_{i=1}^{N} p_i \log p_i \tag{6}$$

where $p_i$ is the probability that $X$ takes the state $i$. This quantity has been used to derive other measures such as joint entropy, which quantifies uncertainty of joint distribution of random variable $X$ and $Y$, or mutual information, which is the expression of the amount of information that can be acquired concerning one random variable by observing another [33].

Once we begin to consider how to predict disease spread, we naturally turn to measuring uncertainties and uncertainty reduction using information theory. Information theory has previously been used to study the uncertainty (and conversely, the predictability) of disease spreading dynamics using the permutation entropy of the total number of infections in the population [34]. Another related study analysed the dynamics of an infectious disease spread by formulating the maximum entropy solutions of the SIS and SIR stochastic models, exploiting the advantage offered by the Principle of Maximum Entropy in introducing the minimum additional information beyond the information implied by the constraints on the conserved quantities [35].

Our investigation will focus on measuring information dynamics of an epidemic process within the population, seeking to relate the spread of the disease to information-processing (computational) primitives such as information storage and transfer. Moreover, while these quantities are defined for studying averages over all observations, the corresponding local measures—specific to each sample of a state $i$ of $X$—can provide us with more suitable tools to study phase transitions in finite-size systems (e.g., [36]).

The information dynamics framework quantifies information transfer, storage and modification, and focusses on their local dynamics at each sample. Information storage is defined as "the amount of information in [an agent's] past that is relevant to predicting its future" while the local active information storage [19] is the local mutual information between an agent's next state ($x_{n+1}$) and its semi-infinite past $x_n^{(k)}$ expressed as:

$$a_x(n+1) = \lim_{k\to\infty} \log \frac{p(x_n^{(k)}, x_{n+1})}{p(x_n^{(k)})p(x_{n+1})}. \tag{7}$$

The average active information storage is the expectation value over the ensemble:

$$A_X = \langle a_x(n+1) \rangle. \tag{8}$$

On the other hand, information transfer is defined via the transfer entropy [37] as the information provided by the source about the destination's next state in the context of the past of the destination [23], whereas local information transfer from a source $Y$ to a destination $X$ is "the local mutual information between the previous state of the source and the next state of the destination conditioned on the semi-infinite past of the destination", expressed as:

$$t_{y\to x}(n+1) = \lim_{k\to\infty} \log \frac{p(x_{n+1}|x_n^{(k)}, y_n)}{p(x_{n+1}|x_n^{(k)})} \tag{9}$$

following [17]. The (average) transfer entropy is the expectation value of the local term over the ensemble:

$$T_{Y\to X} = \langle t_{y\to x}(n+1) \rangle. \tag{10}$$

Transfer entropy was recently analysed on SIS dynamics (generated on brain network structures) in order to investigate information flows on different temporal scales [38].

In order to determine which embedding length $k$ is most suitable for our analysis, we seek to set $k$ so as to maximise the average active information storage, as per the criteria presented in [39]. Importantly though, for this criteria to work we need to maximise the *bias-corrected* active information storage rather than its raw value. Bias-correction pertains to removing the bias in our estimation of $A_X$, i.e., the systematic over- or under-estimation of that quantity as compared to the true value. Typically, as a mutual information quantity, $A_X$ will be overestimated from a finite amount of data, particularly when our embedding length $k$ starts to increase the dimensionality of our state space beyond a point that we have adequately sampled. We can estimate this bias by computing the mutual information between surrogate variables with the same distribution as those we originally consider, but without their original temporal relationship to each other [40]. For $A_X$, one surrogate measurement $A_X^s$ is made with a shuffled version of the $x_{n+1}$ samples (but keeping the $x_n^{(k)}$ samples fixed), and then repeating to produce a population of surrogate measurements. We label the mean of these surrogate measurements $\overline{A_X^s}$, and our effective or bias-corrected active information storage as:

$$A_X' = A_X - \overline{A_X^s}. \tag{11}$$

Subtracting out bias using surrogates was proposed earlier for the transfer entropy as the "effective transfer entropy" [41], simply referred to as "bias-corrected transfer entropy" here. Computing the bias-corrected transfer entropy $T'_{Y \to X}$ is performed in a similar fashion to $A'_X$: first, surrogates $T^s_{Y \to X}$ are computed using a shuffled version of the source samples $y_n$ while holding the destination time series fixed (to retain the destination past–next relationship via $p(x_{n+1}|x_n^{(k)})$), then the mean of the surrogate measurements $\overline{T^s_{Y \to X}}$ is computed, before computing the effective or bias-corrected transfer entropy as:

$$T'_{Y \to X} = T_{Y \to X} - \overline{T^s_{Y \to X}}. \tag{12}$$

To perform the information dynamics calculations in this study we used the Java Information Dynamics Toolkit (JIDT) [40].

### 2.3. Measuring Information Dynamics in the SIS Model

We analysed three simulation runs of the SIS model for each parameter combination and required these runs to be at least 14 time steps of length for $R_0 \leq 1.0$ (as these runs did not have sustained transmission of infection for $10^3$ time steps, 14 being the number of time steps in the third longest run for $R_0 = 0.7$).

We used the past and current status of individuals (1 if infected, 0 if susceptible) and that of their neighbours at a given time to determine $x_{n+1}$, $y_n$, and the $x_n^{(k)}$ vectors for calculations of active information storage and transfer entropy. For instance, if the focal individual is infected and its neighbours are all susceptible at a given time, then $x_n$ was equal to one and $y_n$ was equal to zero for all the neighbours. We then calculated the individual's local transfer entropy by averaging the pairwise transfer entropy (with a given $k$) between itself and each of its neighbours. We then averaged the local transfer entropy across the population to determine the average transfer entropy. For active information storage, we calculated the local values for each individual (with a given embedding length $k$) and averaged these across all the individuals in the population.

To determine the value of $k$ to use, we calculated the bias-corrected active information storage as per Equation (11) for each run, and then averaged the values across runs for each $R_0$. Subsequently, we calculated the mean for each embedding length $k$ across all $R_0$ values. The bias-corrected active information storage maximises for $k = 7$ (Figure 2) and decreases sharply after $k = 8$. As such, applying the criteria discussed above, we select $k = 7$ for the embedding length to be used.

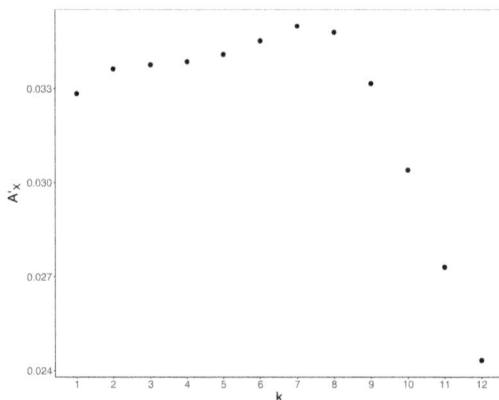

**Figure 2.** Bias-corrected active information storage $A'_X$ in our simulations as a function of embedding length $k$. $A'_X$ was calculated and then averaged for all three replicates for each $R_0$. The mean value (shown in $y$-axis) was then determined for each $k$ (shown in $x$-axis) across the $R_0$ values. The difference increases as the $k$ increases, maximising at $k = 7$, and decreasing subsequently.

## 3. Results and Discussion

In our simulations, the epidemic dies out without any sustained transmission when $R_0 < 1.0$, whereas the number of infected individuals reaches an equilibrium when $R_0 > 1.0$, and the epidemic becomes endemic in the population. We use the mean number of infected individuals throughout the simulation runs to calculate the prevalence for each $R_0$ value, shown in Figure 3.

The average transfer entropy is highest after the critical transition (in the supercritical regime), as shown in Figure 3, reaching its peak at $R_0 = 1.8$ for $k = 7$. This result aligns well with the peak in (collective) transfer entropy slightly in the super-critical regime in the Ising model [25]. In alignment with results in the Ising model, here once the disease dynamics reach criticality, we observe strong effects of one individual on a connected neighbour (measured by the transfer entropy). However, as the dynamics become supercritical, the target neighbour becomes more strongly bound to all of its neighbours collectively, and it becomes more difficult to predict its dynamics based on a single source neighbour alone; as such, the transfer entropy begins to decrease. We also note that the peak in average transfer entropy shifts toward lower $R_0$ values when the embedding time is shorter (not shown).

We see that (raw or non-bias-corrected) average active information storage $A_X$ increases after the critical transition, reaching to its peak at $R_0 = 1.3$ (Figure 3).

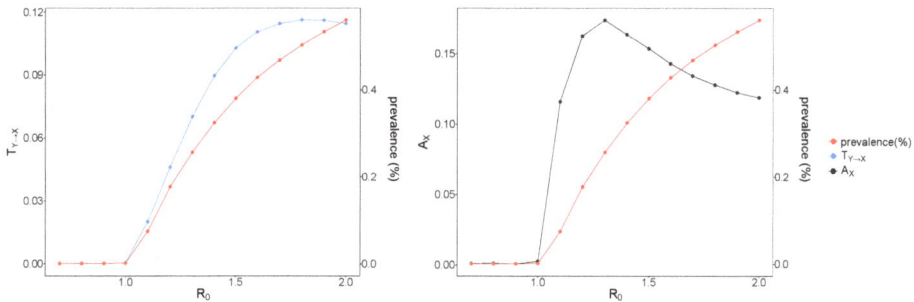

**Figure 3.** Raw average transfer entropy and average active information storage versus $R_0$. Transfer entropy (**left**) calculated by averaging local transfer entropy for each individual across the network and active information storage (**right**) calculated by averaging local active information storage for each individual across the network. For both measures, the embedding time is $k = 7$. The average transfer entropy ($T_{Y \to X}$) is shown in blue, the average active information storage ($A_X$) is shown in gray, and prevalence is shown in red (note the different y-axes). $R_0$ is shown on the x-axis. After the critical transition both $T_{Y \to X}$ and $A_X$ increase and reach to a peak (at $R_0 = 1.8$ and $R_0 = 1.3$, respectively), and subsequently lower down.

To check whether what we observe for $T_{Y \to X}$ and $A_X$ was a real effect or due to increased bias as the time-series activity increased (with $R_0$), we examined the bias-corrected average transfer entropy $T'_{Y \to X}$ and average active information storage $A'_X$. This is shown in Figure 4 for embedding length $k = 7$. The bias-corrected average active information storage shows a similar pattern to $A_X$, however with a sharper peak closer to the phase transition ($R_0 = 1.2$). This shows that most of what was measured as $A_X$ at larger $R_0$ values was indeed due to increased bias. This is even more striking for the bias-corrected transfer entropy, as we observe a sharp peak at $R_0 = 1.2$, similar to $A'_X$ and much earlier than $T_{Y \to X}$ ($R_0 = 1.8$). Therefore, these results suggest that once the disease dynamics reach criticality, the state of each individual first exhibits a large amount of self-predictability from its past (information storage). However, as the dynamics become supercritical, the increasingly chaotic nature of the interactions are reflected in the subsequent decrease in self-predictability.

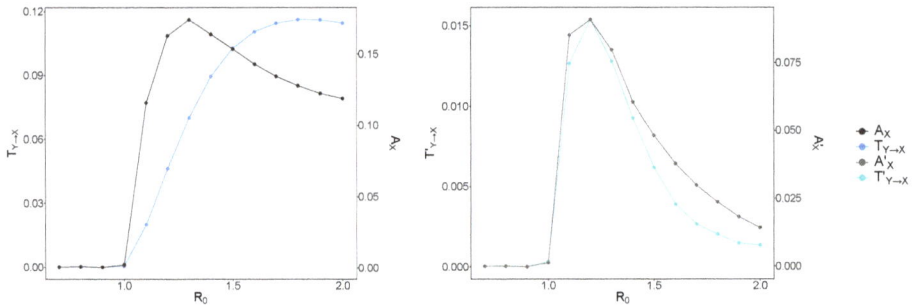

**Figure 4.** Raw and bias-corrected average transfer entropy and average active information storage versus $R_0$. Raw average transfer entropy $T_{Y \to X}$ and average active information storage $A_X$ are shown in dark blue and black, respectively (left panel); bias-corrected average transfer entropy $T'_{Y \to X}$ and average active information storage $A'_X$ are shown in light blue and gray, respectively (right panel). Note the different y-axes for both graphs. $R_0$ is shown on the x-axis. Both $A_X$ and $A'_X$ increase and reach a peak right after the critical transition, and subsequently decrease. $T'_{Y \to X}$ also increases at the same $R_0$ value ($R_0 = 1.2$) as $A'_X$ and plummets thereafter, whereas $T_{Y \to X}$ reaches its highest value later, at $R_0 = 1.8$

We argue that the transfer entropy captures the extent of the distributed communications of the network-wide computation underlying the epidemic spread, while the active information storage corresponds to its distributed memory. Crucially, the peak of both these information-processing operations (measured with the bias correction) occurs at $R_0 = 1.2$, rather than the canonical $R_0 = 1.0$. As mentioned earlier, previous studies of distributed computation and its information-processing operations [23–25,27], concluded that the active information storage peaks just on the ordered side, while transfer entropy maximises on the disordered side of the critical threshold. Therefore, in our case, it may be argued that the concurrence of both bias-corrected peaks, as detected by the maximal information-processing "capacity" of the underlying computation, at $R_0 = 1.2$, indicates an upper bound for the critical threshold in the studied finite-size system.

In the proper thermodynamic limit, as the size of the system goes to infinity, the canonical threshold may well be re-established, but in finite-size systems an additional care may be needed to forecast epidemic spread for intermediate values of the basic reproductive ratio, for instance, $1.0 \leq R_0 \leq 1.2$ as in the presented study. In other words, in finite-size systems one may consider a critical *interval* rather than an exact critical threshold.

Furthermore, in addition to identifying the peak of information-processing capacity of the underlying computation, which pinpointed an upper bound on the critical basic reproductive ratio $R_0 = 1.2$, we studied patterns of coherent information structure, via state-space diagrams formed by transfer entropy and active information storage shown in Figure 5. It is evident that both information-processing operations (communications and memory) are tightly interlinked in the underlying computation, suggesting that the studied epidemic process is strongly coherent.

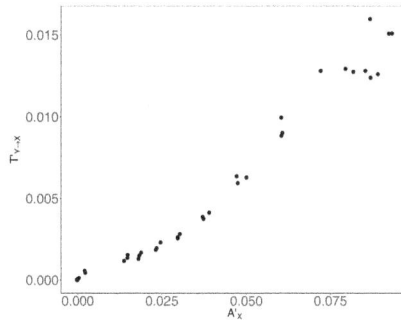

**Figure 5.** Bias-corrected average transfer entropy $T'_{Y \to X}$ versus bias-corrected average active information storage $A'_X$. Bias-corrected transfer entropy $T'_{Y \to X}$ (shown in the $y$-axis) and average active information storage $A'_X$ (shown in the $x$-axis) are calculated separately for three replicates.

## 4. Conclusions

In this paper, we studied the criticality in an SIS epidemic within an information dynamics framework. We argued that the transfer entropy captures the extent of the distributed communications of the network-wide computation and showed that it peaks in the super-critical regime. Similarly, we considered the active information storage as a measure of the distributed memory, observing that its maximum is also attained after the canonical critical transition ($R_0 > 1.0$). To our knowledge, this is the first study to use information dynamics concepts to characterise critical behaviour in epidemics. Crucially, the concurrence of both peaks, which reflect the maximal information-processing capacity of the underlying coherent computation, at $R_0 = 1.2$, indicates an upper bound for the critical threshold in the studied finite-size system. This supports a conjecture that in finite-size systems a critical interval (rather than an exact critical threshold) may be a relevant notion.

At the time of our study, continuous-time measures of information dynamics were not available. Recently, transfer entropy in continuous-time was formalised with a novel approach [42]. One future avenue for improving our analysis would be using continuous-time measures of information dynamics. Our continuous-time simulation results were binned into discrete time steps in order to conduct an information-dynamic analysis. Therefore, using continuous time measure could reveal novel insights that we missed by the discretisation.

We used a very simplistic network topology in this study, where each individual had the same number of undirected connections that were assigned randomly. However, in real-life, a disease can spread in interaction networks that are heterogeneous in terms of the number of contacts (e.g., [43]) or structured differently (e.g., [5,44]). The heterogeneity and the type of the network can influence not only epidemic dynamics but also disease emergence [4]. Therefore, in future, it would be interesting to expand our analysis to different network topologies and relate the insights from information-dynamic analysis to epidemic dynamics and disease emergence probabilities in these networks.

Finally, we note the wider interest in critical dynamics on complex networks in general. This is particularly the case in complex systems and network approaches to computational neuroscience, where it is conjectured that the brain is in or near a critical state so as to advantageously use maximised computational properties here [26,45–48]. (Indeed, as noted earlier, SIS dynamics have been used to model dynamics on brain networks [38]). Our results in this paper regarding SIS dynamics continue to add to the quantitative evidence regarding the maximisation of information storage and transfer as intrinsic computational properties at or near critical states, as previously found in a diverse range of dynamics and network structures including the Ising model [25], recurrent neural networks [26], gene regulatory network (GRN) models [23], and regular–small-world–random transitions in structure [24]. In this way, we have provided another important link for epidemic spreading models to complex networks, criticality and information dynamics.

*Entropy* **2017**, *19*, 194

**Acknowledgments:** Joseph T. Lizier, Mahendra Piraveenan and Mikhail Prokopenko were supported through the Australian Research Council Discovery grant DP160102742, and Joseph T. Lizier was supported through the Australian Research Council DECRA grant DE160100630.

**Author Contributions:** E.Y.E., Ma.P. and Mi.P. conceived and designed the model; E.Y.E. implemented the model and conducted the simulations; J.T.L. contributed analysis tools; E.Y.E., Ma.P., J.T.L. and Mi.P. analysed the data; E.Y.E., Ma.P., J.T.L., and Mi.P. wrote the paper.

**Conflicts of Interest:** The authors declare no conflict of interest.

## References

1. Bernoulli, D. Essai d'une nouvelle analyse de la mortalité causée par la petite vérole et des avantages de l'inoculation pour la prévenir. *Hist. Acad. R. Sci. Mém. Math. Phys.* **1766**, 1–45. (In French)
2. Kermack, W.O.; McKendrick, A.G. A contribution to the mathematical theory of epidemics. *Proc. R. Soc. Lond. A Math. Phys. Eng. Sci.* **1927**, *115*, 700–721.
3. Keeling, M.J.; Rohani, P. *Modeling Infectious Diseases in Humans and Animals*; Princeton University Press: Princeton, NJ, USA, 2008.
4. Leventhal, G.E.; Hill, A.L.; Nowak, M.A.; Bonhoeffer, S. Evolution and emergence of infectious diseases in theoretical and real-world networks. *Nat. Commun.* **2015**, *6*, 6101.
5. Bauer, F.; Lizier, J.T. Identifying influential spreaders and efficiently estimating infection numbers in epidemic models: A walk counting approach. *Europhys. Lett.* **2012**, *99*, 68007.
6. Anderson, R.M.; May, R.M.; Anderson, B. *Infectious Diseases of Humans: Dynamics and Control*; Oxford University Press: Oxford, UK, 1992; Volume 28.
7. Heesterbeek, J.; Dietz, K. The concept of Ro in epidemic theory. *Stat. Neerl.* **1996**, *50*, 89–110.
8. Artalejo, J.; Lopez-Herrero, M. Stochastic epidemic models: New behavioral indicators of the disease spreading. *Appl. Math. Model.* **2014**, *38*, 4371–4387.
9. Heffernan, J.; Smith, R.; Wahl, L. Perspectives on the basic reproductive ratio. *J. R. Soc. Interface* **2005**, *2*, 281–293.
10. Artalejo, J.R.; Lopez-Herrero, M.J. On the Exact Measure of Disease Spread in Stochastic Epidemic Models. *Bull. Math. Biol.* **2013**, *75*, 1031–1050.
11. Pastor-Satorras, R.; Castellano, C.; Van Mieghem, P.; Vespignani, A. Epidemic processes in complex networks. *Rev. Mod. Phys.* **2015**, *87*, 925–979.
12. Yeomans, J.M. *Statistical Mechanics of Phase Transitions*; Oxford University Press: Oxford, UK, 1992.
13. Antia, R.; Regoes, R.R.; Koella, J.C.; Bergstrom, C.T. The role of evolution in the emergence of infectious diseases. *Nature* **2003**, *426*, 658–661.
14. O'Regan, S.M.; Drake, J.M. Theory of early warning signals of disease emergenceand leading indicators of elimination. *Theor. Ecol.* **2013**, *6*, 333–357.
15. Wang, X.R.; Lizier, J.T.; Prokopenko, M. Fisher Information at the Edge of Chaos in Random Boolean Networks. *Artif. Life* **2011**, *17*, 315–329.
16. Prokopenko, M.; Lizier, J.T.; Obst, O.; Wang, X.R. Relating Fisher information to order parameters. *Phys. Rev. E* **2011**, *84*, 041116.
17. Lizier, J.T.; Prokopenko, M.; Zomaya, A.Y. Local information transfer as a spatiotemporal filter for complex systems. *Phys. Rev. E* **2008**, *77*, 026110.
18. Lizier, J.T.; Prokopenko, M.; Zomaya, A.Y. Information modification and particle collisions in distributed computation. *Chaos* **2010**, *20*, 037109.
19. Lizier, J.T.; Prokopenko, M.; Zomaya, A.Y. Local measures of information storage in complex distributed computation. *Inf. Sci.* **2012**, *208*, 39–54.
20. Lizier, J.T. *The Local Information Dynamics of Distributed Computation in Complex Systems*; Springer: Berlin/Heidelberg, Germany, 2013.
21. Lizier, J.T.; Prokopenko, M.; Zomaya, A.Y. A Framework for the Local Information Dynamics of Distributed Computation in Complex Systems. In *Guided Self-Organization: Inception*; Prokopenko, M., Ed.; Springer: Berlin/Heidelberg, Germany, 2014; Volume 9, pp. 115–158.
22. Shannon, C.E. A mathematical theory of communication. *Bell Syst. Tech. J.* **1948**, *27*, 379–423.
23. Lizier, J.T.; Prokopenko, M.; Zomaya, A.Y. The Information Dynamics of Phase Transitions in Random Boolean Networks. *Artif. Life* **2008**, *11*, 374–381.

24. Lizier, J.T.; Pritam, S.; Prokopenko, M. Information dynamics in small-world Boolean networks. *Artif. Life* **2011**, *17*, 293–314.

25. Barnett, L.; Harré, M.; Lizier, J.; Seth, A.K.; Bossomaier, T. Information Flow in a Kinetic Ising Model Peaks in the Disordered Phase. *Phys. Rev. Lett.* **2013**, *111*, 177203.

26. Boedecker, J.; Obst, O.; Lizier, J.T.; Mayer, N.M.; Asada, M. Information processing in echo state networks at the edge of chaos. *Theory Biosci.* **2012**, *131*, 205–213.

27. Lizier, J.T.; Prokopenko, M.; Zomaya, A.Y. Coherent information structure in complex computation. *Theory Biosci.* **2012**, *131*, 193–203.

28. Cliff, O.M.; Lizier, J.T.; Wang, P.; Wang, X.R.; Obst, O.; Prokopenko, M. Quantifying Long-Range Interactions and Coherent Structure in Multi-Agent Dynamics. *Artif. Life* **2017**, *23*, 34–57.

29. Lizier, J.T.; Prokopenko, M.; Cornforth, D.J. The information dynamics of cascading failures in energy networks. In Proceedings of the European Conference on Complex Systems (ECCS), Warwick, UK, 21–25 September 2009; p. 54.

30. Amador, J.; Artalejo, J.R. Stochastic modeling of computer virus spreading with warning signals. *J. Frankl. Inst.* **2013**, *350*, 1112–1138.

31. Anderson, R.M.; May, R.M. *Infectious Diseases of Humans*; Oxford University Press: Oxford, UK, 1991; Volume 1.

32. Gillespie, D.T. Exact stochastic simulation of coupled chemical reactions. *J. Phys. Chem.* **1977**, *81*, 2340–2361.

33. Cover, T.M.; Thomas, J.A. *Elements of Information Theory*; Wiley: New York, NY, USA, 1991.

34. Scarpino, S.V.; Petri, G. On the predictability of infectious disease outbreaks. *arXiv* **2017**, arXiv:1703.07317.

35. Artalejo, J.; Lopez-Herrero, M. The SIS and SIR stochastic epidemic models: A maximum entropy approach. *Theor. Popul. Biol.* **2011**, *80*, 256–264.

36. Lizier, J.T. Measuring the Dynamics of Information Processing on a Local Scale in Time and Space. In *Directed Information Measures in Neuroscience*; Wibral, M., Vicente, R., Lizier, J.T., Eds.; Understanding Complex Systems; Springer: Berlin/Heidelberg, Germany, 2014; pp. 161–193.

37. Schreiber, T. Measuring Information Transfer. *Phys. Rev. Lett.* **2000**, *85*, 461–464.

38. Meier, J.; Zhou, X.; Hillebrand, A.; Tewarie, P.; Stam, C.J.; Mieghem, P.V. The Epidemic Spreading Model and the Direction of Information Flow in Brain Networks. *NeuroImage* **2017**, *152*, 639–646.

39. Garland, J.; James, R.G.; Bradley, E. Leveraging information storage to select forecast-optimal parameters for delay-coordinate reconstructions. *Phys. Rev. E* **2016**, *93*, 022221.

40. Lizier, J.T. JIDT: An Information-Theoretic Toolkit for Studying the Dynamics of Complex Systems. *arXiv* **2014**, arXiv:1408.3270.

41. Marschinski, R.; Kantz, H. Analysing the information flow between financial time series. *Eur. Phys. J. B* **2002**, *30*, 275–281.

42. Spinney, R.E.; Prokopenko, M.; Lizier, J.T. Transfer entropy in continuous time, with applications to jump and neural spiking processes. *Phys. Rev. E* **2017**, *95*, 032319.

43. Lloyd-Smith, J.O.; Schreiber, S.J.; Kopp, P.E.; Getz, W.M. Superspreading and the effect of individual variation on disease emergence. *Nature* **2005**, *438*, 355–359.

44. Schneeberger, A.; Mercer, C.H.; Gregson, S.A.; Ferguson, N.M.; Nyamukapa, C.A.; Anderson, R.M.; Johnson, A.M.; Garnett, G.P. Scale-free networks and sexually transmitted diseases: A description of observed patterns of sexual contacts in Britain and Zimbabwe. *Sex. Transm. Dis.* **2004**, *31*, 380–387.

45. Beggs, J.M.; Plenz, D. Neuronal avalanches in neocortical circuits. *J. Neurosci.* **2003**, *23*, 11167–11177.

46. Priesemann, V.; Munk, M.; Wibral, M. Subsampling effects in neuronal avalanche distributions recorded in vivo. *BMC Neurosci.* **2009**, *10*, 40.

47. Priesemann, V.; Wibral, M.; Valderrama, M.; Pröpper, R.; Le Van Quyen, M.; Geisel, T.; Triesch, J.; Nikolić, D.; Munk, M.H.J. Spike avalanches in vivo suggest a driven, slightly subcritical brain state. *Front. Syst. Neurosci.* **2014**, *8*, 108.

48. Rubinov, M.; Sporns, O.; Thivierge, J.P.; Breakspear, M. Neurobiologically Realistic Determinants of Self-Organized Criticality in Networks of Spiking Neurons. *PLoS Comput. Biol.* **2011**, *7*, e1002038.

*Article*

# Identifying Critical States through the Relevance Index

**Andrea Roli [1,\*], Marco Villani [2,3], Riccardo Caprari [2] and Roberto Serra [2,3]**

[1] Department of Computer Science and Engineering, *Alma Mater Studiorum* Università di Bologna, Campus of Cesena, Cesena I-47521, Italy
[2] Department of Physics, Informatics and Mathematics, Università di Modena e Reggio Emilia, Modena I-41125, Italy; marco.villani@unimore.it (M.V.); 191549@studenti.unimore.it (R.C.); roberto.serra@unimore.it (R.S.)
[3] European Centre for Living Technology, Venezia I-30124, Italy
\* Correspondence: andrea.roli@unibo.it; Tel.:+39-0547-338804

Academic Editor: Mikhail Prokopenko
Received: 7 January 2017; Accepted: 13 February 2017; Published: 16 February 2017

**Abstract:** The identification of critical states is a major task in complex systems, and the availability of measures to detect such conditions is of utmost importance. In general, criticality refers to the existence of two qualitatively different behaviors that the same system can exhibit, depending on the values of some parameters. In this paper, we show that the relevance index may be effectively used to identify critical states in complex systems. The relevance index was originally developed to identify relevant sets of variables in dynamical systems, but in this paper, we show that it is also able to capture features of criticality. The index is applied to two prominent examples showing slightly different meanings of criticality, namely the Ising model and random Boolean networks. Results show that this index is maximized at critical states and is robust with respect to system size and sampling effort. It can therefore be used to detect criticality.

**Keywords:** critical states; relevance index; Ising model; random Boolean networks; complex systems

---

## 1. Introduction

In this paper, the relevance index (RI) is applied to the task of identifying critical states in complex systems (more precisely, we identify regions near critical points; however, in order not to overload the writing, in the following, we use the expression "critical states"). This index had been originally introduced for a different purpose, i.e., as a way to identify key features of the organization of complex dynamical systems, and it has proven able to provide useful results in various kinds of models, including, e.g., those of gene regulatory networks and protein-protein interactions.

Moreover, the method can be applied directly to data, without any need to resort to models, possibly helping to uncover some non-obvious or hidden features of the underlying dynamical organization. As an example, let us mention the discovery of coordinated behaviors of different social and economic agents from the analysis of time series alone, without any a priori knowledge of their interactions.

Essentially, the RI is based on Shannon entropies and can be used to identify groups of variables that change in a coordinated fashion, while they are less integrated with the rest of the system. These groups of integrated variables may form the basis for an aggregate description of the system, at levels higher than that of the single variables. Since the RI allows a variable to belong to more than one group, it can be applied also to "tangled" organizations, which are widespread in complex biological and social systems and which do not have the clean tree-like topology of pure hierarchies.

The RI will be reviewed in detail in Section 2.

The availability of quantitative variables that allow one to identify critical states in complex systems is of utmost importance, and we have found that the RI can also be used to locate critical states with good results, as will be shown below. Moreover, since the RI is affected by the distance of the present state from critical states, it can also be used to identify situations that approach criticality, thus providing early warning signals that can be extremely useful for controlling the behavior of a system.

The use of information-theoretical measures for studying criticality has already been documented in previous studies, such as [1,2], in which Fisher information is used to identify the critical state in both the Ising model and Boolean networks, and [3,4]. However, we remark that the aim of the paper is to show that the RI can be effectively used to identify critical states, rather than to compare different measures of criticality (we also deliberately avoid discussing the tight and intricate relation between criticality and complexity, as it is out of the scope of the paper).

The RI has been applied in two different kinds of systems, where the word "criticality" takes somewhat different meanings. In general, criticality refers to the existence of two qualitatively different behaviors that the same system can show, depending on the values of some parameters. Criticality is then associated with parameter values that separate the qualitatively different behaviors. However, slightly different meanings of the word can be found in the literature, two major cases being (i) the one related to phase transitions and (ii) dynamical criticality, sometimes called the "edge of chaos". In the former case, the two different behaviors refer to equilibrium states that can be observed by varying the value of, e.g., temperature, or of other macroscopic external parameters. In the latter case, the two different behaviors are characterized by their dynamical properties: the attractors that describe the asymptotic behavior of the system can be ordered states, like, e.g., fixed points or limit cycles, or chaotic states. These two meanings are related, but not identical (see [5] for a more detailed discussion).

It is therefore important to understand which kind of critical states can be identified by the RI. That is why we have examined two important models that exhibit the two different meanings of criticality: the Ising model for phase transitions and the random Boolean network model for dynamical criticality. Both are well known, and it will suffice to recall their main properties and notations (in Section 3).

It is also important to stress that the two cases above do not differ only for the different kinds of criticality they show, but also for other important physical and mathematical properties: the Ising model is an ergodic system close to equilibrium, while the random Boolean networks (RBNs) are dissipative, non-ergodic systems. Moreover, the Ising model is inherently stochastic, while the RBN model is deterministic. It is remarkable that the RI is able to satisfactorily locate the critical points in both cases, notwithstanding their differences, as shown in Sections 4 and 5.

Finally, the main results will be summarized in the final section, alongside with indications for further work.

## 2. The Relevance Index

The roots of the RI can be traced back to the work on biological neural networks by Edelman and Tononi [6], who introduced several system-level measures, based on recordings of neural activity, among them the cluster index. The RI is an extension of this latter measure, which can be applied to dynamical systems, while the cluster index had been conceived of for fluctuations around a steady state.

The reasons why the RI has originally been introduced were related to the difficulty in understanding the actual organization of dynamical systems, which requires (i) a proper identification of meaningful organizational "levels" that emerge from the interactions of lower level entities (and possibly also of higher-level entities, such as groups of interacting chemical species inprotocells [7,8]) and (ii) a mapping of the interactions between these meso-levels. In some cases, they can be properly described by a quite familiar tree-like hierarchical structure, as happens in several physical systems where the levels can be identified with the space-time scales of the phenomena (microscopic and macroscopic or micro-meso-macro), in inclusion hierarchies (e.g., a cell that comprises a nucleus that comprises chromosomes that comprise...), in social organizations and others. However, one sometimes finds cases where the interactions among the high levels are of the network type, and their organization cannot be satisfactorily described by a hierarchical structure.

The first step towards understanding these complex organizations is the identification of the meso-level structures, a process that can be far from trivial whenever the interactions are unknown or only partly known. Think for example of a gene regulatory network, where some genes may be known to regulate the expression of a particular gene, but several other interactions are unknown; or of a social or economic organization, where some activities can be observed, but the complete pattern of reciprocal influences cannot (some may be even deliberately hidden, due, e.g., to economic interests).

The purpose of the RI is that of identifying sets of variables that behave in a somehow coordinated way in a dynamical system; the variables that belong to the set are integrated with the other variables of the set, much more than with the others. Since these subsets are possible candidates as higher-level entities, to be used to describe the system organization, they will be called relevant subsets (omitting the specification that they are candidates). A quantitative measure, well suited for identifying them, is defined as follows (the presentation below follows the one given in [9]).

Let us consider a system $U$ whose elements are discrete variables that change in time, and let us suppose that the time series of their values are known. According to information theory, the Shannon entropy of an element $x_i$ is defined as:

$$H(x_i) = - \sum_{v \in V_i} p(v) \, log \, p(v) \tag{1}$$

where $V_i$ is the set of the possible values of $x_i$ and $p(v)$ the probability of occurrence of symbol $v$. In this work, dealing with observational data, probabilities will be estimated by relative frequencies.

The entropy of a pair of elements $x_i$ and $x_j$ is defined by means of their joint probabilities:

$$H(x_i, x_j) = - \sum_{v \in V_i} \sum_{w \in V_j} p(v,w) \, log \, p(v,w) \tag{2}$$

Equation (2) can obviously be extended to sets composed of more than two elements.

Let us now consider a subset $S$ of $U$ composed of $k$ elements. Its integration $I(S)$ is defined as (the integration is also known as intrinsic information or multi-information):

$$I(S) = \sum_{x \in S} H(x) - H(S) \tag{3}$$

$I(S)$ represents the deviation from statistical independence of the $k$ elements in $S$. The integration alone could be used to try to identify the relevant subsets, but it turns out that a more accurate identification requires considering the ratio between $I(S)$ and the mutual information between $S$ and the rest of the universe. The mutual information $M(S; U \setminus S)$ between $S$ and the rest of the system $U \setminus S$ is defined as usual as:

$$M(S; U \setminus S) \equiv H(S) + H(S|U \setminus S) = H(S) + H(U \setminus S) - H(S, U \setminus S) \tag{4}$$

where $H(A|B)$ is the conditional entropy and $H(A, B)$ the joint entropy.

Finally, the relevance index $r(S)$ is defined as:

$$r(S) = \frac{I(S)}{M(S; U \setminus S)} \tag{5}$$

This is the measure that will be used below to identify critical states. Note that it is undefined in all of those cases where $M(S; U \setminus S)$ vanishes. In these cases, however, the subset $S$ is statistically independent from the rest of the system, and it should therefore be analyzed separately (these cases should be screened out in advance).

In our first papers on this subject, following the terminology of Tononi [6], we called the quantity $r(S)$ the dynamical cluster index. However, when applied to time series, this term may be misleading: think for example of two variables that take constant and equal values at all times; one usually tends to think that they (can) belong to the same cluster, but their $r(\cdot)$ value vanishes, since it is based on Shannon entropy, that is zero for states whose probability is one (remember that we estimate probabilities with relative frequencies). Therefore, the measure defined in Equation (5) misses some quite obvious clusters, and in order to avoid ambiguities, we prefer to avoid using that word and refer to it as the relevance index (since it can be used to identify relevant subsets of variables).

When the RI is applied to identify relevant subsets, it is necessary to compare sets of different sizes. However, entropies scale with system size, so this requires considerable ingenuity. Following the original work of Tononi, an "RI method" has been developed for this purpose, where the variable is first normalized with respect to a reference case, and a statistical index is computed that allows meaningful comparisons of sets of different sizes [6,9,10]. However, quite often, these sets overlap, so the actual organization of the system remains opaque; for example, a variable may belong to a set of three variables and also to one of its four-element supersets, both endowed with fairly high values of the statistical index. The RI has been developed to decide which set to consider as a basis for deciphering the system structure, as described in [8,11].

However, in order to identify critical states, it turns out that the comparisons among subsystems of different sizes are not required, and one can directly use the RI as defined in Equation (5), as will be done in Sections 4 and 5.

## 3. Models

As has been discussed in Section 1, one can find in the literature slightly different meanings of the word criticality; critical states are always located in-between different regimes, but the nature of these regimes might differ. Here, we will consider two among the most important cases, namely phase transitions and dynamical criticality. To this end, we will apply the RI to the Ising model (for phase transitions) and the random Boolean network (RBN) model of gene regulation for dynamical criticality.

### 3.1. Ising Model

The Ising model (according to Brush [12], the model was first proposed by Lenz in 1920, as also pointed out by Ising in 1925 in its seminal paper) was originally presented with the aim of reproducing simplified ferromagnetic phenomena in materials, but was then recognized as a notable example of a system that can undergo a phase transition as a function of a control parameter. In this section, we briefly recall its main properties; detailed descriptions of the model may be found in the survey paper by Brush [12] and statistical physics books, such as [13,14].

Let us consider a $d$-dimensional lattice of $N$ atoms characterized by a spin, which can be either up ($+1$) or down ($-1$). The atoms exert short-range forces on each other, and each atom tends to align its spin according to the values of its first neighbors. An external field may also be considered, which biases the orientation of the atoms. The energy of the system is defined as follows:

$$E = -\frac{1}{2} \sum_{\langle i,j \rangle} J \, s_i \, s_j + B \sum_i s_i \tag{6}$$

where $s_i$ is the spin of atom $i$, $J > 0$ is a parameter accounting for the coupling between atoms, $\langle i,j \rangle$ denotes the set of all neighboring pairs and $B$ is a parameter playing the role of an external field.

The system can be studied by means of usual statistical mechanics methods, and it can be assessed whether it undergoes a phase transition. Ising proved that the $d = 1$ case does not have phase transitions, while the $d = 2$ model can undergo a phase transition, as proven by Onsager [15] under the hypothesis that $B = 0$.

In this work, we consider the two-dimensional model, with $B = 0$. We performed Monte Carlo simulations at constant temperature $T$; in this case, the system tends to minimize the value of the free energy $A = E - TS$ (where $S$ is the entropy), and there is competition between the energy term, which tends to align the spins, and the entropy term that accounts for thermal disorder. The Monte Carlo algorithm used is a classical Metropolis algorithm with a Boltzmann distribution:

---

**Algorithm 1** Monte Carlo simulation of a 2D Ising model. Adapted from [16].

**while** maximum number of iterations not reached **do**

Choose a random atom $s_i$
Compute the energy change $\Delta E$ associated to the flip $s_i \leftarrow -s_i$
Generate a random number $r$ in [0,1] with uniform distribution
**if** $r < e^{-\frac{\Delta E}{k_B T}}$ **then**

$s_i \leftarrow -s_i$
**end if**
**end while**

---

The temperature is the control parameter, while the order parameter is the so-called magnetization:

$$\mu = \frac{1}{N} \sum_i s_i \tag{7}$$

For low values of $T$, the steady state of the system will be composed of atoms mostly frozen at the same spin, and the time average of the magnetization $\langle \mu(T) \rangle$ will be close either to one or $-1$; for high values of $T$, the spins will randomly flip, and it will be $\langle \mu(T) \rangle \approx 0$. For values close to the critical temperature $T_c$, a phase transition occurs: the system magnetization undergoes a change in its possible steady state values, as depicted in Figure 1.

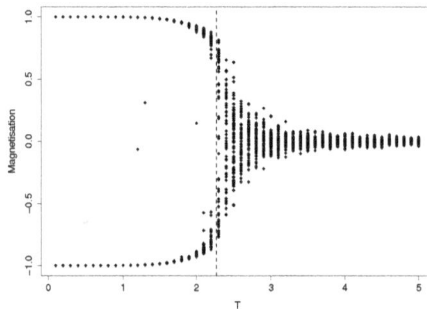

**Figure 1.** Bifurcation diagram for the 2D Ising model obtained by a Monte Carlo simulation. Each dot represents the magnetization of a specific run at temperature $T$; initial conditions are sampled with spin bias in the range $[0.2, 0.8]$ for a total amount of 3000 simulations. The dashed vertical line is located at $T = T_c$.

*3.2. Random Boolean Networks*

Let us now describe RBNs in a very synthetic way, referring the reader to [17–20] for a more detailed account. Several variants of the model have been presented and discussed, but we will restrict our attention here to the "classical" model. A classical RBN is a dynamical system composed of $N$ genes, or nodes, which can take either the value zero (inactive) or one (active). Let $x_i(t) \in \{0, 1\}$ be the activation value of node $i$ at time t, and let $X(t) = [x_1(t), x_2(t) \ldots x_N(t)]$ be the vector of activation values of all of the genes.

The activation of a gene may affect that of other genes; these relationships are represented by directed links and Boolean functions, which model the response of each node to the values of its input nodes. In a classical RBN, each node has the same number of incoming connections $K$, and its $K$ input nodes are chosen at random with uniform probability among the remaining $N - 1$ nodes. The Boolean functions can be chosen in different ways: in this paper, we will only consider the case where they are chosen by assigning, to each combination of $K$ input values, an output value one or zero according to a Bernoulli distribution with parameter $p$. The various outputs are chosen independently from each other, and the probability that the output is one (i.e., the bias $p$) is the same for all of the inputs and for all of the nodes.

In the so-called quenched model, neither the topology nor the Boolean function associated with each node change in time. The network dynamics is discrete and synchronous, so fixed points and cycles are the only possible asymptotic states in finite networks (a single RBN can have, and usually has, more than one attractor). The model shows two main dynamical regimes, ordered and disordered, depending on the degree of connectivity and on the Boolean functions: typically, the average cycle length grows as a power law with the number of nodes $N$ in the ordered region and increases exponentially in the disordered region [17]. The dynamically-disordered region also shows sensitive dependence on the initial conditions (that is why disordered states are often called "chaotic", although the asymptotic states are cycles of finite length in the case of finite networks) not observed in the ordered one.

A well-known method to determine the dynamical regime of an RBN directly measures the spreading of perturbations through the network, by comparing two parallel runs of the same system, whose initial states differ for only a small fraction of the units. This difference is measured by the Hamming distance $h(t)$, defined as the number of units that have different activations on the two runs at the same time step (the measure is performed on many different initial condition realizations, so one actually considers the average value $\langle h(t) \rangle$, but we will omit below the somewhat pedantic brackets). If the two runs converge to the same state, i.e., $h(t) \approx 0$, then the dynamics of the system are robust with respect to small perturbations (a signature of the ordered regime), while if $h(t)$ grows in time (at least initially), then the system is in a disordered state. The critical states are those where $h(t)$ remains initially constant. If a single node is perturbed, the average number of different nodes that differ in the two cases at the following time step is sometimes called the Derrida parameter $\lambda$, so $\lambda > 1$ characterizes disordered states, $\lambda < 1$ ordered states, and $\lambda = 1$ identifies critical states.

In a $p$–$K$ diagram, ordered regimes are separated from those where the dynamics is chaotic by a critical line, whose equation can be shown [19] by:

$$K_c = \frac{1}{2p(1-p)} \tag{8}$$

where $K_c$ is the critical value of the connectivity corresponding to a given value of the bias $p$ (see Figure 2).

The knowledge of the value of $K_c$ is extremely important (Equation (8) can be rigorously derived in the so-called annealed approximation, that is able to provide analytical estimates of the behavior of some variables in RBNs), as it allows us to precisely locate the critical states and to verify how close the RI comes to that value.

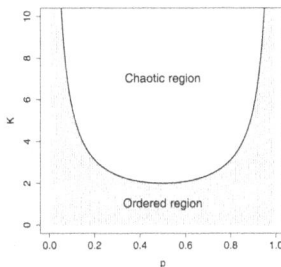

**Figure 2.** The critical line in RBNs. The bold line separates the ordered region (shaded) from the chaotic one.

## 4. Results on the Ising Model

We run Monte Carlo simulations of Ising models with toroidal boundary connections. The neighbors of an atom are its horizontal and vertical first neighbors, i.e., the four adjacent cells in the lattice. We set $J = 1$ and $k_B = 2$, for which $T_c \approx 2.269 \frac{J}{k_B} \approx 1.13$ (the first analytical result on the phase transition in the 2D Ising model has been presented by Onsager [15]). The values of $T$ span the range $[1, 2]$ at steps of 0.05. For each value of $T$, 10 runs are performed starting from random initial conditions chosen with $-1$ spin probability equal to 0.25, so as to start with an intermediate condition between $\mu = -1$ and $\mu = 0$ (we also ran experiments with different biases in the initial condition and did not observe any difference in the results). We run experiments for lattices of size $L \times L$, with $L \in \{10, 20\}$. For each run, $t_{max}$ steps were executed, with $t_{max} = 10^4 \times L^2$; we skipped the first steps so as to reach a steady state, so only the last $t_{max} - 200 \times L^2$ steps were considered and recorded every $L^2$ steps.

In finite-sized Ising models, the critical value of temperature is expected to deviate from the theoretical value. Therefore, the actual critical temperature value was estimated by computing the susceptibility [21], defined as:

$$\chi = \frac{1}{TN} \left( \langle \mu^2 \rangle - \langle \mu \rangle^2 \right) \tag{9}$$

where $T$ is the temperature, $N$ the number of atoms, $\mu$ the magnetization of the system at a given time step and angular brackets denote the time average. The peak of $\chi$ may be used to identify the actual critical temperature value for finite instances. In Figure 3, the median values of susceptibility of ten replicas are plotted against the temperature value. As we can observe, the critical values are around $T = 1.25$ for both lattice sizes considered, which is slightly higher than the theoretical one. This discrepancy is due to the finite size of the systems. This specific value will be taken as the critical one in the Ising models of our experiments.

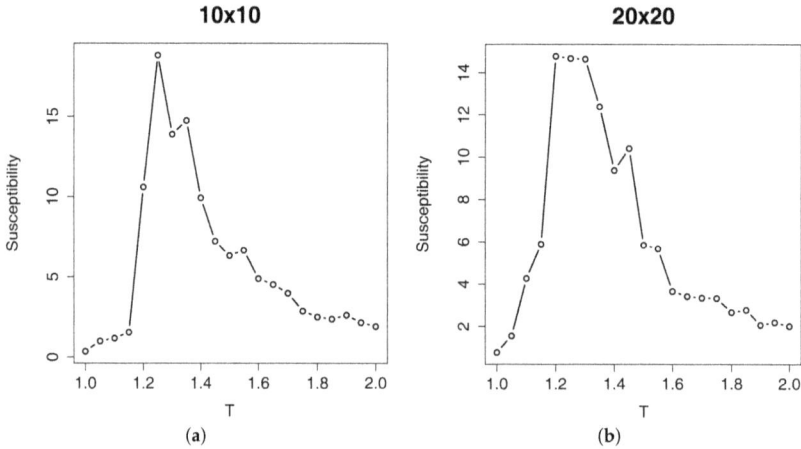

**Figure 3.** Plots of susceptibility values for (**a**) $10 \times 10$ and (**b**) $20 \times 20$ lattices. The median of 10 independent replicas for each temperature value is plotted.

*Relevance Index in the 2D-Ising Model*

For each simulation of the Ising model, we computed the RI for 1000 random subsets of the atoms of size $k_s \in \{2, 3, \ldots, 10\}$. Subsets of larger cardinality would require an impractical number of samples to avoid undersampling in the evaluation of the entropies. However, this is not a limitation of the method, as we will show in the following.

In Figure 4, the median values of RI (averaged over the 1000 samples for each replica) are plotted, against temperature values. For the sake of space, we plotted all of the curves corresponding to all group sizes in the same plot. As $k_s$ increases, the curves shift towards the upper part of the plot, because the (non-normalized) RI values increase with group size. We can observe that the maximum value of RI is attained at a temperature value that corresponds with high precision to the empirically-derived critical value. The effect of undersampling starts to be visible for group sizes approximately greater than 10; indeed, as the cardinality of the groups evaluated increases, the RI peak tends to flatten. However, small group sizes are enough to locate the critical point.

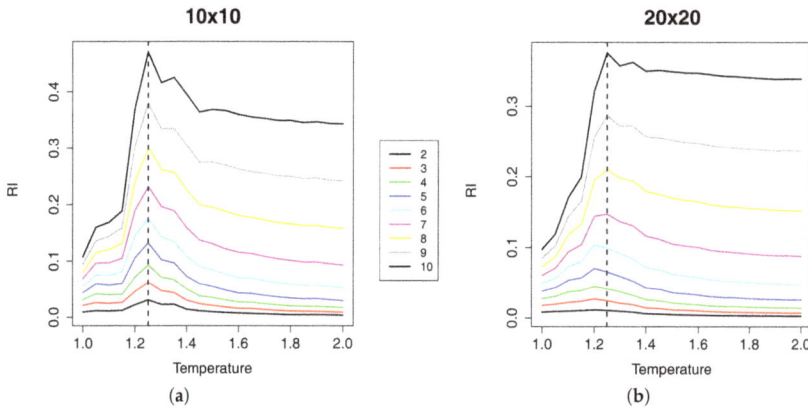

**Figure 4.** Plots of the relevance index (RI) for (**a**) $10 \times 10$ and (**b**) $20 \times 20$ lattices. The median of the average integration values for groups of size two to 10 is plotted against $T$. The curves shift up with group size. Note that in the $20 \times 20$ case, for small group sizes, the index peaks slightly before the dashed line: this discrepancy is ascribed to the small plateau around the maximal value of susceptibility, as can be observed in Figure 3.

A question may arise as to what extent the individual contribution of the integration might impact the overall results. To assess this, we also considered the statistics of the integration only, which are summarized in the plots in Figure 5.

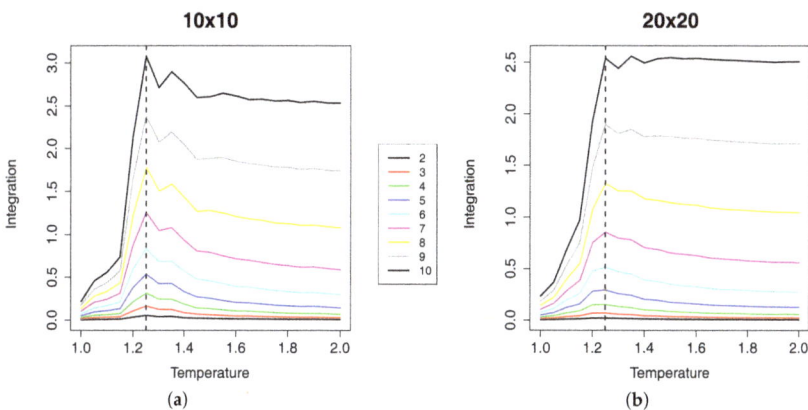

**Figure 5.** Plots of integration for (**a**) $10 \times 10$ and (**b**) $20 \times 20$ lattices. The median of the average integration values for groups of size two to 10 is plotted against $T$. The curves shift up with group size.

From the plots, we can observe that the integration profiles across temperature values show a peak in correspondence with the critical temperature. Nevertheless, the larger the group considered, the less sharp the peak, especially in the $20 \times 20$ instances. Therefore, even if integration alone could provide useful indications to locate the critical temperature, its combination with the mutual information into the RI makes it possible to detect the phase transition with higher precision. The reason for this phenomenon is that the mutual information still moderately grows after $T_c$ (see Figure 6), thus reducing the RI value after the critical point.

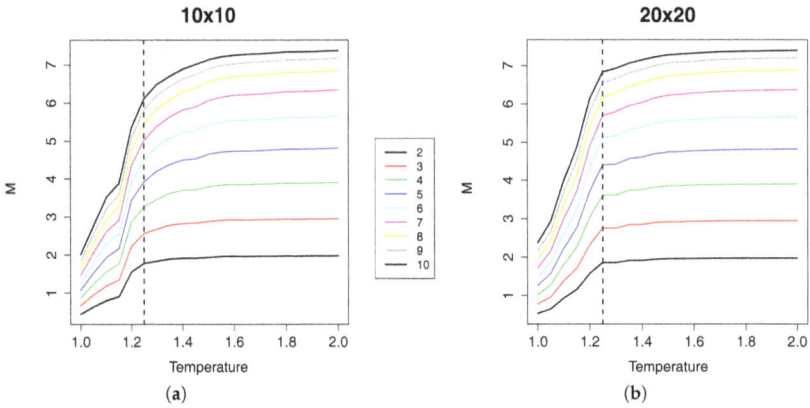

**Figure 6.** Plots of mutual information for (**a**) $10 \times 10$ and (**b**) $20 \times 20$ lattices. The median of the average mutual information values for groups of size two to 10 is plotted against $T$. The curves shift up with group size.

## 5. Results on RBNs

In this section, we show the results of the application of the RI to the task of identifying critical states in RBNs, which have been described in Section 3. There, it was shown that the dynamical behavior of RBNs depends mainly on the values of two parameters, which are (i) their connectivity $K$ and (ii) the bias $p$ of the Boolean functions that rule the responses of their nodes. Since individual network realizations can show different behaviors, the study of the RBN dynamics is based on averages computed on ensembles of networks sharing the same values of connectivity and bias.

The computation of the RI requires a collection of the states of each variable; these have been obtained by collecting in a single series all of the states encountered by a specific RBN starting from $N_{ic}$ random initial conditions. Each time point corresponds to an $N$-dimensional Boolean vector of simultaneous values of the $N$ nodes, and each run continues until an attractor is found or up to a maximum of 1000 steps. The raw data series contain all of the time points, while the attractors series contains only those states that belong to an attractor, and the transients series contains only the states that do not belong to the attractor. In order to avoid excessive computing times, for each series, the analysis is limited to a subset of 1000 randomly-chosen states. The possibility to link together the various time series is directly related to the fact that the RI is based only on relative frequencies of equal-time values and does not depend on states at different times. Note that the choice of random initial conditions implies that the various attractors contribute in a way that is proportional to their basin of attraction. In order to characterize the behavior of families of networks sharing the same parameter values, different random realizations are analyzed; details can be found in Table 1, which summarizes the parameters of the series that have been considered. Since we are interested in their typical dynamical behaviors at different scales, we compute the average of RI for different group sizes; then, the median is taken among all of the systems with the same connectivity and bias.

**Table 1.** In the table, the different combinations of parameters for every experiment are shown. In every experiment, the trajectories are computed until an attractor is found or to a maximum of 1000 steps (obtaining in such a way series of different sizes). In order to avoid excessively long computations in calculating the RI, each index value is computed on 1000 states randomly extracted from the analyzed series. This value is enough to give a suitable approximation of the RI indexes, as discussed in more detail in [22]. For each network and for each group size, 500 random samples are taken starting from 200 initial conditions. RBN, random Boolean network.

| Series Name | Number of Nodes | Number of RBNs |
|---|---|---|
| RBN_20 | 20 | 500 |
| RBN_40 | 40 | 100 |
| RBN_60 | 60 | 100 |
| RBN_100 | 100 | 500 |
| RBN_500 | 500 | 100 |

## 5.1. A Bird's Eye View of the Dynamical Behavior of Families of RBNs

Let us first of all consider the behavior of the RI on a wide set of values of network parameters. As described in Section 3, in a $p$–$K$ diagram, the critical curve is U-shaped (Figure 2); this curve will be called here the Kauffman–Aldana curve [17,19]. Interestingly, the same U-shaped behavior can be observed in Figure 7, where the value of the RI is shown for different $p$–$K$ points. It is also remarkable that the same behavior is observed for different group sizes.

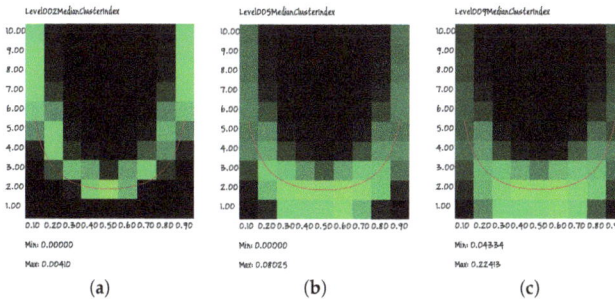

(a)  (b)  (c)

**Figure 7.** Heat maps of the $p$–$K$ diagram of RI indexes computed for groups having different sizes: respectively, groups of (a) size 2, (b) size 5 and (c) size 9. The superimposed red line denotes the position of the edge of chaos curve. This wide region has been sampled in 90 points by combining nine different biases and ten different connectivities; for each point, we tested 100 different RBNs, each RBN being represented by the RI obtained by sampling the states of 200 different trajectories (RBN_40 series, raw data). Each pixel represents the median of the RI of 100 different RBNs sharing the same values of bias and connectivity.

The heat maps indeed show a cloud that surrounds the Kauffman–Aldana curve, indicated by the U-shaped red line superimposed on the plots. Moreover, the position and dispersion of the clouds are similar in all of the group sizes (for clarity, in this section, we present only the results regarding groups of sizes 2, 5 and 9; similar results have been obtained for all groups having sizes from 3 to 10), the only exception being that of the size of two group: actually, as will be shown in the following section, this visual impression is an effect of the low granularity of the wide-range data rather than the signature of a truly different behavior.

Indeed, due to computational limitations and to the attempt to cover a wide range of parameter values, the resolution of the plots in Figure 7 is quite low. However, if we perform a higher resolution search for the peaks of the RI, we find that they actually approach the correct theoretical values (see Equation (8)). An example of such a high-resolution analysis is described in the following section.

Note also that the extremal values of the RI are not the same for all of the critical states, as shown in Figure 8. While this might seem surprising, it should be recalled that critical states share some properties, but there may be differences. Indeed, it has already been shown elsewhere [23] that RBNs can show heterogeneous behaviors in different positions along the critical line (even maintaining critical dynamics).

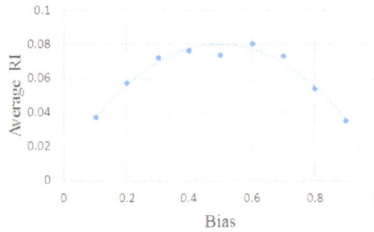

**Figure 8.** The peaks of the RI values shown in Figure 7b (the interpolating curve has been obtained by fitting a quadratic function to the measured points, and it is only a visual aid).

Besides studying the RI, it is interesting to observe the separate contributions of integration ($I$) and mutual information ($M$): as shown in Figure 9, RI and $I$ are both close to the edge of chaos region, but the RI cloud provides a better estimation of the critical curve, especially in the zone of large biases. Moreover, the integration tends to worsen in identifying the chaotic region as the size of the groups increase; it seems indeed that the RI, i.e., the ratio between integration and mutual information, allows a better identification of the chaotic region.

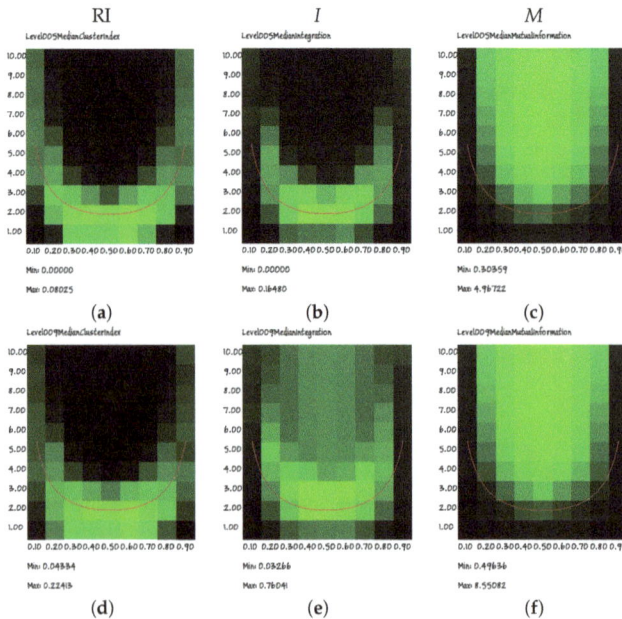

**Figure 9.** Heat maps of the connectivity-bias diagrams of RI (left) and for integration $I$ (center) and mutual information $M$ (right) for (**a–c**) size five groups and (**d–f**) size nine groups (second row). It is possible to note that RI is closer to the critical region (identified by the superimposed red line) than the integration alone, especially in the regions of high biases.

## 5.2. A High-Resolution Analysis

As anticipated, in this section, we show small sections of the whole bias-connectivity diagrams, with a finer sampling of the values of the independent variables and with larger RBNs.

Figure 10 shows the RI results obtained from a section of the *p–K* diagram at $k = 3$, in nets having 100 and 500 nodes: it is possible to note that the RI maximum value is attained for values that are very close to those predicted by the Kauffman–Aldana curve, for all of the investigated group sizes. This observation supports the idea that a similar distribution of organizations is present at many scales (represented here by different group sizes). Moreover, series involving larger RBNs identify more precisely the critical point (a fact already observed in the literature, as for example in [24]) and have narrower RI distributions. Similar observations can be made for other *p–K* diagram sections (data not shown).

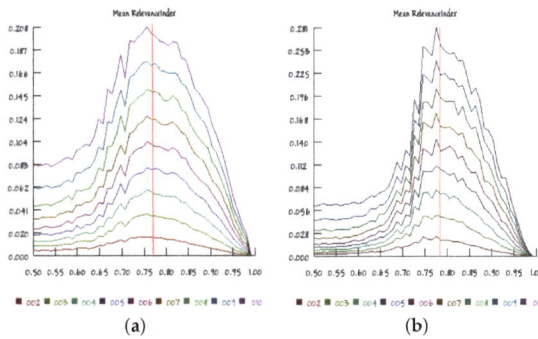

**Figure 10.** The median values of the RI index in RBN having $K = 3$ and respectively (**a**) 100 and (**b**) 500 nodes, for group sizes in the range [2,10]. Bias varies from 0.5 to 1.0, with steps of 0.01. The vertical red line identifies the experimentally-determined edge of chaos position.

The high-resolution analysis confirms that the RI locates the critical region more precisely than integration alone, as shown in Figure 11. Moreover, in the same figure, a measure $\lambda'$, closely related to the Derrida parameter and described in the legend, is also shown. It is interesting to observe that $\lambda'$ has a maximum close to the theoretical critical value of Equation 8, but that its distribution is not sharply peaked. On the other hand, the plot of the RI has also a very close maximum, but the width of its distribution is narrower, so that it can better help to localize the critical value.

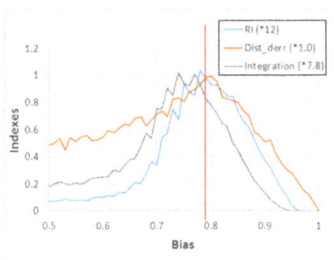

**Figure 11.** The plot shows the median values of the RI and *I* indexes in $K = 3$ RBNs having 500 nodes. Bias varies from 0.5 to 1.0, with steps of 0.01. Moreover, the median values of the index $\lambda'$ are also shown, defined as $\lambda' = 1 - |1 - \lambda|$. This variable is used instead of $\lambda$ itself, which would grow in the chaotic region; in this way, a better visual recognition of the critical point is possible. In order to have similar vertical scales, the RI and *I* values are respectively multiplied by the constants 12.0 and 7.8. The vertical red line identifies the theoretical edge of chaos position.

Transients and Attractors

Differently from ergodic systems (like those represented by the Ising model), RBNs are highly dissipative systems; their asymptotic states are a very small subset of the ensemble of all of the possible states. Therefore, an interesting question regards the dynamical organization of these systems when they reach their attractors: is it significantly different from the organization expressed when the systems are traveling along their transients?

In order to address this question, we split the low-resolution data of 20-node networks and the high-resolution data of 100-node networks into series containing respectively only the states belonging to transients and those that belong to attractors, and we compute the RI separately (while the results of the previous sections refer to complete raw data series). Interestingly, the results are quite different, as shown in Figure 12: the transients of critical systems show RI values close to their maxima, as was the case in the raw data series, although with a relatively wider dispersion, but the attractors' series show peaks of the RI distributions significantly far from the theoretical expectations. In spite of these quantitative difference, the general U-shape of the cloud on the $p$–$K$ diagram is still observed (as shown in Figure 13).

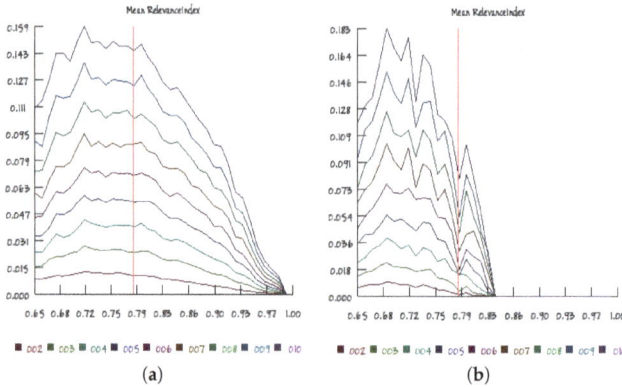

**Figure 12.** The median values of the RI index in RBN having $K = 3$ and 100 nodes obtained (**a**) by using only the states belonging to the transients or (**b**) by using only the states belonging to the attractors. Bias varies from 0.65 to 1.0, with steps of 0.01; the group sizes are in the range [2,10]. The vertical red line identifies the theoretical edge of chaos position.

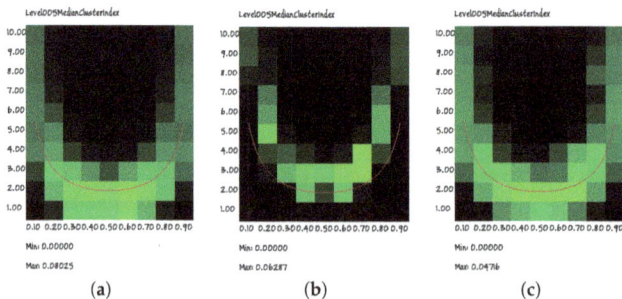

**Figure 13.** The median values of the RI index in an RBN having 20 nodes for groups of size five, obtained (**a**) by using the states belonging to the whole trajectories; (**b**) by using only the states belonging to the attractors and (**c**) by using only the states belonging to the transients. The superimposed red line identifies the theoretical edge of chaos position (computed by assuming the ergodicity of the systems).

One might guess that the difference may be due to the fact that many feedbacks strongly constrain the attractor states, while transients should be able to explore a broader set of conditions. Indeed, the theoretical curve of Equation (8) had been rigorously obtained in the so-called annealed model, where at every time step, connections and Boolean functions are redrawn at random. Physical systems like gene regulatory networks are however much better described by the quenched model, where the links and the Boolean functions are constant in time, and it is this model that has been used in the simulations described here. While the annealed model may provide useful estimates of some properties of the quenched one, it is likely to fail when the analysis considers only dynamical attractors that do not exist in the annealed case; and this seems to happen here, looking at the plots in Figure 12.

It has also been observed elsewhere that the Derrida plots computed by perturbing a subset of all of the possible initial states (for example, those belonging to some attractor) can be very different from the theoretical ones [23]. The phenomena related to the peculiarities of restricting the set of states of an RBN require further studies, as the differences shown here confirm.

## 6. Conclusions and Future Work

The results discussed in the previous sections support the statement that the RI can be effectively used to identify critical states in different kinds of complex systems, both in terms of phase transition and dynamical criticality. Our experimental analysis concerned two prominent classes of complex systems, which stands in favor of the hypothesis that the results may hold in general. The results show also that the RI behaves robustly against sampling effort and system size. Results on both the Ising model and RBN ensembles show that the RI identifies the criticality profile with high precision. In addition, a detailed analysis shows also that, while still detecting critical states, the RI varies across RBN critical ensembles, providing further evidence to the observation that critical RBNs exhibit a spectrum of different behaviors. Moreover, our analysis also supports the statement that the information provided by the dynamics of an RBN along its transients might considerably differ from that of its attractors. Preliminary results in this direction were obtained in [23,25].

It is worth emphasizing that, whilst the index was originally proposed with a different aim, it has been proven able to detect features of criticality. One can imagine possible reasons for this interesting phenomenon, but further investigation is required.

Finally, we plan to study the possibility of applying this method to detect early warning signals of change in complex systems, with the aim of identifying in advance dynamical changes toward or away from criticality.

**Acknowledgments:** We thank Marco Zattoni who made the first investigations on the use of information-theoretical measures on threshold networks and Alessandro Filisetti who contributed to the development of the RI method.

**Author Contributions:** Andrea Roli, Marco Villani and Roberto Serra conceived of and designed the experiments. Andrea Roli, Riccardo Caprari and Marco Villani performed the experiments. Andrea Roli, Marco Villani, Roberto Serra and Riccardo Caprari analyzed the data. Andrea Roli, Marco Villani and Roberto Serra wrote the paper. All authors have read and approved the final manuscript.

**Conflicts of Interest:** The authors declare no conflict of interest.

## References

1. Prokopenko, M.; Lizier, J.; Obst, O.; Wang, R. Relating Fisher information to order parameters. *Phys. Rev. E* **2011**, *84*, 041116.
2. Wang, X.; Lizier, J.; Prokopenko, M. Fisher information at the edge of chaos in random Boolean networks. *Artif. Life* **2011**, *17*, 315–329.
3. Bossomaier, T.; Barnett, L.; Harré, M. Information and phase transitions in socio-economic systems. *Complex Adapt. Syst. Model.* **2013**, *1*, doi:10.1186/2194-3206-1-9.
4. Zubillaga, D.; Cruz, G.; Aguilar, L.; Zapotécatl, J.; Fernández, N.; Aguilar, J.; Rosenblueth, D.; Gershenson, C. Measuring the Complexity of Self-Organizing Traffic Lights. *Entropy* **2014**, *5*, 2384–2407.

5. Roli, A.; Villani, M.; Filisetti, A.; Serra, R. Dynamical criticality: overview and open questions. *arXiv* **2015**, arXiv:1512.05259v2.

6. Tononi, G.; McIntosh, A.; Russel, D.; Edelman, G. Functional clustering: Identifying strongly interactive brain regions in neuroimaging data. *Neuroimage* **1998**, *7*, 133–149.

7. Villani, M.; Filisetti, A.; Graudenzi, A.; Damiani, C.; Carletti, T.; Serra, R. Growth and Division in a Dynamic Protocell Model. *Life* **2014**, *4*, 837–864.

8. Villani, M.; Filisetti, A.; Nadini, M.; Serra, R. On the Dynamics of Autocatalytic Cycles in Protocell Models. In *Advances in Artificial Life, Evolutionary Computation and Systems Chemistry*; Rossi, F., Mavelli, F., Stano, P., Caivano, D., Eds.; Springer: Cham, Switzerland, 2015; Volume 587, pp. 92–105.

9. Villani, M.; Roli, A.; Filisetti, A.; Fiorucci, M.; Poli, I.; Serra, R. The search for candidate relevant subsets of variables in complex systems. *Artif. Life* **2015**, *21*, 412–431.

10. Villani, M.; Filisetti, A.; Benedettini, S.; Roli, A.; Lane, D.; Serra, R. The detection of intermediate-level emergent structures and patterns. In *Advances in Artificial Life, ECAL 2013*; Liò, P., Miglino, O., Nicosia, G., Nolfi, S., Pavone, M., Eds.; The MIT Press: Cambridge, MA, USA, 2013; pp. 372–378.

11. Filisetti, A.; Villani, M.; Roli, A.; Fiorucci, M.; Serra, R. Exploring the organisation of complex systems through the dynamical interactions among their relevant subsets. In *Proceedings of the European Conference on Artificial Life 2015, ECAL 2015*; Andrews, P., Caves, L., Doursat, R., Hickinbotham, S., Polack, F., Stepney, S., Taylor, T., Timmis, J., Eds.; The MIT Press: Cambridge, MA, USA, pp. 286–293.

12. Brush, S. History of the Lenz-Ising Model. *Rev. Mod. Phys.* **1967**, *39*, 883–895.

13. Stanley, H. *Introduction to Phase Transitions and Critical Phenomena*; Oxford University Press: Oxford, UK, 1971.

14. Binney, J.; Dowrick, N.; Fisher, A.; Newman, M. *The Theory of Critical Phenomena*; Oxford University Press: Oxford, UK, 1992.

15. Onsager, L. Crystal Statistics. I. A Two-Dimensional Model with an Order-Disorder Transition. *Phys. Rev.* **1944**, *65*, 117–149.

16. Solé, R. *Phase Transitions*; Princeton University Press: Princeton, NJ, USA, 2011.

17. Kauffman, S. *The Origins of Order: Self-Organization and Selection in Evolution*; Oxford University Press: Oxford, UK, 1993.

18. Kauffman, S. *At Home in the Universe*; Oxford University Press: Oxford, UK, 1996.

19. Aldana, M.; Coppersmith, S.; Kadanoff, L. Boolean dynamics with random couplings. In *Perspectives and Problems in Nolinear Science*; Springer: New York, NY, USA, 2003; pp. 23–89.

20. Serra, R.; Villani, M.; Semeria, A. Genetic network models and statistical properties of gene expression data in knock-out experiments. *J. Theor. Biol.* **2004**, *227*, 149–157.

21. Christensen, K.; Moloney, R. *Complexity and Criticality*; Imperial College Press: London, UK, 2005.

22. Caprari, R. Applicazione Della Teoria Dell'informazione Allo Studio di Regimi Critici. Bachelor's Thesis, Università di Modena e Reggio Emilia, Modena, Italy, 2016. (In Italian)

23. Villani, M.; Campioli, D.; Damiani, C.; Roli, A.; Filisetti, A.; Serra, R. Dynamical regimes in non-ergodic random Boolean networks. *Nat. Comput.* **2016**, doi:10.1007/s11047-016-9552-7.

24. Ribeiro, A.; Kauffman, S.; Lloyd-Price, J.; Samuelsson, B.; Socolar, J. Mutual information in random Boolean models of regulatory networks. *Phys. Rev. E* **2008**, *77*, 011901.

25. Zattoni, M. Threshold Networks: Simulazione Progettazione e Analisi. Master's Thesis, Università di Bologna, Bologna, Italy, 2014. (In Italian)

![entropy logo] *entropy*

MDPI

*Article*

# Emergence of Distinct Spatial Patterns in Cellular Automata with Inertia: A Phase Transition-Like Behavior

**Klaus Kramer** [1,2], **Marlus Koehler** [2], **Carlos E. Fiore** [3] and **Marcos G.E. da Luz** [2,*]

[1]  CEPLAN, Universidade do Estado de Santa Catarina, 89283-081 São Bento do Sul-SC, Brazil; glausians@gmail.com

[2]  Departamento de Física, Universidade Federal do Paraná, C.P. 19044, 81531-980 Curitiba-PR, Brazil; koehler@fisica.ufpr.br

[3]  Instituto de Física, Universidade de São Paulo, C.P. 66318, 05315-970 São Paulo-SP, Brazil; fiore@if.usp.br

*  Correspondence: luz@fisica.ufpr.br; Tel.: +55-41-3361-3664

Academic Editor: Mikhail Prokopenko
Received: 24 December 2016; Accepted: 28 February 2017; Published: 7 March 2017

**Abstract:** We propose a Cellular Automata (CA) model in which three ubiquitous and relevant processes in nature are present, namely, spatial competition, distinction between dynamically stronger and weaker agents and the existence of an inner resistance to changes in the actual state $S_n$ (=−1,0,+1) of each CA lattice cell $n$ (which we call inertia). Considering ensembles of initial lattices, we study the average properties of the CA final stationary configuration structures resulting from the system time evolution. Assuming the inertia a (proper) control parameter, we identify qualitative changes in the CA spatial patterns resembling usual phase transitions. Interestingly, some of the observed features may be associated with continuous transitions (critical phenomena). However, certain quantities seem to present jumps, typical of discontinuous transitions. We argue that these apparent contradictory findings can be attributed to the inertia parameter's discrete character. Along the work, we also briefly discuss a few potential applications for the present CA formulation.

**Keywords:** cellular automata; spatial-temporal patterns; complexity; phase transitions; emergent behavior; phase segregation; ecotones

## 1. Introduction

Spatial-temporal pattern structures are ubiquitous [1,2], especially in biological and human phenomena [3–6]. They commonly originate from (nonlinear) driving forces acting locally on the elements of a spatially-extended system. Often, this type of global emergent behavior cannot be "guessed" simply from a direct qualitative inspection of the interactions between the individual constituents (i.e., at the microscopic level). The full dynamics, the macroscopic description [6], can be understood only as a collective effect. Such a scenario frames what is frequently termed in the literature complex systems.

Likewise, phase transitions do constitute an extremely relevant class of processes. In particular, the so-called critical phenomena lead to a very rich range of distinct comportment [7,8]. Briefly, not too close to a critical point $\lambda_c$, the system microscopic organization varies continuously with a given control parameter $\lambda$ (for instance, temperature, chemical potential, etc.), maintaining a certain main characteristic, e.g., a non-null magnetization or some degree of ordered aggregation, quantified by a macroscopic order parameter $\Gamma$. However, by crossing $\lambda_c$ (with $\Gamma(\lambda_c)$ usually vanishing), the system qualitatively changes, going through a phase transition, e.g., becoming non-magnetic or complete disordered. Moreover, around the critical point [7], there is the development of infinite correlation lengths and the rise of universal features (like detail-independent critical exponents). A non-trivial association between complexity and criticality is generally believed to exist [9].

The wide purpose of Cellular Automata (CA) (the acronym CA will mean either Cellular Automaton or its plural form Cellular Automata; which one is actually being used should be clear from each sentence's specific context) models [10,11] couples very well with the idea of emergent complexity. From a pure computational (algorithmic) point of view, CA is an extremely versatile framework: in principle CA would be able to realize an universal Turing machine [12–14], hence to perform complete general information processing. Thus, putting aside the controversial discussion if CA could (or not) describe any natural process [15], the fact is that CA are extremely useful in simulating distinct complex systems [16]. For example, the quite hard task of mimicking life and/or live organisms (individually or collectively) was the initial motivation for John von Neumann, with the help of Stanislaw Ulam, to propose CA in the 1940s, an approach reborn in the 1970s, in part due to the general interest in the John H. Conway's "Game of Life" CA. Nowadays, CA have become a frequently employed theoretical tool in biological and ecological studies (refer, e.g., to [17–19]).

The key aspects underlying CA can be summarized as the following. Suppose we cross-grade the system pertinent configurational (spatial, phase, etc.) space, portraying it in terms of cells (or elements). Then, to each cell $n = 1, \ldots, N$, we can ascribe a state variable $S_n(t)$ (function of a discrete time $t = 0, 1, 2, \ldots$), whose numerical (usually also discrete) values indicate, say, particles' density, pigment color, action states (active/inactive, life/death, infected/non-infected), energy content, etc. As the system dynamically evolves [20], the set $\{S_n(t)\}$ may give rise to intricate tile-like arrangements. This can be so even though at the elements' level, the local responses to the interactions are simple. In this way, the full CA behavior reflects global complexity [21].

Due to the CA inherent discrete formulation, contrary to the already mentioned close connection between CA and complex systems, it may be much more difficult to establish an appropriate link between deterministic CA and phase transitions. This is particularly true with respect to continuous transitions, i.e., critical phenomena (for probabilistic CA, phase transitions are more easily observed; see, for instance, [22,23]). Actually, in most CA constructions, there are no "good" $\lambda$'s, which could play the role of control parameters (eventually, they might exist, but then demanding certain restrictions to act as such, like requiring the number of possible states $S$ to be very high [24]).

Recently, a deterministic CA with a new ingredient, inertia, has been proposed [25]. The original goal in [25] was to find a minimal, not dedicated model capable of describing the basic aspects of ecotones [26–28]: zones between geographic regions of distinct biomes. In such transition areas, there is the coexistence of groups of species coming from different ecosystems. In many concrete situations, it is still not completely understood how less fitted (e.g., to the nearby biomes conditions) exogenous animals and plants can survive [29,30]. They eventually would perish in a more homogeneous environment if competing with the same stronger (better adapted) local endogenic species [31,32]. To address the problem, some of us have introduced an intrinsic (in opposition to certain probabilistic proposals in the literature; see, e.g., [33–35]) resistance, the inertia $I$, to the system's natural rules of evolution [25]. As a consequence, depending on $I_n$ (which assumes discrete values), it becomes more difficult for the interactions to change the state $S_n$ of each cell $n$. By playing with the initial conditions and the $I_n$ distribution, it is possible to generate spatial patterns of meta-communities qualitatively similar to those in ecotones [21,26–28].

In the present contribution, we shall explore CA with inertia in very general terms, showing that $I$ can act as a proper control parameter. Considering analysis procedures typical from statistical physics, we study the evolution of ensembles of CA with different random initial configurations. We thus are able to characterize average properties for the CA final spatial patterns' structures, clearly identifying behavior similar to phase transitions. Although, because of their discrete nature, CA cannot lead to "true" critical phenomena, certain features of usual continuous phase transitions can be observed as we change $\{I_n\}$. In fact, by assuming different distribution of inertia values among the CA cells, we observe distinct and intricate heterogeneities in the emergence of spatial patterns.

Intriguingly, for certain quantities, we even can see indications of discontinuous phase transitions. These apparent contradictory results eventually can be explained in terms of the finite character of the CA model and the discreteness of the inertia parameter $I$. Along the work, potential applications for our model are also briefly mentioned. Finally, in the last discussion and conclusion section, we make final considerations about the rise of phase transitions in CA and the important role, in this regard, represented by inertia-like quantities (using ecotones as an illustration).

## 2. The Model: CA with Inertia

### 2.1. Basic Definitions

In our CA construction, the space is represented by a square lattice of $N \times N$ elements (or cells). At the discrete time $t \, (= 0, 1, 2, \ldots)$, the state of cell $(i, j)$ (hereafter labeled $n \, (= 1, 2, \ldots, N \times N)$) is given by $S_n(t)$, whose possible numerical values are $-1$, $0$ and $+1$. We assume that only cells in the (active) states $+1$ and $-1$ have the eventual power of altering the $S$'s of their neighbors. Therefore, the state $0$ is neutral in terms of any competition dynamics. We consider a further internal parameter associated with each cell $n$, $I_n$, which is an integer between zero and eight. We call such a quantity "inertia" since it characterizes the cells' intrinsic resistance to changes in their actual state values. In the more general case, $I_n$ could be a function of time, but for our purposes, here, we restrict the analysis only to time-independent $I$'s.

At time $t$, let us define $V_n(t) = V_n^{(+)}(t) - V_n^{(-)}(t)$, with $V_n^{(S)}(t)$ the population of the state $S \, (=-1, 0, +1)$ in the neighborhood of $n$. The cell $n$ has $\mathcal{N}_n = 8, 5, 3$ contiguous neighbors if it is, respectively, in the bulk, border or corner of the square lattice (we are supposing hard wall boundary conditions (HWBD)). This is known as Moore's neighborhood of radius one. We assume HWBD because many systems that potentially can be described by our model (like the already mentioned ecotones landscape, as well as fragmentation patterns in ecosystems [36]) require limited boundaries as their spatial arena. The time evolution of each $S_n(t)$ follows from two deterministic rules:

(i)  The inertial rule: If $I_n < |V_n(t)|$ the dynamical rule below is applied, otherwise $S_n(t+1) = S_n(t)$, i.e., the state of $n$ remains unchanged.
(ii)  The dynamical rule: $S_n(t+1) = \text{sign}[V_n(t)]$, with $\text{sign}[x]$ the signal of $x$.

Two aspects of (i)–(ii) should be highlighted. First, any neighbor of $n$ in the state $0$ does not contribute to make $V_n(t) \neq 0$, a necessary condition to change $S_n$ (if $V_n(t) = 0$, the dynamical rule (ii) is not applied). Therefore, cells in state $0$ cannot modify their neighborhoods. Second, cells in the active state $S$ (either $-1$ or $+1$) belonging to the neighborhood of $n$ can alter $S_n \neq S$ only if $V_n^{(S)} - V_n^{(-S)}$ is greater than the cell $n$ resistance to switch (given by the inertia parameter $I_n$).

From now on, by 'one step' of evolution ($t \to t+1$), we will mean that we have considered Rules (i)–(ii) for all of the $N \times N$ elements in the CA lattice, obtaining the full set $\{S_n(t+1)\}$ from $\{S_n(t)\}$.

To characterize certain system features, we define two groups of quantities calculated at each time $t$. The first relates to the population of a state $S$, given by $p_S(t)$. For our $N \times N$ lattice, the total population is $p = N \times N = p_-(t) + p_0(t) + p_+(t)$. The second represents the degree of clusterization of the CA lattice spatial pattern, either of the whole system, $c$, or only of the cells in state $S$, $c_S$. The clusterization measures the amount of "agglomeration" of the CA elements, i.e., the number and size of clusters formed by cells in the same state. For their definition, suppose $S_{major}$ the state corresponding to $\text{Max}\{V_n^{(-1)}(t); V_n^{(0)}(t); V_n^{(+1)}(t)\}$ (obviously, making sense only if there is a unique most populated state). If there is a $S_{major}$, we set $\mathcal{N}_n^{(S_{major})}(t) = V_n^{(S_{major})}(t)$, otherwise $\mathcal{N}_n^{(S_{major})}(t) = 0$. Then, $c(t)$ and $c_S(t)$ read:

$$c(t) = \frac{1}{p} \sum_n [c]_n(t), \qquad c_S(t) = \frac{1}{p_S(t)} \sum_n [c_S]_n(t), \qquad (1)$$

with:

$$[c]_n(t) = \mathcal{N}_n^{(S_{major})}(t) / \mathcal{N}_n,$$
$$[c_S]_n(t) = [c]_n(t) \text{ if } S = S_n(t) = S_{major} \text{ and } 0 \text{ otherwise.} \qquad (2)$$

In Equation (2), $[c]_n$ is the local clusterization around the site $n$ regardless of the state of $n$. On the other hand, $[c_S]_n$ estimates the accumulation of the state $S$ around a cell $n$ if such a cell is also in the state $S$. These functions have been proven very useful in typifying certain spatial patterns in biology [25]. However, as we have explicitly verified, the use of other methods (like Hoshen–Kopelman [37]) to gauge the CA degree of clusterization yields the same qualitative results obtained in the present work.

Finally, we specify a third parameter, $\tau$, representing the the minimal number of full iterations (i.e., the number of time steps; see above) for a specific initial lattice to reach a stationary configuration. In other words, for $t > \tau$, the CA remains unchanged for any further application of the evolution rules. As we are going to discuss, only in relatively few cases we will not have a finite $\tau$ for the CA initial lattices considered.

### 2.2. A Statistical Physics-Like Analysis: Averaging over Ensembles of Initial Configurations

Since we shall identify typical properties of the proposed CA, we assume a statistical physics point of view and consider ensembles of initial configurations for the CA lattice. Then, we calculate the quantities described in Section 2.1 by performing averages over a large number of time evolved lattices (from such ensembles), obtaining "mean characteristics" of the system.

To standardize the analysis, we always take lattices having initially $p_+(0) \geq p_-(0)$ and the same fixed number of elements in the 0 state, with $p_0(0)$ being equal to the integer closest to $N^2/3$. Unless otherwise explicitly mentioned, we set $N = 22$ (a value already large enough to illustrate the CA main aspects and also allowing relatively fast simulations). We commonly generate about $N_L \sim 7 \times 10^4$ lattices (in the numerically harder cases $N_L \sim 3.5 \times 10^4$), resulting in very good means for any situation studied. For $N = 22$ (total population $p = 484$ and $p_0(0) = 161$), we discuss two distinct ensembles of lattices: (a) one where $p_+(0)$ is homogeneously distributed in the range $162 \leq p_+(0) \leq 182$; and (b) another in which the clusterization $c_+(0)$ is homogeneously distributed in the interval $0.12 \leq c_+(0) \leq 0.22$. (we observe that such an interval, appropriate for our purposes here, is consistent with $p_+(0) > p_-(0)$ (but hard to meet for $p_+(0) \gg p_-(0)$), also allowing one to generate a large number of replicas for each $c_+(0)$ value).

Observing these restrictions, the spatial distribution of initial states in the cells is random.

We mention that comparatively few initial CA lattices either are not able to converge to a final stationary structure or may take a too long time to do so. Thus, from all of the initial lattices created, we have used only those with $\tau \leq 200$ (typically corresponding to 97%–99% of $N_L$). In the Appendix A, we give simple examples of end patterns that are not stationary because they oscillate between very similar (but not the same) configurations.

Our procedure is therefore: (a) to dynamically evolve each replica in the ensembles according to Rules (i)–(ii) until achieving a steady condition, and then; (b) to perform the pertinent averages over the resulting CA configurations.

## 3. Results

In Section 3, we use the following notational convention. Since any $q(t)$ will be a constant for $t \geq \tau$, the final stationary value of $q(t \geq \tau)$ will be denoted simply as $q$. Furthermore, any quantity $q$ should be understood as the resulting average over the corresponding ensemble.

### 3.1. CA with Zero Inertia

We start by briefly presenting conventional results for a CA without inertia ($I_n = I = 0, \forall n$), the most usual context in the literature. This case will serve as a reference to discuss the behavior of CA with $I \neq 0$.

First, we note that as expected (so, we do not show any plot for such situation here), the final population and clusterization of elements in the $+1$ state increase fairly linearly with the initial value of the $+1$ population, $p_+(0)$. This linearity comes from the fact that the average distribution of a state $S$ in the neighborhood of a cell $n$ is $\overline{f}_S \sim p_S/p$. If $p_S$ increases, $\overline{f}_S$ increases accordingly. Since the dynamical "pressure" to change the state of $n$ to an active $S$ goes with $\overline{f}_S$, it follows that the final number of elements in state $S$ is basically proportional to $p_S(0)$.

An interesting behavior emerges when we calculate $p_+$ and $c_+$ as a function of $c_+(0)$. Examples are shown in Figures 1 and 2. Observe that both $p_+$ and $c_+$ decrease with increasing $c_+(0)$, so $c_+(0)$ plays an inverse role to that of $p_+(0)$. This is apparently counter-intuitive: one could expect higher initial agglomerations of an active $S$ (acting as compact "source" regions at $t = 0$) to enhance the spread of $S$ throughout the lattice, leading to a dominance of $S$ over the other states.

To comprehend the above, it is important to recall that the parameter $c_S(0)$, examined alone, in principle gives just the average degree of agglomeration of a particular $S$, but not any detailed information about the spatial distribution of the $S$ cells across the whole lattice. However, the construction of ensembles with the $c_+(0)$'s in the range of values used here results only in mild variations of $\Delta = p_+(0) - p_-(0)$. Hence, in this case, a higher $c_+(0)$ necessarily implies that the $+1$ cells are more localized in specific regions of the CA lattice. On the opposite, lower $c_+(0)$'s yield more uniformly distributed $+1$ along it. Thus, around the initially localized $+1$ clusters (whose sizes increase and number decreases with an increasing in $c_+(0)$), certainly there is a tendency for the growth of $+1$. However, in the remainder of the lattice, $-1$ is mostly competing with the neutral state 0, so very quickly, $-1$ will dominate. As a consequence, the more concentrated is $+1$ originally, the higher the final overall preeminence of $-1$, explaining the trends in Figures 1 and 2.

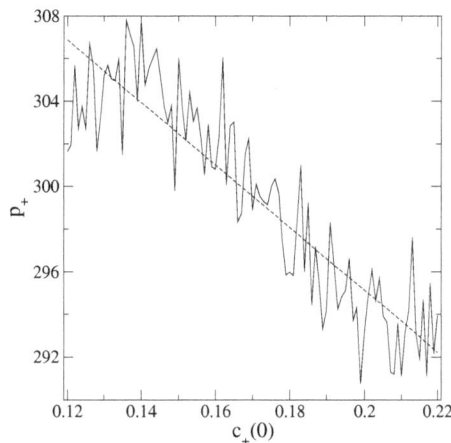

**Figure 1.** Final (i.e., after reaching the stationary configuration) mean population $p_+$ as a function of the initial clusterization $c_+(0)$ for the CA model without inertia. For each $c_+(0)$, $p_+$ is obtained as an average over an ensemble of $N_L$ initial lattices (all having the same $c_+(0)$ value, but different $p_+(0)$'s; see the main text). Even then, the curve $p_+ \times c_+(0)$ fluctuates around a decreasing linear trend represented by the dashed line (the best linear fit). By further generating for each $c_+(0)$ a number $N_S \sim 300$ of ensembles (of $N_L$ replicas in each), one obtains a good straight line $p_+ \times c_+(0)$ with a well-defined slope $\theta$ (see Sections 3.2 and 3.3.3).

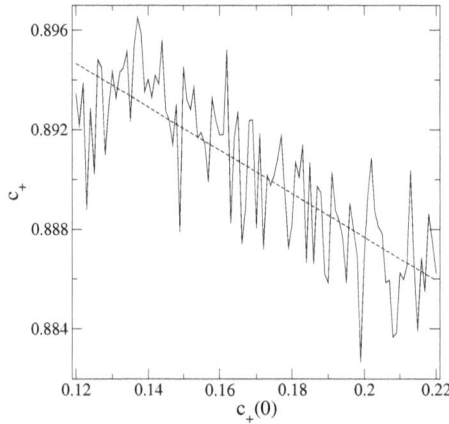

**Figure 2.** The same as in Figure 1, but for $c_+ \times c_+(0)$.

We finish this brief analysis of the $I = 0$ case considering how $\tau$ changes with $p_+(0)$ and $c_+(0)$. From simulations, one finds that $\tau$ increases with $p_+(0)$ and decreases with $c_+(0)$ (recall that we are assuming an interval for $p_+(0)$, such that $\Delta$ is not too large). Regarding $p_+(0)$, if originally $\Delta = 0$ with both $\pm 1$ states equally dispersed, rapidly, there will be a situation of equilibrium between them, leading to shorter $\tau$'s. If $\Delta > 0$, typically at the stationary condition ($t \geq \tau$), $p_+ - p_- > \Delta$. Nonetheless, this relative gain of $p_+$ will require slightly longer times to be achieved as $p_+(0)$ moderately increases. On the other hand, high $c_+(0)$ means that the $+1$ cells are agglomerated in certain regions of the lattice (see the previous discussion). Therefore, quickly, these regions will be populated by $+1$ and the others by the $-1$, resulting in an expeditious convergence to the stationary configuration.

### 3.2. CA with Homogeneous Inertia

The simplest (homogeneous) case of a CA with inertia is that in which $I_n = I$ constant $\forall\, n$ ($1 \leq I \leq 8$). Note that the extreme value $I = 8$ leads to no dynamics (i.e., $\tau = 0$) once the evolution rules (with $I = 8$) cannot modify the system, because any cell has at most eight first neighbors.

First, consider $p_+$ and $c_+$ in terms of $p_+(0)$. Similar to $I = 0$ in Section 3.1, on average, a rather linear dependence of $p_+$ on $p_+(0)$ is observed. Furthermore, as expected, when $I > 0$, both the magnitude of $p_+$, as well as the slopes of the resulting straight lines $p_+ \times p_+(0)$ decrease steadily with increasing $I$. This reflects the natural fact that the cells tend to remain in their initial state for higher $I$'s, producing a final pattern, which does not substantially differ from the initial one. Likewise, on average, $c_+$ displays a linear variation with $p_+(0)$. However, differently from $p_+$, the angular coefficient of the $c_+ \times p_+(0)$ straight line does not display a simple monotonic behavior with $I$. This is clear in Figure 3a, where we plot the slope (i.e., the angle $\theta$ (in degrees)) of the straight line $c_+ \times p_+(0)$ as a function of $I$. There is a clear peak for $\theta$: for $I \leq 4$ the (positive) correlation between $p_+(0)$, and $c_+$ increases with $I$, whereas when $I > 4$, such a correlation decreases with $I$. To understand such a result, which indicates some sort of dynamical transition, we recall that when $I \neq 0$, only a neighborhood of $n$ with $I + 1$ or more cells in the same active $S \neq S_n$ would be able to eventually alter $S_n$. Then, as $I$ grows, the system evolution becomes more sensitive to the initial population because high values of $p_+(0)$ (and, consequently, larger $+1$ state concentrations) are fundamental to trigger state changes. Nevertheless, when $I > 4$ and if $p_+(0)$ is not overwhelming, $c_+$ starts to decouple from $p_+(0)$ since the dynamics passes to be strongly controlled by the cells' internal resistance, and only huge initial clusterization of $+1$ (not the case for $p_+(0)$ not very large) could significantly modify the CA lattice original configuration.

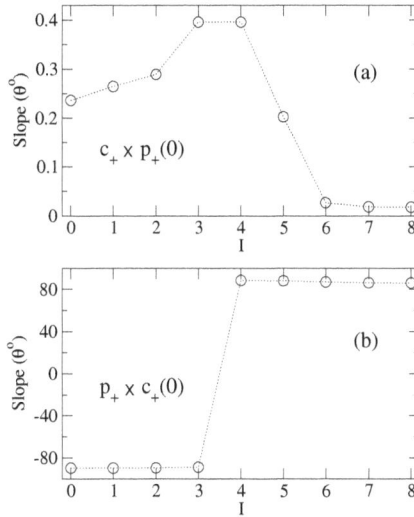

**Figure 3.** The slope, angle $\theta$ in degrees (circles), of the straight lines corresponding to (**a**) $c_+ \times p_+(0)$ and (**b**) $p_+ \times c_+(0)$ (for the calculation details, see the caption of Figure 1) as a function of the inertia $I$. The dotted line is just a guide for the eye.

Second, we observe that even stronger evidence of transition is given by the behavior of $p_+$ versus $c_+(0)$. The slopes (again, the corresponding straight lines' angles $\theta$) are shown in Figure 3b. Similarly to the $I = 0$ case in Section 3.1, for $I \leq 3$, one finds that $p_+$ decreases with increasing $c_+(0)$ (negative $\theta$). Therefore, the mechanism relating $c_+(0)$ with $p_+$ discussed in Section 3.1 is still dominant if the inertia is low. However, when $I \geq 4$, the final $p_+$ increases with $c_+(0)$. Indeed, then, only very aggregated $+1$'s around a cell $n$ (of high inertia) will have the ability to modify $S_n$. Further, in such a context of greater $I$'s, the more randomly-distributed population of $-1$'s across the lattice barely will increase (contrasting with the dynamics of the equivalent situation, but with $I = 0$ in Section 3.1).

Third, regarding $c_+ \times c_+(0)$, for the $0 \leq I < 3$, interval, $c_+$ displays only a weak dependence on $c_+(0)$, with the magnitude of $c_+$ slightly decreasing as $I$ increases (see the explanation in Section 3.1). For $I \geq 3$, the cells' enhanced resistance induces a stronger correlation between $c_+$ and $c_+(0)$, so that higher final will demand higher initial clusterizations. This is so because for large inertia, changes in a given $S_n \neq S$ are possible only if $n$ is in the vicinity of a large cluster of the active $S$. As a consequence, an active state $S$ will spread out only if "supported" by already existing clusters of $S$.

Lastly, the cells tend to remain in their original states as the inertia grows. Therefore, the number of accessible intermediary configurations until stationarity usually reduces with $I$, shortening $\tau$. As an illustration, in Figure 4, we show a typical CA evolution. We consider the same initial lattice, for which $c_+(0) = 0.169864$ and $p_+(0) = 172$ and different $I$'s. The black, white and grey colors represent, respectively, states $+1$, $0$ and $-1$. In this particular example, when $I = 0$, $\tau = 9$ iteration steps are necessary to achieve the steady condition. For the other $I$'s, we have: $\tau = 7$ for $I = 1, 2$; $\tau = 6$ for $I = 3$; and $\tau = 2$ for $I = 4, 5$. In the first six cases, the final CA patterns are all distinct from each other. For $I \geq 6$, the initial configuration cannot be changed from the evolution rules.

Some other discussed features of the present CA system can also be identified in Figure 4. For instance, we see that the final clusterization becomes strongly correlated to the initial clusterization for high $I$'s. However, perhaps the most interesting property observed in Figure 4 is that the number of cells in the neutral (or dynamically passive) $S = 0$ state in the final configuration increases with $I$. In this particular example, when $I = 0$, all the initial 0 elements are transformed into the active $\pm 1$

states. However, as $I$ increases, a fraction of the initial 0 cells are able to survive up to the stationary configuration, with $p_0$ growing with $I$ (obviously $p_0 = p_0(0)$ for $I = 8$).

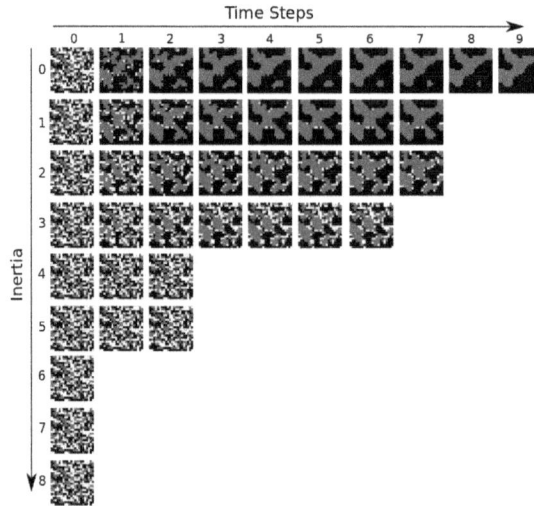

**Figure 4.** Evolution until stationarity of the same initial lattice (of $p_+(0) = 172$ and $c_+(0) = 0.169864$) for increasing inertia $I$ values. Each $t$ corresponds to a full iterated time step ($t = 0$ is the initial lattice). The black, white and gray colors represent, respectively, the $+1$, 0 and $-1$ states. For this particular case, the initial configuration cannot be changed by the CA rules (i)–(ii) when $I \geq 6$.

Thus, it is clear that the presence of an internal resistance variable, inertia, is essential for the endurance of states that are dynamically neutral (or somehow "fragile") during the time evolution. This property has been used to model ecotones (i.e., biome transition regions) in a previous work [25]. Nonetheless, here, we shall briefly mention a further potential application for our CA formulation. As nicely put in [38], the phenomenon of frustration is a very common behavior in complex systems, arising when a local minimum of energy cannot be achieved due to opposite force mechanisms. For example, three 1/2-spins located at the vertices of a triangle cannot simultaneously be at the lowest energy level if their mutual interaction is antiferromagnetic. Frustration-induced phase transitions have already been described by CA [39], but through stochastic implementations (briefly, we define two deterministic set of rules, $R_1$ and $R_2$, and at each time $t$, choose probabilistically one of them to be applied to the CA elements). In magnetic lattices displaying frustration, if the coupling extends over larger neighborhoods and the interactions are anisotropic, the system may develop regions of regular ordering, separated by irregular lines of frustrated magnets [40]. With a proper extension of the rules in Section 2.1, the evolution of our deterministic CA with inertia can simulate the formation of such regions, where the state 0 could then represent the frustrated magnets (this is presently an ongoing study) (cf., in Figure 4, see the configurations for $I = 2$ at $t = 7$ and $I = 3$ at $t = 6$).

### 3.3. CA with Inertia Following Spatial Patterns

So far, we have discussed the case of the same inertia value for the whole CA lattice. However, one can imagine complete arbitrary distributions of inertia among the CA cells, conceivably leading to very diverse and rich dynamics. Therefore, just to give a flavor of more general possibilities, we next consider three relatively simple illustrative examples, which we call Patterns I, II and III for the inertia spatial distribution.

### 3.3.1. Pattern I: Regional Block Distribution of Inertia

The inertia Pattern I is depicted in Figure 5. There are in total nine regional blocks of inertia, each having a fixed value of $I = 0, 1, \ldots, 8$. Successive adjacent horizontal blocks (from the top-left to the bottom-right) have $I$ incremented by one unit.

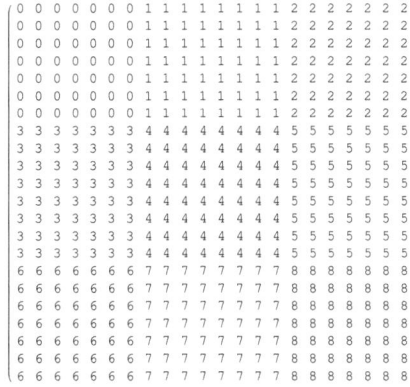

$$
\begin{pmatrix}
0 & 0 & 0 & 0 & 0 & 0 & 0 & 1 & 1 & 1 & 1 & 1 & 1 & 1 & 2 & 2 & 2 & 2 & 2 & 2 \\
0 & 0 & 0 & 0 & 0 & 0 & 0 & 1 & 1 & 1 & 1 & 1 & 1 & 1 & 2 & 2 & 2 & 2 & 2 & 2 \\
0 & 0 & 0 & 0 & 0 & 0 & 0 & 1 & 1 & 1 & 1 & 1 & 1 & 1 & 2 & 2 & 2 & 2 & 2 & 2 \\
0 & 0 & 0 & 0 & 0 & 0 & 0 & 1 & 1 & 1 & 1 & 1 & 1 & 1 & 2 & 2 & 2 & 2 & 2 & 2 \\
0 & 0 & 0 & 0 & 0 & 0 & 0 & 1 & 1 & 1 & 1 & 1 & 1 & 1 & 2 & 2 & 2 & 2 & 2 & 2 \\
0 & 0 & 0 & 0 & 0 & 0 & 0 & 1 & 1 & 1 & 1 & 1 & 1 & 1 & 2 & 2 & 2 & 2 & 2 & 2 \\
0 & 0 & 0 & 0 & 0 & 0 & 0 & 1 & 1 & 1 & 1 & 1 & 1 & 1 & 2 & 2 & 2 & 2 & 2 & 2 \\
3 & 3 & 3 & 3 & 3 & 3 & 3 & 4 & 4 & 4 & 4 & 4 & 4 & 4 & 5 & 5 & 5 & 5 & 5 & 5 \\
3 & 3 & 3 & 3 & 3 & 3 & 3 & 4 & 4 & 4 & 4 & 4 & 4 & 4 & 5 & 5 & 5 & 5 & 5 & 5 \\
3 & 3 & 3 & 3 & 3 & 3 & 3 & 4 & 4 & 4 & 4 & 4 & 4 & 4 & 5 & 5 & 5 & 5 & 5 & 5 \\
3 & 3 & 3 & 3 & 3 & 3 & 3 & 4 & 4 & 4 & 4 & 4 & 4 & 4 & 5 & 5 & 5 & 5 & 5 & 5 \\
3 & 3 & 3 & 3 & 3 & 3 & 3 & 4 & 4 & 4 & 4 & 4 & 4 & 4 & 5 & 5 & 5 & 5 & 5 & 5 \\
3 & 3 & 3 & 3 & 3 & 3 & 3 & 4 & 4 & 4 & 4 & 4 & 4 & 4 & 5 & 5 & 5 & 5 & 5 & 5 \\
3 & 3 & 3 & 3 & 3 & 3 & 3 & 4 & 4 & 4 & 4 & 4 & 4 & 4 & 5 & 5 & 5 & 5 & 5 & 5 \\
6 & 6 & 6 & 6 & 6 & 6 & 6 & 7 & 7 & 7 & 7 & 7 & 7 & 7 & 8 & 8 & 8 & 8 & 8 & 8 \\
6 & 6 & 6 & 6 & 6 & 6 & 6 & 7 & 7 & 7 & 7 & 7 & 7 & 7 & 8 & 8 & 8 & 8 & 8 & 8 \\
6 & 6 & 6 & 6 & 6 & 6 & 6 & 7 & 7 & 7 & 7 & 7 & 7 & 7 & 8 & 8 & 8 & 8 & 8 & 8 \\
6 & 6 & 6 & 6 & 6 & 6 & 6 & 7 & 7 & 7 & 7 & 7 & 7 & 7 & 8 & 8 & 8 & 8 & 8 & 8 \\
6 & 6 & 6 & 6 & 6 & 6 & 6 & 7 & 7 & 7 & 7 & 7 & 7 & 7 & 8 & 8 & 8 & 8 & 8 & 8 \\
6 & 6 & 6 & 6 & 6 & 6 & 6 & 7 & 7 & 7 & 7 & 7 & 7 & 7 & 8 & 8 & 8 & 8 & 8 & 8 \\
6 & 6 & 6 & 6 & 6 & 6 & 6 & 7 & 7 & 7 & 7 & 7 & 7 & 7 & 8 & 8 & 8 & 8 & 8 & 8
\end{pmatrix}
$$

**Figure 5.** The Pattern I spatial distribution of inertia.

Qualitatively, the dependence of $p_+$ and $c_+$ on $p_+(0)$ is similar to that for homogeneous constant $I$'s. There is a linear positive correlation of these quantities with the initial $+1$ population, so that $p_+$ and $c_+$ grow linearly with $p_+(0)$. Furthermore, $p_+$ decreases with increasing $c_+(0)$, which is the trend in Section 3.2 for $I$ small. On the other hand, Figure 6 shows that (up to fluctuations, cf. Figures 1 and 2) $c_+$ tends to increase with $c_+(0)$. This is the case in Section 3.2, but only for greater $I$'s. Such results point to an interesting propensity for gradual block-like inertia distributions. The behavior of $p_+$ is always more strongly determined by the lower $I$'s, whereas $c_+$ as a function of $c_+(0)$ ($p_+(0)$) is more influenced by the higher (lower) $I$'s. Regarding $\tau$, in the ranges considered for the initial conditions, the inertia Pattern I is akin to $I = 0$, with larger $p_+(0)$'s ($c_+(0)$'s) giving rise to longer (shorter) $\tau$'s. Nevertheless, there is an important difference: the average number of iterations is usually lower for Pattern I than for $I = 0$. This is so because regions with high $I$ quickly converge to a local stationary configuration, thus slightly decreasing $\tau$ compared to CA without inertia (Section 3.1).

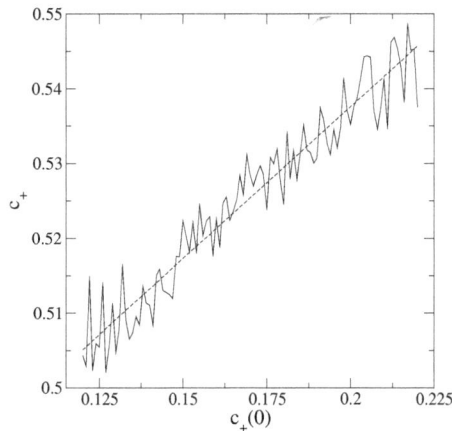

**Figure 6.** Similar to Figure 2, but for the spatial distribution of inertia given by Pattern I.

Figure 7 shows a typical time evolution of a CA with the inertia Pattern I (in this example, $\tau = 6$). Observe that the cells belonging to blocks with low inertia are able to form clusters of active $\pm 1$ states (only very few cells in the 0 state survive). In contrast, elements in the blocks with higher inertia are more evenly distributed among the three possible $S$ values. In such blocks, there is no prevalence of a major state, and the final $p_0$ increases as $I$ (regionally) increases.

Time Steps

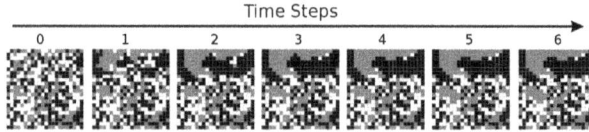

**Figure 7.** Evolution until stationarity (here, $\tau = 6$) of the same initial lattice (of $p_+(0) = 172$ and $c_+(0) = 0.16531$) with the inertia Pattern I. The color convention is the same as in Figure 4.

CA are very useful to model phase separation processes [41], e.g., as occurring in binary mixtures [42]. In this regard, we observe that the blocks in Figure 7, where $S = \pm 1$ have become very clusterized (for $t \geq 6$), could be interpreted as the regions of the final segregated phases. On the other hand, certain lattice areas do not allow a strong domination by a single type of state (say, representing a specific molecule). Therefore, a cluster distribution of inertia acts like a heterogeneous medium for the distinct states (or particles) diffusion, thus engendering complex morphological partition [43].

### 3.3.2. Pattern II: A Unique Central Region with a Non-Null Constant Inertia

The inertia spatial Pattern II is represented in Figure 8. Just the elements of a central square block $B$ have non-null inertia ($I_n = i \neq 0$ for any $n$ in $B$ and $I_n = 0$ otherwise). Note that the number of $I = 0$ neighbors to the $B$ border (corners) cells is three (five). Thus, usually, $i \geq 5$ would lead to no dynamics for the $B$ elements in the case of a random distribution of states for the initial CA lattice. Therefore, here, we only assume $i = 1, 2, 3, 4$.

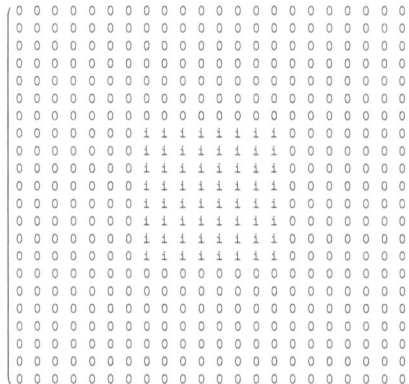

**Figure 8.** The Pattern II spatial distribution of inertia.

This kind of inertia pattern is interesting to test how an initial compact group with a certain opposition to changes (but with a random distribution of $S$'s) can influence the full system evolution. For applications of CA to problems where there are resistant agents or refractory periods, during which there is no response to external signals from certain elements, see, e.g., models of resistance to market innovations [33] and rippling arrangements in biological cells [44].

A first distinct feature of Pattern II (with respect to the case of $I = 0$ everywhere, i.e., $i = 0$) is that the number of replicas in the ensembles (Section 2.2), which does not converge to stationarity, slightly decreases (increases) when $i = 1$ ($i > 1$). It can be understood as the following. For null inertia, there are some lattices that oscillate between few final, similar, but not equal, configurations (Appendix A). For Pattern II with $i \geq 1$, a certain number $N_e(i)$ of these oscillations are eliminated (see Appendix A), and the corresponding lattices stabilize. For $i = 1$ (still a low inertia value), unless for these eliminations, other aspects of the CA dynamics are not drastically altered. However, for greater $i$'s, the border cells of $B$ start to act as a contour of fixed $S$'s, generating a geometric constraint in the evolution of the remaining lattice cells (especially those in the immediate vicinity of $B$). Hence, a given number $N_c(i)$ of initial lattices will not converge to a steady condition due to the appearance of border oscillating structures (see the discussion in the Appendix A) around $B$. For higher $i$'s, $N_c(i)/N_e(i) > 1$, if $B$ is big enough, the situation here. As a consequence, compared to Section 3.1, the average $\tau$ for Pattern II very mildly increases (decreases) if $i = 1$ ($i \geq 2$). In fact, for $i = 1$, the extra lattices ($N_e(1)$), which are now included in the calculations, often have their individual $\tau$'s longer than the average $\tau$ in Section 3.1. For $i \geq 2$, the block $B$ generally converges more rapidly to a steady configuration, thus lowering the lattices overall $\tau$.

Qualitatively, the time evolutions of the cells with zero inertia in both Pattern II and $I = 0$ everywhere are akin. Therefore, Pattern II leads to $p_+$ and $c_+$ with a fair linear growth with $p_+(0)$. Furthermore, $p_+$ and $c_+$ decrease with increasing $i$, following the same trend of a fixed $I$ for the whole lattice (Section 3.2). As for $p_+$ and $c_+$ as a function of $c_+(0)$, they have the same behavior shown in Figures 1 and 2, with the values of $p_+$ and $c_+$ decreasing as $i$ increases.

Figure 9 illustrates the dynamics of a specific CA with the inertia Pattern II. Here, $\tau$ is 9, 11, 9, 10, 9, respectively, for $i$ equal to 0, 1, 2, 3, 4. From the plots, we observe that basically the cells in and around $B$ are the most affected by the values of $i$. The remaining lattice elements evolve following almost the same sequence of configurations seen in the case without inertia (Figure 9 first row, in which $i = 0$). These results contrast with previously-mentioned models in the literature [33,44], whose resistive elements tend to change the final configuration of the entire system. The main difference is that the initial states of the $B$ cells (of $I = i \neq 0$) are randomly distributed. Hence, they do not share a common $S$ (say, a strong opinion about some specific subject), which in the long term could influence the global distribution of states, representing voting intention, language, cultural preferences, etc., and eventually giving rise to phase transition-like phenomena [45]. Actually, the CA with inertia Pattern II is an appropriate construction to describe systems presenting regions of homogeneous phases coexisting with regions of very heterogeneous structures (cf., Figure 9).

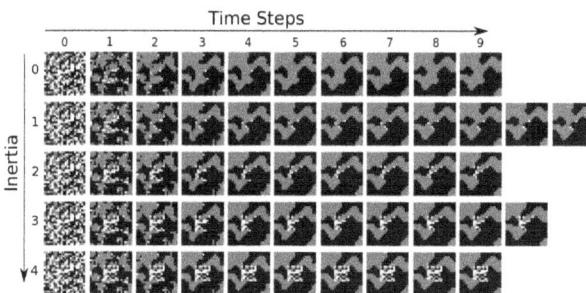

**Figure 9.** Evolution until stationarity of the same initial lattice (of $p_+(0) = 172$ and $c_+(0) = 0.167878$) with the inertia Pattern II. The inertia values are those for $i$ in the central block $B$ ($i = 0$ corresponds to $I = 0$ everywhere; Section 3.1). The color convention is the same as in Figure 4.

### 3.3.3. Pattern III: Non-Null Inertia Initially Only for the Neutral $S = 0$ State

The $S = 0$ state is passive under the evolutionary rules described in Section 2.1. Hence, it is often completely annihilated by the active states when $I = 0$ everywhere. This is no longer the case if all of the cells have a same non-null inertia (Section 3.2). However, then one could ask which other inertia distributions can preserve $S = 0$. The intuitive answer is to assume that only the elements starting in the neutral state can withstand changes. To investigate such situation, we finally consider the inertia Pattern III, where we suppose that initially the cells $n$ at $S = 0$ ($S = \pm 1$) have an inertia value $I_n = i \neq 0$ ($I_n = 0$).

Here, again, $p_+$ and $c_+$ depend linearly on $p_+(0)$. The $p_+$ and $c_+$ magnitudes strongly decrease with $I$, the same trend seen in Section 3.2. Further, the always positive slopes (the angles $\theta$) of the straight lines $p_+ \times p_+(0)$ and $c_+ \times p_+(0)$ monotonically diminish with $I$, however with a very mild variation (especially for $c_+ \times p_+(0)$) when $I < 4$. Hence, contrary to Figure 3a, the slope $\theta$ of the $c_+ \times p_+(0)$ curve versus $I$ does not present a peak, which would characterize a qualitative dynamical change (at least regarding the final $+1$ clusterization in terms of $p_+(0)$). This distinct behavior from that in Section 3.2 occurs because initially, only the $S = 0$ cells (so with $I \neq 0$) can act like "buffer" elements to prevent the growth of $+1$ agglomerations. But dynamically $S = 0$ is a passive state. Therefore, for $I$ small, the formation of $+1$ clusters at the end of the evolution is not so critically dependent on $p_+(0)$, as would be the case if the $-1$ active state at the beginning had also a non-null inertia.

On the other hand, an important transition is observed when we analyze $p_+$ versus $c_+(0)$. For $I \leq 4$ ($I \geq 5$), the final $p_+$ decreases (increases) with $c_+(0)$. This is illustrated in Figure 10, where one sees that the slope of $p_+ \times c_+(0)$ versus $I$ has the same behavior of Figure 3b (but notice a different onset). Such similarity indicates that the strongest influence on the final value of $p_+$ in terms of $c_+(0)$ is played by the susceptibility of neutral elements to become active $S = \pm 1$ states (a mechanism that depends on $I$) and not due to the competition between $+1$ and $-1$ to transform each other.

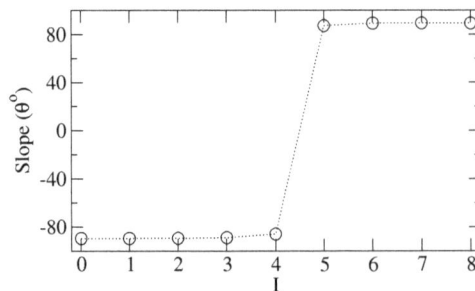

**Figure 10.** The same as in Figure 3b, the slope (angle $\theta$) of the straight line $p_+ \times c_+(0)$ versus $I$, but for the inertia Pattern III.

For Pattern III, the average $\tau$ drops quickly with increasing $I$. Indeed, as $I$ grows, the $S = 0$ cells tend to become "immune" and then excluded from evolution. With a reduced number of elements participating in the dynamics, naturally, the convergence to a stationary configuration is quicker.

A typical evolution is illustrated in Figure 11. As $I$ increases, $S = 0$ can survive until the stationary configuration is reached, especially if they are located in the borders that divide clusters of the $+1$ and $-1$ states (this is just the mechanism allowing the ecotones' formation [25]). For larger $I$'s, the elements in the zero state are able to distribute equally in all regions of the lattice. As already discussed, once the $S = 0$ cells with $I \neq 0$ play the role of "buffers", the spatial aggregation of $\pm 1$ becomes more difficult as $I$ gets larger. Nevertheless, the final clusterization degree of $\pm 1$ is still considerably higher when confronted with the case of the same $I$ for all cells (compare, e.g., Figures 4 and 11). This can be understood from the fact that during the very first time steps, in small regions $R$ where there exists a prevalence of $S = \pm 1$, it will become easier for $S$ to rapidly dominate $R$ over $-S$ (whose $I = 0$). Such "nucleation" guarantees a certain minimum level of clusterization at the end of the evolution process.

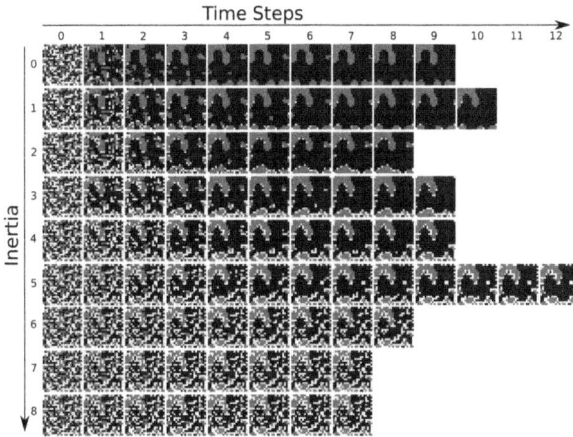

**Figure 11.** Evolution until stationarity of the same initial lattice (of $p_+(0) = 172$ and $c_+ = 0.169574$) with the inertia Pattern III and distinct values of $I$. Since initially only the neutral $S = 0$ has non-null inertia, even for $I = 8$, the system needs a finite number of steps to reach a steady configuration. The color convention is the same as in Figure 4.

## 4. Discussion and Conclusions

The understanding of the processes involved in spatio-temporal patterns' formation (especially in biological systems) is a relatively old [3], but still very fascinating, subject in science. The corresponding underlying mechanisms are closely related to the idea of emergent behavior in complex systems [46], as well as to some concepts of critical phenomena [47]. In particular, resistance of some elements to a specific driving force might be one of the key factors to explain different rich structures observed in nature [38,42,48,49]. Nevertheless, the investigation of this latter dynamics using CA (especially deterministic ones) is still not very well explored in the literature.

Here, we have examined a simple CA of straightforward evolution rules. The idea was to propose a minimalist model displaying phase transition-like behavior rather than to describe a concrete particular problem (but see below). However, in the CA formulation, we have considered three very common ingredients in real systems (deterministically implemented): distinction between dynamically-active (in a way "stronger") and passive (in a way "weaker") agents, spatial competition and inertia, i.e., intrinsic opposition to modifications in the actual CA cells' states. By assuming distinct distributions and values for the $I_n$'s, we have been able to identify qualitative modifications in the final stationary configurations of the CA lattice. This clearly demonstrates that such an innate quantity, inertia, can adjust the emergent spatial patterns resulting from the CA evolution. In fact, the observed pattern changes with the inertia distribution display a striking resemblance with phase transitions, indicating a more fundamental reorganization of the CA final steady configurations as a function of a control parameter, in the present case $I$.

The first evidence of a macroscopic change in the system, due to the increase of inertia, can be grasped from a detailed inspection of Figure 4, characterized by well-defined islands (domains) of cells in the states $+1$ and $-1$ when $I \leq 3$ (for $I = 3$, this is not so prominent, but $S = \pm 1$ islands still can be spotted). On the contrary, for $I \geq 4$, the domains are practically absent. These qualitative distinctions points to a phase transition controlled by the inertia. Moreover, the structures in Figure 4 have a close parallel with ferromagnetic-paramagnetic critical transitions, in which the existence of domains with well-defined magnetization is continuously suppressed as the temperature progressively rises. Figure 3a,b reinforces that the global changes induced by the inertia are related to a phase transition occurring around $3 \leq I \leq 4$. Figure 3a shows that the slope of $c_+ \times p_+(0)$ has a peak just in this $I$ interval, whereas a jump of the slope $p_+ \times c_+(0)$ (Figure 3b) is observed in the same $I$ range. It is worth mentioning that the curve in Figure 3a could be related to a kind of "susceptibility", which usually presents accentuated peaks at the onset of a phase transition [50].

We emphasize that the quantities we have chosen to characterize the transitions, as they should be, are very sensitive to the actual distribution of the $I_n$'s. Indeed, since the inertia is attached to cells, not to cell states, the CA dynamics is completely determined by the initial lattice and the specific $\{I_n\}$. In this work, we have made particular options for such distributions. Besides the same $I$ for all cells (Section 3.2), the inertia Patterns I, II and III were motivated by phenomena associated with, respectively, segregation of binary mixtures, opinion formation and the existence of buffer-like states (e.g., important in some condensed matter problems). As an example, for the inertia Pattern III of Section 3.3.3, for which initially only the cells $n$ with $S = 0$ have $I_n \neq 0$, the slope of $c_+ \times p_+(0)$ as a function of $I$ no longer displays a peak (although the slope of $p_+ \times c_+(0)$ versus $I$ still presents a jump (Figure 10), following the same behavior of Figure 3b). Such a distinction, as discussed in Section 3.3.3, reflects the fact that the final stationary patterns in this latter case tend to have the states $\pm 1$ more clusterized than those in Section 3.2, even for larger $I$'s.

Surely, a very pertinent question relates to the exact type of transition (either discontinuous or continuous) we are observing in the present model. We first remark that even for more conventional statistical physics systems, in different situations, a proper identification is not a trivial or a direct task. For instance, in some cases, "weak" discontinuous transitions exhibit just small jumps for the order parameters, becoming hard to distinguish between phase coexistence and critical phenomena [50]. As an illustration, the so-called explosive percolation, in which the transport through a network occurs discontinuously, is a very peculiar process [51–53]. Although some examples of explosive percolation initially pointed to discontinuous phase transitions [54]; afterwards, they were revealed to be continuous [55]. The difficulty of discriminating critical from first-order transitions also appears in the context of absorbing states (AS), in which three adjacent species are required for producing an offspring (recall that our CA has also three states, moreover with the possibility of extinction, usually of $S = 0$). Distinct works have claimed a discontinuous phase transition for AS systems [56,57], but it is strongly believed they belong to the same (continuous transition) universality class of the 1D directed percolation [58,59].

For CA, the situation is not different [60]. Actually, it can be even more tricky. The great distinction is that in traditional problems, for $\lambda \neq \lambda_c$, the thermodynamics quantities are usually continuous and differentiable functions of the control parameter $\lambda$. Therefore, in principle, one can study their analytical properties around $\lambda_c$, so to determine the phase transition character. The big challenge with CA is that often, one cannot perform such a kind of analysis.

In our case, $I$ is not continuous, but we can change it through a set of distinct values. Taking into account the discrete (and finite) nature of the lattices, by inspecting the CA patterns for different $I$'s, Figures 4 and 11 seem to indicate a continuous phase transition (like a magnetic process). On the other hand, the leap for the slope of $p_+ \times c_+(0)$ might imply discontinuous phase transitions. However, observe that this jump can be related to the discreteness of the inertia parameter, masking an eventual sharp, but still continuous variation of $\theta$. Certainly, further investigations would be desirable. As a possibility, we recall that we have discussed only the case of time-independent inertia distributions. The case of $I_n = I_n(t)$ could be related, e.g., to seasonal effects in biological environments [4–6]. Then, by calculating an average $I$ during pertinent time intervals, one can define an effective $I_{eff}$, which does not need to be an integer, assuming a broader range of values. In this way, the analysis could be performed with such $I_{eff}$, allowing a better characterization of the phase transition in terms of the inertia order parameter.

Along Section 3, we have mentioned different potential applications for this kind of CA construction, especially concerning biological structures and phase segregation. However, our main goal has not been to describe a particular situation. Instead, our aim has been to discuss general aspects of how a CA, a paradigmatic tool in modeling complex systems, can generate distinct spatial-temporal patterns by fine tuning an inner variable, inertia, directly associated with resistance to changes in the system microstates (i.e., at the cells' level [6]). Nevertheless, a few comments about a specific situation should help to put our results in a more concrete perspective.

In [25], some of us have used a very similar framework to generate typical patterns in ecotones (see the Introduction section), although not addressing phase transitions. It is important to mention that actual ecotones have been associated with phase transitions [61,62], in fact to second-order ones, i.e., to critical phenomena [63]. Thus, in spite of the great simplicity of our CA model, it readily reproduces two of the most important aspects of ecotones:

(i) The coexistence of alien species (AS) living in the interface region, represented by $S = 0$, between two distinct spatial areas corresponding to biomes $B_1$, $S = +1$ and $B_2$, $S = -1$ (cf., Figures 4 and 11) (note that usually, these AS are not able to survive within either $B_1$ or $B_2$).

(ii) Ecotones' boundary extensions and shapes usually are driven by climate conditions and species competition [26–28,63]. Such boundaries may or may not arise depending on the strength and interplay of these factors (say, quantified by a parameter $\lambda$). The variation of $\lambda$ will simply modify or eventually destroy ecotones. The corresponding transition, as $\lambda$ varies, appears to be akin to continuous phase transitions [63]. Again, qualitatively, this is what we observe from the CA lattices' time evolution considering distinct values of $I$ (see Figures 4 and 11).

The last point is how to interpret inertia in the present context. The AS would be defeated if "clashing" against the species just from $B_1$ or just from $B_2$. However, in the intermediate region, both latter sets of species are not so well adapted as in their original ecosystems. Furthermore, they are competing with each other. This gives a certain contextual advantage to the AS, allowing them to survive, but only when there is tension between the active species. The inertia (at least for the AS) somehow quantifies this emergent (and relative) fitness due to the impairment and competition between the $B_1$ and $B_2$ species.

Finally, we remark on an interesting observation made by one of the anonymous referees. A possible distinct version of our CA is to associate inertia with cell states rather than with cells. In this case, the pattern of inertia would evolve with (and partly determine) the pattern of cell states. More generally, both approaches could be used in a single system. For example, in a CA describing the rich dynamics of land use [64], some inertia values could be attached to cells, so as to represent inherent cell qualities (i.e., heterogeneity in the cell space), while others could be attached to cell states $S$, representing the cost of changing $S$. Such a necessity of a hybrid treatment in certain classes of problems is discussed, for instance, in [64].

We hope the presented work can motivate future studies relating the concept of inertia to complex systems and phase transitions.

**Acknowledgments:** Klaus Kramer acknowledge CAPES for a PhD scholarship, and Marlus Koehler, Carlos E. Fiore and Marcos G.E. da Luz acknowledge CNPq for research grants. CEF also acknowledges the financial support from FAPESP under grant 2015/04451-2.

**Author Contributions:** Klaus Kramer, Marlus Koehler and Marcos G.E. da Luz, conceived of the study. Klaus Kramer and Marcos G.E. da Luz designed the study. Klaus Kramer performed the simulations. All of the authors analyzed and interpreted the results. Marlus Koehler and Marcos G.E. da Luz wrote the manuscript with input from all.

**Conflicts of Interest:** The authors declare no conflict of interest.

## Appendix A. Examples of Periodic Oscillating Structures (Not Converging to Steady Configurations)

In the following, we give two examples of end situations (i.e., after enough time steps), which are not stationary because some small blocks in the CA lattice oscillate between a few different local configurations, whereas the rest of the cells remains unchanged. For simplicity, we just illustrate the case of $I = 0$ for all of the elements. Furthermore, for a better visualization, we consider $N = 30$, but the behavior here is qualitatively the same for other values of $N$ (including $N = 22$ of Section 3).

In Figure A1a, we show the final oscillating pattern to which a certain typical initial CA lattice (in the sense of the construction described in Section 2.2) has evolved after a certain number of steps. We schematically represent in Figure A1b the blocks $B$ (of 25 cells each) indicated in Figure A1a. Note that if somehow the surrounding of $B$ is fixed, so the $B$ border cells cannot change, then a direct analysis shows that for $I = 0$, the evolution Rules (i)–(ii) necessarily make $B$ alternate between the two displayed structures in Figure A1b. This is the case for any lattice, unless for a localized region (in the form of $B$) that has evolved to a stationary condition. On the other hand, setting $I = 1$ for the configurations in Figure A1a suffices to "freeze" the $B$ oscillations into the left shape of Figure A1b, leading thus to a final steady state.

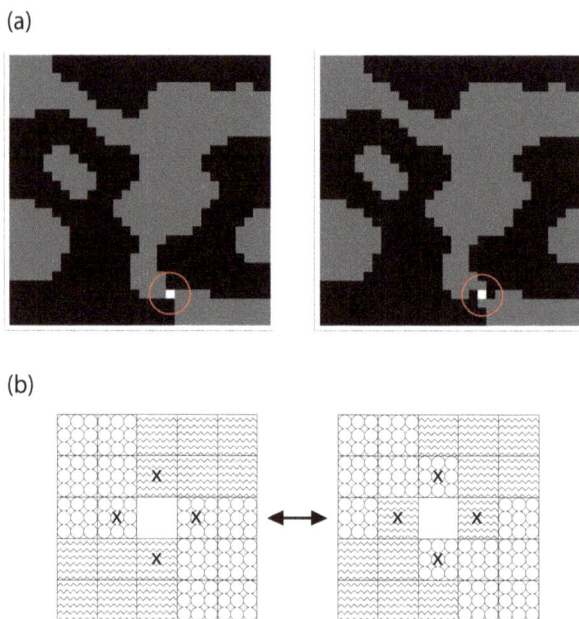

(a)

(b)

**Figure A1.** (**a**) The two oscillating configurations to which an initial $30 \times 30$ CA lattice (of all elements with $I = 0$) has evolved under the (i)–(ii) evolution rules. Here, black, gray and white represent, respectively, the states $+1$, $-1$, and $0$. Interestingly, in this case only one cell with $S = 0$ has remained, belonging to the indicated blocks $B$; (**b**) Schematics of the blocks $B$ (of 25 cells each) responsible for the oscillatory behavior when $I = 0$. The four cells marked by **x** will switch (as illustrated) each time rule (ii) is applied. $I = 1$ is large enough to stop the oscillations, driving $B$ to the left structure.

The basic local blocks $B$ resulting in the oscillatory behavior for final CA lattice patterns can have different shapes. For instance, for the example shown in Figure A2, an initial lattice has evolved to structures with two of such blocks, which moreover are composed only by $S = \pm 1$ states. This contrasts with $B$ of Figure A1, having a single $S = 0$ cell in the center. We observe that if for the lattices in Figure A2 we switch the inertia to $I = 1$, the blocks in the lattices' right side will stop oscillating (stabilizing in their present configurations). However, the blocks in the left side will continue oscillating into two different structures, but of slightly different forms than the ones in Figure A2.

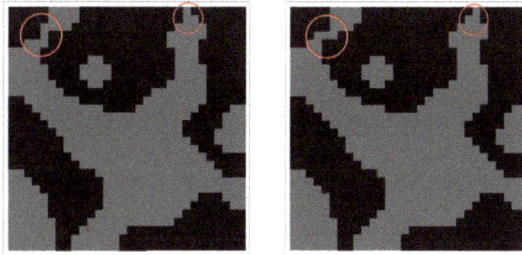

**Figure A2.** Another example of a final non-steady pattern for the CA. Differently from Figure A1, as indicated now there are two blocks where the oscillations take place. Furthermore, for this case only the $S = \pm 1$ states have survived. The $B$ in the right side of the lattice is an emblematic configuration of borders oscillating structures.

Lastly, we comment that oscillating structures also exist for $I = 1$ (as mentioned for the example of Figure A2). However, from exploratory numerical simulations (with many $N$'s), we have never seen similar behavior when $I \geq 2$. Therefore, it is an interesting theoretical question if one can have finite fundamental blocks $B$ in the case of larger than one inertia values.

## References

1. Mandelbrot, B.B. *The Fractal Geometry of Nature*; W.H. Freeman and Company: San Francisco, CA, USA, 1983.
2. Cladis, P.E.; Palffy-Muhoray, P. (Eds.) *Spatio-Temporal Patterns In Nonequilibrium Complex Systems*; Addison-Wesley: Reading, MA, USA, 1995.
3. Thompson, D.W. *On Growth and Form: A New Edition*; Cambridge University Press: Cambridge, UK, 1942.
4. Solé, R.V.; Bascompte, J. *Self-Organization in Complex Ecosystems*; Princeton University Press: Princeton, NJ, USA, 2006.
5. Malchow, H.; Petrovskii, S.V.; Venturio, E. *Spatiotemporal Patterns in Ecology and Epidemiology: Theory, Models, and Simulation*; Chapman and Hall/CRC: Boca Raton, FL, USA, 2007.
6. Anteneodo, C.; da Luz, M.G.E. Complex Dynamics of Life at Different Scales: From Genomic to Global Environmental Issues. *Philos. Trans. R. Soc. A* **2010**, *368*, doi:10.1098/rsta.2010.0286.
7. Sornette, D. *Critical Phenomena in Natural Sciences: Chaos, Fractals, Selforganization and Disorder: Concepts and Tools*, 2nd ed.; Springer: Berlin, Germany, 2006.
8. Bak, P. *How Nature Works: The Science of Self-Organized Criticality*; Copernicus: New York, NY, USA, 1999.
9. Christensen, K.; Moloney, N.R. *Complexity and Criticality*; Imperial College Press: London, UK, 2005.
10. Wolfram, S. *Cellular Automata and Complexity: Collected Papers*; Westview Press: Boulder, CO, USA, 2002.
11. Schiff, J.L. *Cellular Automata: A Discrete View of the World*; Wiley: London, UK, 2008.
12. Wolfram, S. Universality and complexity in cellular automata. *Physica D* **1984**, *10*, 1–35.
13. Wolfram, S. Twenty problems in the theory of cellular automata. *Phys. Scr.* **1985**, *T9*, 170–185.
14. Lindgren, K.; Nordahl, M.T. Universal computation in simple one-dimensional cellular automata. *Complex Syst.* **1990**, *4*, 299–318.
15. Wolfram, S. *A New Kind of Science*; Wolfram Media: Champaign, IL, USA, 2002.
16. Hoekstra, A.G.; Kroc, J.; Sloot, P.M.A. *Simulating Complex Systems by Cellular Automata*; Hoekstra, A.G., Kroc, J., Sloot, P.M.A., Eds.; Springer: Berlin, Germany, 2010.
17. Ermentrout, G.B.; Edelstein-Keshet, L. Cellular automata approaches to biological modeling. *J. Theor. Biol.* **1993**, *160*, 97–133.
18. Rohde, K. Cellular automata and ecology. *Oikos* **2005**, *110*, 203–207.
19. Maini, P.K.; Deutsch, A.; Dormann, S. *Cellular Automaton Modeling of Biological Pattern Formation*; Birkhäuser: Boston, MA, USA, 2005.
20. Haken, H. *Dynamic Patterns in Complex Systems*; World Scientific: Singapore, 1988.
21. Barabasi, A.L. *Bursts: The Hidden Pattern Behind Everything We Do*; Dutton: New York, NY, USA, 2010.

22.  Derrida, B.; Stauffer, D. Phase transitions in two-Dimensional Kauffman cellular automata. *Europhys. Lett.* **1986**, *2*, 739–745.

23.  Boccara, N. Phase transitions in cellular automata. In *Computational Complexity*; Meyers, R.A., Ed.; Springer: New York, NY, USA, 2012; pp. 2157–2167.

24.  Wootters, W.K.; Langton, C.G. Is there a sharp phase transition for deterministic cellular automata? *Physica D* **1990**, *45*, 95–104.

25.  Kramer, K.; Koehler, M.; da Luz, M.G.E. Cellular automata with inertia: Species competition, spatial patterns, and survival in ecotones. *J. Phys. Conf. Ser.* **2010**, *246*, 012040.

26.  Peters, D.P.C. Plant species dominance at a grassland–shrubland ecotone: An individual-based gap dynamics model of herbaceous and woody species. *Ecol. Model.* **2002**, *152*, 5–32.

27.  Favier, C.; Chave, J.; Fabing, A.; Schwartz, D.; Dubois, M.A. Modelling forest–savanna mosaic dynamics in man-influenced environments: Effects of fire, climate and soil heterogeneity. *Ecolog. Model.* **2004**, *171*, 85–102.

28.  Zeng, Y.; Malanson, G.P. Endogenous fractal dynamics at alpine treeline ecotones. *Geogr. Anal.* **2006**, *38*, 271–287.

29.  Ngai, J.T.; Jefferies, R.L. Nutrient limitation of plant growth and forage quality in Arctic coastal marshes. *J. Ecol.* **2004**, *92*, 1001–1010.

30.  Chaneton, E.J.; Mazia, C.N.; Kitzberger, T. Facilitation vs. apparent competition: Insect herbivory alters tree seedling recruitment under nurse shrubs in a steppe–woodland ecotone. *J. Ecol.* **2010**, *98*, 488–497.

31.  Freeman, D.C.; Wang, H.; Sanderson, S.; McArthur, E.D. Characterization of a narrow hybrid zone between two subspecies of big sagebrush (*Artemisia tridentata*, Asteraceae). VII. Community and demographhic analyses. *Evol. Ecol. Res.* **1999**, *15*, 487–502.

32.  Scarano, F.R. Plant communities at the periphery of the Atlantic rain forest: Rare-species bias and its risks for conservation. *Biol. Conserv.* **2009**, *142*, 1201–1208.

33.  Moldovan, S.; Goldenberg, J. Cellular automata modeling of resistance to innovations: Effects and solutions. *Technol. Forecast. Soc. Chang.* **2004**, *71*, 425–442.

34.  Zhang, T.; Xuan, H.; Gao, B. Modeling diffusion of innovation with cellular automata. In Proceedings of the 2005 International Conference on Services Systems and Services Management (ICSSSM'05), Chongqing, China, 13–15 June 2005.

35.  Zupan, N. Using cellular automata to simulate electronic commerce receptivity in small organisations. *Technol. Forecast. Soc. Chang.* **2007**, *74*, 798–819.

36.  Young, T.P. Restoration ecology and conservation biology. *Biol. Conserv.* **2000**, *2*, 73–83.

37.  Hoshen, J.; Kopelman, R. Percolation and cluster distribution. I. Cluster multiple labeling technique and critical concentration algorithm. *Phys. Rev. B* **1976**, *14*, 3438–3445.

38.  Binder, P.M. Frustration in Complexity. *Science* **2008**, *320*, 322–323.

39.  Mousseau, N. Frustration induced phase transition in high-dimensional deterministic cellular automata. *Europhys. Lett.* **1994**, *28*, 551–556.

40.  Ke, X.; Li, J.; Zhang, S.; Nisoli, C.; Crespi, V.H.; Schiffer, P. Tuning magnetic frustration of nanomagnets in triangular-lattice geometry. *Appl. Phys. Lett.* **2008**, *93*, 252504.

41.  Rothman, D.H.; Zaleski, S. Lattice-gas models of phase separation: Interfaces, phase transitions, and multiphase flow. *Rev. Mod. Phys.* **1994**, *66*, 1417–1481.

42.  Vannozzi, C.; Fiorentino, D.; D'Amore, M.; Rumshitzki, D.S.; Dress, A.; Mauri, R. Cellular automata model of phase transition in binary mixtures. *Ind. Eng. Chem. Res.* **2006**, *45*, 2892–2896.

43.  Jiao, Y.; Torquato, S. Emergent behaviors from a cellular automaton model for invasive tumor growth in heterogeneous microenvironments. *PLoS Comput. Biol.* **2011**, *7*, e1002314.

44.  Alber, M.S.; Kiskowski, M.A.; Glazier, J.A.; Jiang, Y. On cellular automaton approaches to modeling biological cells. In *Mathematical Systems Theory in Biology, Communications, Computation, and Finance*; Rosenthal, J., Gilliam, D.S., Eds.; Springer: New York, NY, USA, 2003; pp. 1–39.

45.  Kacperski, K.; Holyst, J.A. Formation of opinions under the influence of competing agents—A mean field approach. In *Traffic and Granular Flow'99*; Helbing, D., Herrmann, H.J., Schreckenberg, M., Wolf, D.E., Eds.; Springer: Berlin, Germany, 2000; pp. 69–80.

46.  Darley, V. Emergent phenomena and complexity. In *Artificial Life IV: Proceedings of the Fourth International Workshop on the Synthesis and Simulation of Living Systems*; Brooks, R., Maes, P., Eds.; MIT Press: Cambridge, UK, 1995; pp. 411–416.

47.  Scheffer, M.; Bascompte, J.; Brock, W.A.; Brovkin, V.; Carpenter, S.R.; Dakos, V.; Held, H.; van Nes, E.H.; Rietkerk, M.; Sugihara, G. Early-warning signals for critical transitions. *Nature* **2009**, *461*, 53–59.
48.  Dumont, S.; Prakash, M. Emergent mechanics of biological structures. *MBoC* **2014**, *25*, 3461–3465.
49.  Laine, A.-L.; Burdon, J.J.; Dodds, P.N.; Thrall, P.H. Spatial variation in disease resistance: From molecules to metapopulations. *J. Ecol.* **2011**, *99*, 96–112.
50.  Landau, D.P.; Binder, K. *A Guide to Monte Carlo Simulation in Statistical Physics*, 2nd ed.; Cambridge University Press: Cambridge, UK, 2005.
51.  Araujo, N.A.M.; Herrmann, H.J. Explosive percolation via control of the largest cluster. *Phys. Rev. Lett.* **2010**, *105*, 035701.
52.  Araujo, N.A.M.; Andrade, J.S., Jr.; Ziff, R.M.; Herrmann, H.J. Tricritical point in explosive percolation. *Phys. Rev. Lett.* **2011**, *106*, 095703.
53.  Boettcher, S.; Singh, V.; Ziff, R.M. Ordinary percolation with discontinuous transitions. *Nat. Commun.* **2012**, *3*, 787.
54.  Achliopatas, D.; D'Souza, R.M.; Spencer, J. Explosive percolation in random networks. *Science* **2009**, *323*, 1453–1455.
55.  Riordan, O.; Warnke, L. Explosive percolation is continuous. *Science* **2011**, *333*, 322–324.
56.  Dickman, R.; Tomé, T. First-order phase transition in a one-dimensional nonequilibrium model. *Phys. Rev. A* **1991**, *44*, 4833–4838.
57.  Fiore, C.E.; de Oliveira, M.J. Phase transition in conservative diffusive contact processes. *Phys. Rev. E* **2004**, *70*, 46131.
58.  Odor, G.; Dickman, R. On the absorbing-state phase transition in the one-dimensional triplet creation model. *J. Stat. Mech.* **2009**, *2009*, P08024.
59.  Park, S.-C. Absence of the discontinuous transition in the one-dimensional triplet creation model. *Phys. Rev. E* **2009**, *80*, 061103.
60.  Wentian, L.I.; Packard, N.H.; Langton, C.G. Transition phenomena in cellular automata rule space. *J. Phys. D* **1998**, *31*, 2751–2753.
61.  Loehle, C.; Li, B.-L.; Sundell, R.C. Forest spread and phase transitions at forest-prairie ecotones in Kansas, U.S.A. *Landsc. Ecol.* **1996**, *11*, 225–235.
62.  Gastner, M.T.; Oborny, B.; Zimmermann, D.K.; Pruessner, G. Transition from connected to fragmented vegetation across an environment gradient: Scaling laws in ecotone geometry. *Am. Nat.* **2009**, *174*, E23–E39.
63.  Ivanova, Y.; Soukhovolsky, V. Modeling the boundaries of plant ecotones of mountain ecosystems. *Forest* **2016**, *7*, 271.
64.  White, R.; Engelen, G.; Uljee, I. *Modeling Cities and Regions as Complex Systems*; MIT Press: Cambridge, UK, 2015.

*entropy*

MDPI

*Article*
# Echo State Condition at the Critical Point

Norbert Michael Mayer

Department of Electrical Engineering and Advanced Institute of Manufacturing with
High-Tech Innovations (AIM-HI), National Chung Cheng University, Chia-Yi 62102, Taiwan;
mikemayer@ccu.edu.tw; Tel.: +886-5-272-0411 (ext. 33219)

Academic Editor: Mikhail Prokopenko
Received: 29 October 2016; Accepted: 15 December 2016; Published: 23 December 2016

**Abstract:** Recurrent networks with transfer functions that fulfil the Lipschitz continuity with $K = 1$ may be echo state networks if certain limitations on the recurrent connectivity are applied. It has been shown that it is sufficient if the largest singular value of the recurrent connectivity is smaller than 1. The main achievement of this paper is a proof under which conditions the network is an echo state network even if the largest singular value is one. It turns out that in this critical case the exact shape of the transfer function plays a decisive role in determining whether the network still fulfills the echo state condition. In addition, several examples with one-neuron networks are outlined to illustrate effects of critical connectivity. Moreover, within the manuscript a mathematical definition for a critical echo state network is suggested.

**Keywords:** reservoir computing; uniformly state contracting networks; power law

## 1. Introduction

Classic approaches of recurrent neural networks (RNNs), such as back-propagation through time [1], have been considered difficult to handle. In particular learning in the recurrent layer is slow and problematic due to potential instabilities. About 15 years ago, reservoir computing [2] was suggested as an alternative approach for RNNs. Here, it is not necessary to train connectivity in the recurrent layer. Instead, constant, usually random, connectivity weights are used in the recurrent layer. The supervised learning can be done by training the output layer using linear regression. Two types of reservoir computing are well established in the literature. The first is called liquid state machines (LSMs, [3]), which are usually based on a network of spiking neurons. The second type is called an echo state nework (ESN, [4]), which uses real valued neurons that initially use a sigmoid as a transfer function. Although a random recurrent connectivity pattern can be used, heuristically it has been found that typically the performance of the network depends strongly on the statistical features of this random connectivity (cf. for example [5] for ESNs).

Thus, what is a good reservoir with regard to particular stationary input statistics? This has been a fundamental question for research in this field since the invention of reservoir computing. One fundamental idea is that a reservoir can only infer training output from this window of the input history of which traces still can be found inside the reservoir dynamics. However, if the necessary inference from time series in order to learn the training output is far in the past, it may happen that no traces of this input remain inside the reservoir. So, the answer seems to be that a good reservoir is a reservoir from whose states it is possible to reconstruct an input history with a time span that is as long as possible. More precisely, they should be reconstructed in a way that is sufficiently accurate in order to predict the training output. In other words, a good reservoir is a reservoir that has a good memory of the input history.

There have been efforts to quantify the quality of the memory of the reservoir. Most common is the *"memory capacity"* (MC) according to Jaeger's definition [4]. However, MC has several drawbacks. For example, it is not directly compatible to a Shannon information based measure. Still, it illustrates that ESNs are relatively tightly restricted in the way that the upper limit of the MC is equal to the number of hidden layer neurons. So the capabilities of the network increase with the number of neurons.

One more important limiting factor with regard to the reservoir memory is the strength of the recurrent connectivity. According to the echo state condition, the nature of the reservoir requires that the maximum $|\lambda|_{max}$ of its eigenvalues in modulus is smaller than 1, which is called the echo state property (ESP). This seems always to result in a exponential forgetting of previous states. Thus, forgetting is independent from the input statistics but instead has to be pre-determined and is due to the design of the reservoir dynamics.

In order to proceed, there are several important aspects. First, it is necessary to get rid of the intrinsic time scale of forgetting that is induced by $|\lambda|_{max} < 1$. More precisely, the remaining activity of inputs to the reservoir that date back earlier than $\Delta t$ is a fraction smaller than $|\lambda|_{max}^{\Delta t}$. Networks where the largest eigenvalue is larger than 1 cannot be used as reservoirs anymore, a point which is detailed below. One can try $|\lambda|_{max} = 1$ and see if this improves the network performance and how this impacts the memory of the reservoir on earlier events. Steps toward this direction have been made by going near the "edge of chaos" [5] or even further where the network may not be an echo state network for all possible input sequences but instead just around some permissible inputs [6]. Presumably, these approaches still all forget exponentially fast.

Strictly, networks with $|\lambda|_{max} = 1$ are not covered by the initial proof of Jaeger for the echo state condition. One important purpose of this paper is to close this gap and to complete Jaeger's proof in this sense. The other purpose is to motivate the principles of [7] in as simple as possible examples and thus to increase the resulting insight.

The intentions of the following sections of the paper are to motivate the concept of critical neural networks and explain how they are related to memory compression. These intentions comprise a relatively large part of the paper because it seems important to argue for the principle value of critical ESNs. Section 2 introduces the concept of reservoir computing and also defines important variables for the following sections. An important feature is that Lyapunov coefficients are reviewed in order to suggest a clear definition for critical reservoirs that can be used analytically on candidate reservoirs. Section 3 describes how critical one-neuron reservoirs can be designed and also introduces the concept of extending to large networks. Section 4 explains why the critical ESNs are not covered by Jaeger's proof. The actual proof for the echo state condition can be found in Section 5. Certain aspects of the proof have been transferred to the appendix.

## 2. Motivation

The simplest way to train with data in a supervised learning paradigm is to interpolate data (cf. for example [8]). Thus, for a time series of input data $\mathbf{u}_t \in \mathbb{R}^n$ that forms an input sequence $\bar{\mathbf{u}}^\infty$ and a corresponding output data $\mathbf{o}_t \in \mathbb{R}^m$ one can choose a vector of non-linear, linearly independent functions $\mathbf{F}(\mathbf{u}_t) : \mathbb{R}^n \to \mathbb{R}^k$ and a transfer matrix $\mathbf{w}^{out} : \mathbb{R}^k \to \mathbb{R}^m$. Then, one can define

$$\mathbf{x}_t = \mathbf{F}(\mathbf{u}_t)$$
$$\bar{\mathbf{o}}_t = \mathbf{w}^{out}\mathbf{x}_t.$$

$\mathbf{w}^{out}$ can be calculated by linear regression, i.e.,

$$\mathbf{w}^{out} = (AA')^{-1}(AB), \tag{1}$$

where the rectangular matrices $A = [\mathbf{x}_0, \mathbf{x}_1, \dots, \mathbf{x}_t]$ and $B = [\mathbf{o}_0, \mathbf{o}_1, \dots, \mathbf{o}_t]$ are composed from the data of the training set and $A'$ is the transpose of $A$. Further, one can use a single transcendental function $\theta(.)$ such that

$$\mathbf{x}_t = \mathbf{F}(\mathbf{u}_t) = \theta(\mathbf{w}^{in}\mathbf{u}_t), \tag{2}$$

where $\mathbf{w}^{in} : \mathbb{R}^n \to \mathbb{R}^k$ is a matrix in which each line consists of a unique vector and $\theta(.)$ is defined in the Matlab fashion; so the function is applied to each entry of the vector separately. Linear independence of the components of $\mathbf{F}$ can then be guaranteed if the column vectors of $\mathbf{w}^{in}$ are linearly independent. Practically, linear independence can be assumed if the entries of $\mathbf{w}^{in}$ are chosen randomly from a continuous set and $k \geq n$.

The disadvantage of the pure interpolation method with regard to time series is that the input history, that is $\mathbf{u}_{t-1}, \mathbf{u}_{t-2}, \dots \mathbf{u}_0$, has no impact on training the current output $\tilde{\mathbf{o}}_t$. Thus, if a relation between previous inputs and current outputs exists, that relation cannot be learned.

Different from Equation (1), a reservoir in the sense of reservoir computing [2,9,10] can be defined as

$$\mathbf{x}_t = \mathbf{F}(\mathbf{x}_{t-1}, \mathbf{u}_t). \tag{3}$$

The recursive update function adds several new aspects to the interpolation that is outlined in Equation (1):

1. The new function turns the interpolation into a dynamical system:

$$\mathbf{x}_t = \mathbf{F}(\mathbf{x}_{t-1}, \mathbf{u}_t) = \mathcal{F}_{\tilde{\mathbf{u}}^\infty}(\mathbf{x}_{t-1}), \tag{4}$$

where the notation $\mathcal{F}_{\tilde{\mathbf{u}}^\infty}(\mathbf{x}_{t-1})$ is intended to illustrate the character of the particular discrete time, deterministic dynamical system and the fact that each possible time series $\tilde{\mathbf{u}}^\infty$ defines a specific set of dynamics (For obvious reasons one may call $\mathbf{F}$ an input driven system.).

The superset $\mathcal{F}$ over all possible $\tilde{\mathbf{u}}^\infty$,

$$\mathcal{F} = \bigcup_{\tilde{\mathbf{u}}^\infty} \mathcal{F}_{\tilde{\mathbf{u}}^\infty},$$

may be called the reservoir dynamics. Thus, $\mathcal{F}$ covers all possible dynamics of a particular reservoir with regard to any time series of input vectors in $\mathbb{R}^n$. Note that this way of looking into the dynamics of reservoirs is non-standard. Rather the standard approach is to interpret the reservoir as a non-autonomous dynamical system and then to formalize the system accordingly [6]. For the present work, the turn towards standard dynamical systems has been chosen because here the relevant methodology is well established and the above mentioned formalization appears sufficient for all purposes of this work.

2. It is now possible to account for information from a time series' past in order to calculate the appropriate output.

One important question is if the regression step in the previous section, and thus the interpolation, works at all for the recursive definition in Equation (3). Jaeger showed ([4], p. 43; [11]) that the regression is applicable, i.e., the echo state property (ESP) is fulfilled, if and only if the network is uniformly state contracting. Uniformly state contraction is defined in the following.

Assume an infinite stimulus sequence $\tilde{\mathbf{u}}^\infty = \{\mathbf{u}_n\}_{n=0}^\infty$ and two random initial internal states of the system $\mathbf{x}_0$ and $\mathbf{y}_0$. To both initial states $\mathbf{x}_0$ and $\mathbf{y}_0$ the sequences $\tilde{\mathbf{x}}^\infty = \{\mathbf{x}_n\}_{n=0}^\infty$ and $\tilde{\mathbf{y}}^\infty = \{\mathbf{y}_n\}_{n=0}^\infty$ can be respectively assigned.

$$
\begin{aligned}
\mathbf{x}_t = \mathbf{F}(\mathbf{x}_{t-1}, \mathbf{u}_t) &= \mathcal{F}_{\tilde{\mathbf{u}}^\infty}(\mathbf{x}_{t-1}) \\
\mathbf{y}_t = \mathbf{F}(\mathbf{y}_{t-1}, \mathbf{u}_t) &= \mathcal{F}_{\tilde{\mathbf{u}}^\infty}(\mathbf{y}_{t-1}) \\
q_{\tilde{\mathbf{u}}^\infty, t} &= d(\mathbf{x}_t, \mathbf{y}_t),
\end{aligned}
\tag{5}
$$

where $q_{\bar{u}^\infty}$ is another series and $d(.,.)$ shall be a distance measure using the square norm.

Then the system $\mathcal{F}$ is uniformly state contracting if it is independent from $\bar{u}^\infty$ and if for any initial state $(x_0, y_0)$ and all real values $\epsilon > 0$ there exists a finite $\tau_\mathcal{F}(\epsilon) < \infty$ for which

$$\max_{\mathcal{F}} q_{\bar{u}^\infty, t} \leq \epsilon \tag{6}$$

for all $t \geq \tau_\mathcal{F}$.

Another way to look at the echo state condition is that the network $\mathcal{F}$ behaves in a time invariant manner, in the way that some finite subsequence in an input time series will roughly result always in the same outcome. In other words

$$x_{\Delta t + t_0} \approx y_{\Delta t + t_0}$$

independent of $t_0$, $x_{t_0}$ and $y_{t_0}$ and if $\Delta t$ is sufficiently large.

Lyapunov analysis is a method to analyze predictability versus instability of a dynamical system (see [12]). More precisely, it measures exponential stability.

In the context of non-autonomous systems, one may define the Lyapunov exponent as

$$\Lambda_{\bar{u}^\infty} = \lim_{|q_{\bar{u}^\infty, t=0}| \to 0} \lim_{t \to \infty} \frac{1}{t} \log \frac{|q_{\bar{u}^\infty, t}|}{|q_{\bar{u}^\infty, 0}|}, \tag{7}$$

Thus, if

$$q_{\bar{u}^\infty, t} \propto \exp(bt),$$

then $\Lambda_{\bar{u}^\infty}$ approximates $b$ and thus measures the exponent of exponential decay. For power law decays, the Lyapunov exponent is always zero (For example, one may try $q_{\bar{u}^\infty, t} = \frac{1}{t+1}$).

In order to define criticality, we use the following definition.

**Definition 1.** *A reservoir that is uniformly state contracting shall be called critical if for at least one input sequence $\bar{u}^\infty$ there is at least one Lyapunov exponent $\Lambda_{\bar{u}^\infty}$ that is zero.*

The echo state network (ESN) is an implementation of reservoir dynamics as outlined in Equation (3). Like other reservoir computing approaches, the system is intended to resemble the dynamics of a biologically inspired recurrent neural network. The dynamics can be described for discrete time-steps $t$, with the following equations:

$$x_{lin,t} = \mathbf{W} x_{t-1} + \mathbf{w}^{in} u_t$$
$$x_t = \theta \left( x_{lin,t} \right) \tag{8}$$
$$\tilde{o}_t = \mathbf{w}^{out} x_t.$$

With regard to the transfer function $\theta(.)$, it shall be assumed that it is continuous, differentiable, transcendental and monotonically increasing with the limit $1 \geq \theta'(.) \geq 0$, which is compatible with the requirement that $\theta(.)$ fulfills the Lipschitz continuity with $K = 1$. Jaeger's approach uses random matrices for $\mathbf{W}$ and $\mathbf{w}^{in}$, learning is restricted to the output layer $\mathbf{w}^{out}$. The learning (i.e., training $o_t$) can be performed by linear regression (cf. Equation (1)).

The ESN fulfills the echo state condition (i.e., it is uniformly state contracting) if certain restrictions on the connectivity of the recurrent layer apply, for which one can name a necessary condition and a sufficient condition:

- C1 A network has echo states only if

$$1 > |\lambda|_{\max} = \max \mathrm{abs}(\lambda(\mathbf{W})), \tag{9}$$

i.e., the absolute value of the biggest eigenvalue of **W** is below 1. The condition means that a network is **not** and ESN if $1 < \max \operatorname{abs}(\lambda(\mathbf{W}))$.

- C2 Jaeger named here initially

$$1 > s_{max} = \max s(\mathbf{W}), \tag{10}$$

where $s$ is the vector of singular values of the matrix **W**. However, a closer sufficient condition has been found in [13]. Thus, it is already sufficient to find a full rank matrix $D$ for which

$$1 > \max s(D\mathbf{W}D^{-1}). \tag{11}$$

The authors of [14] found another formulation of the same constraint: *The network with internal weight matrix* **W** *satisfies the echo state property for any input if W is diagonally Schur stable, i.e., there exists a diagonal P > 0 such that $W^T PW - P$ is negative definite.*

Apart from the requirement that a reservoir has to be uniformly state contracting, the learning process itself is not of interest in the scope of this paper.

## 3. Critical Reservoirs with Regard to the Input Statistics $\bar{u}^\infty$

Various ideas on what types of reservoirs work better than others have been brought up. One can try to keep the memories of the input history in the network as long as possible. The principle idea is to tune the network's recurrent connectivity to a level where the convergence for a subset of $\mathcal{F}_{crit} \in \mathcal{F}$ with regard to Equation (6) is

$$q_{\bar{u}^\infty_{crit},t} \propto t^a \tag{12}$$

rather than

$$q_{\bar{u}^\infty,t} \propto b^t, \tag{13}$$

where $a < 0$ and $0 < b < 1$ are system specific values, i.e., they depend on $\mathcal{F}_{\bar{u}^\infty}$.

A network according to Equation (12) is still an ESN since it fullfils the ESP. Still, forgetting of initial states is not bound to a certain time scale. Remnants of information can —under certain circumstances— remain for virtually infinite long times within the network given that not too much unpredictable new input enters the network. Lyapunov analysis of a time series according to Equation (12) would result in zero, and Lyapunov analysis of Equation (13) yields a nonzero result.

In ESNs forgetting according to the power law of Equation (12) for an input time series $q_{\bar{u}^\infty_{crit},t}$ is achievable if the following constraints are fulfilled:

- The recurrent connectivity **W**, the input connectivity $\mathbf{w}^{in}$ of the ESN and the transfer function $\theta(.)$ have to be arranged in a way that if the ESN is fed with $q_{\bar{u}^\infty_{crit},t}$ one approximates

$$\lim_{t \to \infty} |\dot\theta(\mathbf{x}_{lin,t})| = 1.$$

Thus, the aim of the training is

$$|\dot\theta(\mathbf{x}_{lin,t})| = 1. \tag{14}$$

Since the ESN has to fulfil the Lipschitz continuity with $K = 1$, the points where $\dot\theta = 1$ have to be inflection points of the transfer function. In the following these inflection points shall be called epi-critical points (ECPs).
- The recurrent connectivity of the network is to be designed in a way that the largest absolute eigenvalue and the largest singular value of **W** both are equal to one. This can be done by using normal matrices for **W** (see Section 3.5).

*3.1. Transfer Functions*

The standard transfer function that is widely used in ESNs and other types of neural networks is the sigmoid function, i.e.,

$$\theta(x) = \tanh(x). \tag{15}$$

The sigmoid transfer function has one ECP $\Pi_0 = 0$.
As a second possible transfer function, one may consider

$$\theta(x) = 0.5x - 0.25\sin(2x). \tag{16}$$

Here, one has a infinite set of ECPs at $\Pi_i = (n + 0.5) \times \pi$. It is important to have more than one ECP in the transfer functions because for a network with discrete values it appears necessary that each neuron has at least 2 possible states in order to have any inference from the past to the future. In the case of tanh, the only solution of Equation (14) is

$$\mathbf{x}_{lin,t} = 0.$$

That type of trained network cannot infer information from the past to the present for the expected input, which significantly restricts its capabilities. One can see that from

$$\mathbf{x}_t = \tanh(\mathbf{x}_{lin,t}) = 0.$$

The next iteration yields

$$\mathbf{x}_{t+1} = \tanh(\mathbf{w}^{in}\mathbf{u}_t),$$

which again, after training would require

$$0 = \mathbf{w}^{in}\mathbf{u}_t.$$

Thus, the network can only anticipate the same input at all time steps.

In the case of Equation (16), the maximal derivative $\theta'(x)$ is 1 at $x = \pi(n + 1/2)$, where $n$ is an integer number (confer Figure 1). Here the main advantage is that there exists an infinite set of epi-critical points. However, all these points are positioned along the linear function $y = x/2$. This setting still significantly restricts the training of $\mathcal{F}_{crit}$. Here one can consider the polynomial with the lowest possible rank (cf. the green line in Figure 1, left side) that interpolates between the epi-critical points (in the following called an epi-critical transfer function). In the case of Equation (16) the epi-critical transfer function is the linear function

$$\mathbf{x}_{t+1} = 0.5\mathbf{W}\mathbf{x}_t + 0.5\mathbf{w}^{in}\mathbf{u}_t.$$

Thus, the effective dynamics of the trained reservoir on the expected input time series is -if this is possible- the dynamics of a linear network. This results in a very restricted set of trainable time series.

As an alternative, one could consider also a transcendental function for the interpolation between the points, such as depicted in Figure 2. The true transfer function (blue line in Figure 2) can be constructed in the following way. Around a set of defined epi-critical points $\Pi_i$, define $\theta$ as either

$$\theta(x) = \tanh(x - \Pi_i) + \tanh(\Pi_i)$$

or

$$\theta(x) = \tanh(x).$$

This is one conceptional suggestion for further investigations. The result is a transfer function with the epi-critical points $\Pi_i$ and 0. The epi-critical transfer in this case is a tanh function.

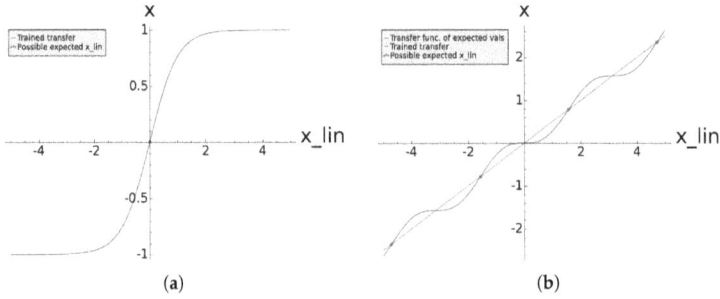

(a)           (b)

**Figure 1.** Variant possible transfer functions with interesting features regarding criticality. On the left side (**a**) tanh has only one epi-critical point. Equation (16) is graphed on the right side (**b**). Here, there is an infinite set of epi-critical points $n\pi + \pi/2$ that are all positioned along the line $y = x/2$. In both graphs green dots indicate epi-critical points, green curves are smooth interpolations between those points, and the blue line indicates the particular transfer function itself.

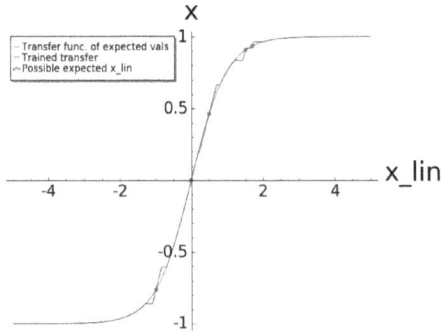

**Figure 2.** Tailored transfer function where the epi-critical points can be organized in an adaptive way and follow a transcendental function (which has certain advantages to the examples depicted in Figure 1, in blue). Green dots indicate epi-critical points, the green curve is a smooth interpolation between those points, and the blue line indicates the particular transfer function itself. Note that the green curve is a tanh and thus a transcendental function.

### 3.2. Examples Using a Single Neuron as a Reservoir

In this and the following sections, practical examples are brought up where a single neuron represents a reservoir. Single neuron reservoirs have been studied in other researches [15,16]. Here the intention is to illustrate the principle benefits and other features of critical ESNs.

First one can consider a neuron with tanh as a transfer function along with a single input unit

$$x_{t+1} = \tanh(bx_t + u_t), \tag{17}$$

where in order to achieve a critical network $|b|$ has to be equal to 1. i.e., the network exactly fulfills the boundary condition. From previous consideration one knows that $\mathcal{F}_{crit}$ has the dynamics that results from $u_t = 0$ as an input. In this case the only fixed-point of the dynamics is $x_t = 0$, which is also the epi-critical point if $|b| = 1$.

Power law forgetting: Starting from the two initial values $x_0 = 0$ and say $y_0 = 0.01$, one can see that the two networks converge in a power law manner to zero. On the other hand, for a linear network

$$x_{t+1} = bx_t + u_t$$

with the same initial conditions the dynamics of the two networks never converge (independently of $u_t$). Instead, the difference between $x_0 = 0$ and say $y_0 = 0.01$ stays the same forever. Thus, the network behavior in the the case of $|b| = 1$ depends on the nature of the transfer function. For all other values of $b$, both transfer functions result qualitatively in the same behavior in dependence on $b$: either they diverge or they converge exponentially. Since $|b| = 1$ is also the border between convergence and divergence and thus the border between uniformly state contracting networks and not uniformly state contracting networks, the case of $|b| = 1$ is a critical point of the dynamical system, in a similar manner as a critical point at the transition from ordered dynamics to instability. In the following it is intended to extend rules for different transfer functions, where different transfer functions result in the critical point in uniformly state contracting networks and where this is not the case.

As a final preliminary remark, it has to be emphasized that a network being uniformly state contracting means that the states are contracting for any kind of input $u_t$. It does not mean that for any kind of input the contraction follows a power law. In fact, for all input settings $u_t \neq 0$ the contraction is exponential for the neuron of Equation (17), even in the critical case ($|b| = 1$).

### 3.3. Single Neuron Network Example with Alternating Input and Power Law Forgetting

As outlined above, for practical purposes the sigmoid function, i.e., $\theta = tanh(.)$, is not useful for critical networks because the only critical state occurs when total activity of such a network is null. In that case it is not possible to transfer information about the input history. The reason is illustrated in the following example, where instead of a sigmoid function other types of transfer functions are used.

So, the one-neuron network

$$x_{t+1} = \theta(-bx_t + (2-b)u_t) \tag{18}$$

with a constant $b$, the transfer function of Equation (16) and the expected alternating input $u_t = (-1)^t \pi/4$ has an attractor state when also $x_t$ is alternating with $x_t = (-1)^t \pi/4$ independently from $b$. Thus, $u_t = (-1)^t \pi/4$ shall be interpreted as the expected input that directs the activity of the network exactly to where $\theta' = 1$ and thus induces a critical dynamic $\mathcal{F}_{\bar{u}^\infty} = \mathcal{F}_{crit}$.

It is now interesting to investigate the convergence behavior for different values of $b$ considering differing starting values for internal states $x_t$ and $y_t$. Figure 3 depicts the resulting different Lyapunov exponents for one-neuron networks with different values of $b$. One can see that—not surprisingly—the Lyapunov exponent for $b = 1$ is zero. This is the critical point that marks the transition from order to instability in the system and at the same time the transition from ESP to non-ESP.

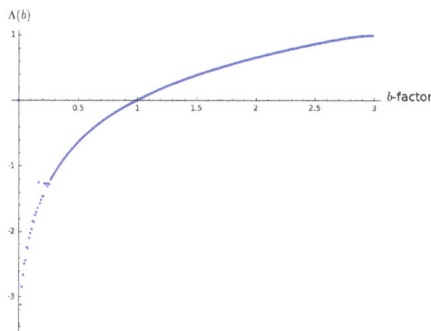

**Figure 3.** Depicted is the Lyapunov exponent for the example system of Equation (18) for different values of $b$. At $b = 1$ the Lyapunov exponent crosses zero. At this point all types of linear analysis fail. The point marks the border between networks that fulfil the echo state property (ESP) and those that do not. So, this point is the called critical point. Further analysis shows that the point itself belongs to the area with ESP. All results from Figures 4 and 5 are drawn from the point where the Lyapunov exponent is zero, that is $b = 1$.

In the following the same network at $b = 1$ is investigated. Some unexpected input, i.e., $u_t \neq (-1)^t \pi/4$, lets the network jump out of the attractor. If the input afterwards continues as expected, the network slowly returns to the attractor state in a power law fashion. Thus,

$$d(x_t, y_t) \propto t^{-a},$$

where $y_t$ represents the undisturbed time series and $a$ is a constant value. Note that $x_t$ contains all information of the network history and that the network was simulated using IEEE 754-2008 double precision variables, which have a memory size of 64 bits on Intel architecture computers. Although floating point variables are organized in a complicated way of three parts, the sign, exponent and mantissa, it is clear that the total reservoir capacity cannot exceed those 64 bits, which means in the limit a reservoir of one-neuron cannot remember more than 64 binary i.i.d. random numbers.

Thus, about 64 iterations after an unexpected input the difference between $x_t$ and $y_t$ should be annihilated. Thus, if both networks $x_t$ and $y_t$ receive the same unexpected input that is of the same magnitude, the difference between the $x_t$ and $y_t$ should reach virtually 0 within 64 iterations. The consideration can be tested by setting the input

$$u_t = \pi/4 \times \text{rand},$$

where rand is an i.i.d. random list of +1s and -1s, that produces a representative of $\mathcal{F}_{\bar{u}^\infty} \neq \mathcal{F}_{crit}$. Figures 4 and 5 depict results from simulations, where after one iteration when two networks receive different inputs, both networks receive again the same input. Depicted is again the development of the difference between both networks $d(x_t, y_t)$ versus the number of iterations. The graphs appear in a double logarithmic fashion, so power law decays appear as straight lines. One can see that the networks that receive alternating, i.e., the expected input, pertains the difference for very long time spans (that exceed 64 iterations). On the other hand, if the network input for both networks is identical but i.i.d random and of the same order of magnitude as the expected input, the difference vanishes within 64 iterations.

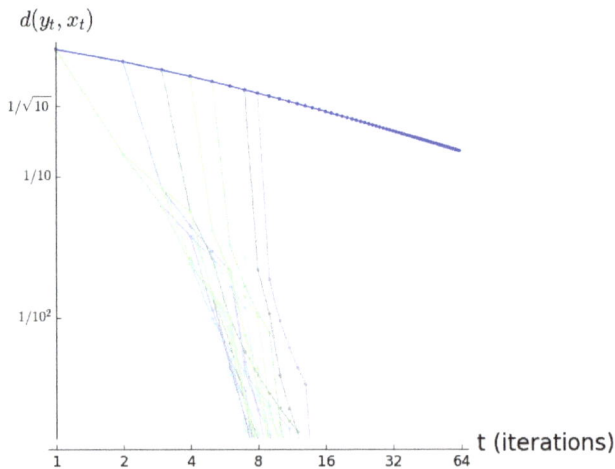

**Figure 4.** Results from two dynamics defined in Equation (18), with $b = 1$ and different types on input $u_t$. One copy of the two initially identical networks has received a variant input at iteration one. Depicted here is the decay of the difference of the state in the recurrent layer if both networks receive the same and expected (alternating) input (in dark blue). The pale curves are data from Figure 5 embedded for comparison. One can see that the difference function (blue) follows a power law and pertains for longer than 64 iterations.

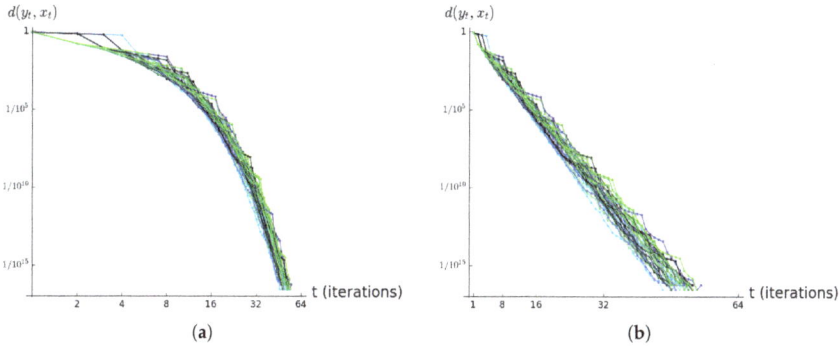

**Figure 5.** Complementary data for Figure 4: (**a**) Decay if both networks receive the same but irregular (i.i.d random) input (different colors are used for different trials). The difference vanishes much faster than in the left case. Finally the difference becomes and stays null, which is out of a logarithmic scale. Thus, later iterations are not depicted anymore; (**b**) Depicted here is the same data as at the left, however in a double-logarithmic plot. The data forms a straight line, which indicates an exponential decay.

The network distinguishes regular input from irregular input. Memories of irregular events pertain for a long time in the network provided that the following input is regular again. How a reservoir can be trained to anticipate certain input statistics has been discussed [7]. Additional solutions to this problem are subject to further investigations.

### 3.4. Relation to "Near Edge of Chaos" Approaches

It is a common experience that—in spite of given theoretical limits for the ESPs—the recurrent connectivity can be significantly higher than $s_{max} = 1$ for many practical input statistics. Those over-tuned ESNs in many cases show a much better performance than those that actually obey Jaeger's initial limit ESP. So, recently researchers came up with theoretical insights with regard to ESPs that are subject to a network and a particular input statistic [6]. In the scope of this work, instead of defining the ESP for a network the ESP is always defined as related to a network and an input statistic. Also, similar efforts have been undertaken in the field of so-called liquid state machines [17,18].

One may assume that those approaches show similar properties as the one that has been presented here. However, for a good reason those approaches all are called *'near edge of chaos'* approaches. In order to illustrate the problems that arise from those approaches, one may consider what happens if those overtuned ESNs are set exactly to the critical point. Here, just for the general understanding one may consider again a one-neuron network and a tanh as a transfer function, so

$$x_{t+1} = \tanh(-bx_t + u_t). \tag{19}$$

Note that the ESP limit outlined above requires that the recurrent connectivity should be $b < 1$. An input time series than one can use is from the previous section

$$u_t = (-1)^t \pi/4 \tag{20}$$

Slightly tedious but basically simple calculus results in a critical value of $b \approx 2.344$ for the input time series, where $x_t \approx (-1)^t \times 0.757$. In this situation one can test for the convergence of two slightly different initial conditions and obtain a power law decay of the difference. However, setting up the amplitude of the input just a tiny bit higher is going to result in two diverging time series $x_t$ and $y_t$. If the conditions of the ESN are chosen to be exactly at the critical point, it is possible that a untrained input sequence very close to the trained input sequence turns the ESN into a state where the ESP is not fulfilled anymore (for a related and more detailed discussion with a numerical analysis, confer [19]).

For this reason all the networks have to be chosen at a significant margin away from the edge of instability. That is very different from the approach in the previous section where although the expected input sequence for the network is exactly at the critical point, all other input sequences result in a stable ESN in most cases with exponential forgetting.

*3.5. How the One-Neuron Example Can Be Extended to Multi-Neuron Networks: Normal Matrices*

A normal matrix **W** commutes with its own transpose, i.e.,

$$\mathbf{W}\mathbf{W}^T = \mathbf{W}^T\mathbf{W}.$$

For a normal matrix **W**, it can easily be shown that

$$\max s(\mathbf{W}) = \max \mathrm{abs}(\lambda(\mathbf{W})).$$

These matrices apply to the spectral theorem; the largest absolute eigenvalue is the same as the largest singular value, which makes a clear theoretical separation between networks that are uniformly state contracting and those that are not compatible to the echo state condition. Still, for normal matrices all previously known ES conditions do not determine to which of those two groups the critical point itself belongs.

Summarizing, all previous works result in theorems for an open set of conditions that are defined by the strict inequalities Equations (9) and (11). In the closest case, the case of normal matrices, when considering the singular condition

$$1 = \max s(\mathbf{W}) = \max \mathrm{abs}(\lambda(\mathbf{W})) \tag{21}$$

there is no statement of the above mentioned theorems if the network is uniformly state contracting.

Some simple, preliminary numerical tests reveal that in the case of networks that satisfy Equation (21) the further development of the network strictly depends on the exact shape of transfer function.

## 4. Echo State Condition Limit with Weak Contraction

In the previous section, it has been shown how power law forgetting may occur in an ESN type neural network. These networks are all tuned to the point where $|\lambda|_{\max} = S = 1$. For this tuning it is still undetermined if the ESP is fulfilled or not even if normal matrices or one-neuron RNNs are used. The current section is dedicated to determining under which conditions Jaeger's ESP can be extended to this boundary condition.

Jaeger's sufficient echo state condition (see [4], App. C, p. 41) has strictly been proven only for non-critical systems (largest singular value $S < 1$) and with $\tanh(.)$ as a transfer function. The original proof is based on the fact that $\tanh$ in combination with $S < 1$ is a contraction. In that case Jaeger shows an exponential convergence.

The core of all considerations of a sufficient condition is to give a close upper estimate of the distance between the next iterations of two different states $y_t$ and $x_t$. The estimate is of the form

$$\max_{\bar{u}^\infty} d(y_{t+1}, x_{t+1}) = \max_{\mathcal{F}} d(\mathcal{F}(y_t), \mathcal{F}(x_t)) \le \phi_1 \cdot d(y_t, x_t),$$

where the parameter $\phi_1$ basically is quantified by the nature of the transfer function and the strength of the connectivity matrix. The estimate has to be good enough that the iterative application of $\phi_1$ should result in a convergence to 0:

$$\lim_{t \to \infty} [\phi_1 \cdot]^t d(y_0, x_0) = 0, \tag{22}$$

This is equivalent to investigating a series $q_t$ with

$$d(y_t, x_t) < q_t = [\phi_1 \cdot]^t d(y_0, x_0)$$

and

$$\lim_{t\to\infty} q_t = 0, \tag{23}$$

which can prove that the requirement of uniformly state contraction (cf. Equation (6)) is fulfilled. For example, consider the case of a reservoir with one-neuron as described in Equation (17). Here the challenge is to find an estimator for $\phi_1$ such that

$$\max_{u_t\in\mathbb{R}} d(\theta(by_t + u_t), \theta(bx_t + u_t)) \leq \phi_1 \cdot d(y_t, x_t),$$

where the chosen $\phi_1$ still holds the limit in Equation (22). For $|b| < 1$, convergence can be proven easily:

$$\max_{u_t\in\mathbb{R}} d(\theta(by_t + u_t), \theta(bx_t + u_t)) \leq$$
$$\max_{u_t\in\mathbb{R}} ||by_t + u_t - bx_t - u_t||_2 =$$
$$|b| \cdot d(y_t, x_t).$$

Thus for $|b| < 1$, one can easily define $\phi_1 = |b|$. So $q_t$ can be defined as

$$q_t = |b|^t \cdot d(y_0, x_0).$$

So Equation (23) is fulfilled. The convergence is exponential. The arguments so far are analogous to the core of Jaeger's proof for the sufficient echo state condition C2 that is restricted to one dimension.

For $|b| = 1$, this argument does not work anymore. Obviously, Jaeger's proof is not valid under these circumstances. However, the initial theorem can be extended. As a pre-requisite, one can replace the constant $\phi_1$ with a function that depends on $q_t$ as an argument.

So one can try

$$\phi_1(q_t) = 1 - \eta q_t^\kappa,$$

where $\eta > 0$ and $\kappa > 1$ have to be defined appropriately. This works for small values of $q_t$. However, it is necessary to name a limit for large $q_t > \gamma$. Define

$$\phi_1(z) := \begin{cases} 1 - \eta z^\kappa & \text{if } z < \gamma \\ 1 - \eta \gamma^\kappa & \text{if } z \geq \gamma. \end{cases} \tag{24}$$

Three things have to be done to check this cover function:

- First of all, one needs to find out if indeed the cover function $\phi_1$ fulfills

$$\max_{u_t\in\mathbb{R}} d(\theta(y_t + u_t), \theta(x_t + u_t)) \leq d(y_t, x_t)\phi_1(d(y_t, x_t)).$$

In order to keep the proof compatible with the proof for multiple neurons for this work, one has to chose a slightly different application for $\phi_1$,

$$\max_{u_t\in\mathbb{R}} d^2(\theta(y_t + u_t), \theta(x_t + u_t)) \leq d^2(y_t, x_t)\phi_1(d^2(y_t, x_t)),$$

which serves the same purpose and is much more convenient for multiple neurons.

In Appendix C one can find a recipe for this check.
- Second, one has to look for the convergence of

$$q_{t+1} = q_t \phi_1(q_t) = q_t(1 - \eta q_t^\kappa),$$

when $q_t \leq \gamma$. The analysis is done in Appendix A.

- Third, one needs to check

$$q_{t+1} = q_t \, \phi_1(q_t) = q_t \, (1 - \eta \gamma^\kappa),$$

as long as $q_t > \gamma$. Since the factor $(1 - \eta \gamma^\kappa)$ is positive, smaller than one and constant, the convergence process is exponential, obviously.

Note that the next section's usage of the cover function differs slightly even through it has the same form as Equation (24).

## 5. Sufficient Condition for a Critical ESN

The content of this section is a replacement of the condition C2 where the validity of the ESP is inferred for $S \leq 1$.

**Theorem 1.** *If hyperbolic tangent or the function of Equation (16) are used as transfer functions, the echo state condition (see Equation (6)) is fulfilled even if $S = 1$.*

**Summary of the Proof.** As an important precondition, the proof requires that both transfer functions fulfill

$$d(\theta(y_t), \theta(x_t)) \leq d(y_t, x_t) \phi_k(d^2(y_t, x_t)),\tag{25}$$

where $\phi_k(z)$ is defined for a network with $k$ hidden neurons as

$$\phi_k(z) := \begin{cases} 1 - \eta z^\kappa & \text{if } z < \gamma \\ 1 - \eta \gamma^\kappa & \text{if } z \geq \gamma \end{cases}.\tag{26}$$

Here, $1 > \gamma > 0, 1 > \eta > 0, \kappa \geq 1$ are constant parameters that are determined by the transfer function and the metric norm $d(.,.) = ||.||_2$ $\square$

In Appendix C it is shown that indeed both transfer functions fulfil that requirement. It then remains to prove that in the slowest case we have a convergence in a process with 2 stages. In the first stage, if $d^2(y_t, x_t) > \gamma$ there is a convergence that is faster or equal to an exponential decay. The second stage is a convergence process that is faster or equal to a power law decay.

**Proof.** Note with regard to the test function $\phi_k$:

$$\phi_k \leq 1,$$
$$\forall z, \forall Z : 0 \leq z \leq Z \quad \leftrightarrow \quad \phi_k(z) \geq \phi_k(Z),$$
$$\text{and } \forall z, \forall Z : 0 \leq z \leq Z \quad \leftrightarrow \quad Z \times \phi_k(Z) \geq z \times \phi_k(z)$$

In analogy to Jaeger, one can check now the contraction between the time step $t$ and $t + 1$:

$$d^2(\mathbf{y}_{t+1}, \mathbf{x}_{t+1}) = d^2(\theta(\mathbf{y}_{lin,t+1}), \theta(\mathbf{x}_{lin,t+1}))$$
$$\leq d^2(\mathbf{y}_{lin,t+1}, \mathbf{x}_{lin,t+1}) \times \phi_k(d^2(\mathbf{y}_{lin,t+1}, \mathbf{x}_{lin,t+1}))\tag{27}$$

One can rewrite

$$d^2(\mathbf{y}_{lin,t+1}, \mathbf{x}_{lin,t+1}) = ||\mathbf{W}\mathbf{y}_t + I - \mathbf{W}\mathbf{x}_t - I||_2^2 = ||\mathbf{W}(\mathbf{y}_t - \mathbf{x}_t)||_2^2,\tag{28}$$

where $I = \mathbf{w}^{in}\mathbf{u}_t$. Next one can consider that one can decompose the recurrent matrix by using singular value decomposition (SVD) and obtain $\mathbf{W} = \mathbf{U} \cdot \mathbf{S} \cdot \mathbf{V}^T$. Note that both $\mathbf{U}$ and $\mathbf{V}$ are orthogonal matrices and that $\mathbf{S}$ is diagonal with positive values $s_i$ on the main diagonal. We consider

$$\mathbf{a} = \mathbf{V}^T(\mathbf{y}_t - \mathbf{x}_t).$$

Because $\mathbf{V}$ is an orthogonal matrix, the left side of the equation above is a rotation of the right side and the length $||\mathbf{a}||$ is the same as $||\mathbf{y}_t - \mathbf{x}_t||$. One can write

$$d^2(\mathbf{y}_t, \mathbf{x}_t) = \sum_i a_i^2,$$

where the $a_i$ are entries of the vector $\mathbf{a}$. Since

$$\mathbf{y}_{lin,t+1} - \mathbf{x}_{lin,t+1} = \mathbf{U}\mathbf{S}\mathbf{a}$$

and $\mathbf{U}$ is again a rotation matrix, one can write

$$d^2(\mathbf{y}_{lin,t+1}, \mathbf{x}_{lin,t+1}) = \sum_i s_i^2 a_i^2,$$

where $s_i$ is the $i$-th component of the diagonal matrix $\mathbf{S}$, i.e., the $i$-th singular value.
In the following we define $s_{max} = \max_i s_i$ and calculate

$$
\begin{aligned}
& d^2(\mathbf{y}_{lin,t+1}, \mathbf{x}_{lin,t+1}) \times \phi_k(d^2(\mathbf{y}_{lin,t+1}, \mathbf{x}_{lin,t+1})) \\
&= (\sum_i s_i^2 a_i^2) \times \phi_k(\sum_i s_i^2 a_i^2) \\
&\leq (s_{max}^2 \sum_i a_i^2) \times \phi_k(s_{max}^2 \sum_i a_i^2) \\
&\leq (s_{max}^2 d^2(\mathbf{y}_t, \mathbf{x}_t)) \times \phi_k(s_{max}^2 d^2(\mathbf{y}_t, \mathbf{x}_t))
\end{aligned}
\tag{29}
$$

Merging Equations (27) and (29) results in the inequality

$$d(\mathbf{y}_{t+1}, \mathbf{x}_{t+1}) \leq s_{max} d(\mathbf{y}_t, \mathbf{x}_t) \times (\phi_k(s_{max}^2 d^2(\mathbf{y}_t, \mathbf{x}_t)))^{0.5}.$$

First, assuming $s_{max} < 1$ and since we know $\phi_k \leq 1$, we get an exponential decay

$$d(y_n, x_n) \leq s_{max}^t d(x_0, y_0).$$

This case is handled by Jaeger's initial proof. With regard to an upper limit of the contraction speed (cf. Equation (6)), one can find

$$\tau(\epsilon) = \frac{\log \epsilon - \log(d(y_0, x_0))}{\log s_{max}}.$$

If the largest singular value $s_{max} > 1$, then for some type of connectivities (i.e., normal matrices) the largest absolute eigenvalue is also larger than 1 due to the spectral theorem. In this case, the echo state condition is not always fulfilled, which has been shown also by Jaeger.
What remains is to check the critical case $s_{max} = 1$. Here again one can discuss two different situations (rather two separate phases of the convergence process) separately:
If $d^2(\mathbf{y}_{lin,t}, \mathbf{x}_{lin,t}) > \gamma$, we can write the update inequality of Equation (30) as:

$$d^2(\mathbf{y}_{t+1}, \mathbf{x}_{t+1}) \leq (1 - \eta\gamma^\kappa)d^2(\mathbf{y}_t, \mathbf{x}_t).$$

Thus, for all $\epsilon^2 \geq \gamma$, the slowest decay process can be covered by

$$\tau(\epsilon) = \frac{2\log\epsilon - 2\log(d(y_0, x_0))}{\log(1 - \eta\gamma^\kappa)}.$$

If $\epsilon^2 < \gamma$, then Equation (30) becomes:

$$d^2(y_{t+1}, x_{t+1}) \leq d^2(y_t, x_t)(1 - \eta d(y_t, x_t)^{2\kappa}).$$

One can replace

$$q_t = d^2(y_t, x_t) \tag{30}$$

and again consider the sequence

$$q_{t+1} = q_t(1 - \eta q_t^\kappa),$$

which is discussed in Appendix A. The result there is that the sequence converges faster than

$$q_*(t) = [\frac{\eta}{\kappa}t + q_0^{-\kappa}]^{-1/\kappa}. \tag{31}$$

Note that, although the Lyapunov exponent (cf. Equation (7)) of $q_*(t)$ is zero, the sequence $q_*(t)$ converges in a power law fashion. Thus,

$$\lim_{t\to\infty} q(t) = d^2(y_t, x_t) = 0 \tag{32}$$

and thus ESP has been proven. $\square$

Moreover, one can calculate the upper time limit $\tau(\epsilon)$:

$$\tau(\epsilon) = \frac{\kappa}{\eta}(\epsilon^{-2\kappa} - a_0^{-\kappa}).$$

## 6. Summary, Discussion and Outlook

The background of this paper is to investigate the limit of recurrent connectivity in ESNs. The preliminary hypothesis towards the main work can be summarized in Figure 6. Initially it is hard to quantify the transition point between uniformly state contracting and non-state contracting ESNs exactly. However, for normal matrices and one-neuron networks the gap between the sufficient condition and the necessary condition collapses in a way that there are two neighboring open sets. The first open set is known to have the ESP, and the other open set evidently does not have the ESP. What remains is the boundary set. The boundary set is interesting to analyze because it can easily be shown that here power law forgetting can occur.

The proof of Section 4 shows a network is an ESN even if the largest eigenvalue of the recurrent connectivity matrix is equal to 1 and if the transfer function is either Equation (15) or (16). The proof is also extensible to other transfer functions. On the other hand it is obvious that some transfer functions result in networks that are not ESNs. For example a linear transfer function ($\theta(x) = x$) is *not* state contracting.

Even if the network is state contracting, it is not necessarily exponentially uniformly state contracting. Its rate of convergence might follow a power law in the slowest case. Several examples for power law forgetting have been shown in the present work. More examples of preliminary learning have been outlined in [7]. One important target of the present research is to allow for a kind of memory compression in the reservoir by letting only the unpredicted input enter the reservoir.

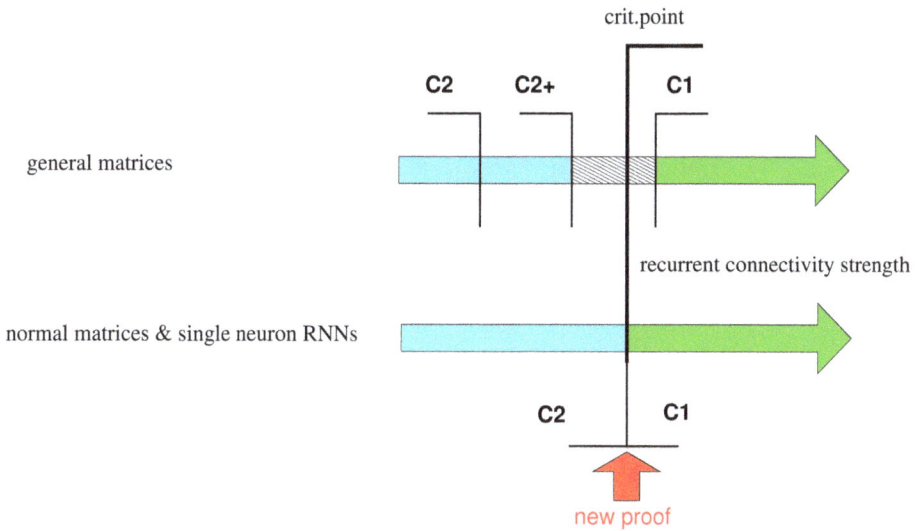

**Figure 6.** The graph illustrates the relation between the different conditions with regard to echo state property (ESP). C1 and C2 are the necessary and sufficient condition according to Jäger [4]. The large arrows represent the connectivity strength of the recurrent synaptic weight matrix **W**. For a small connectivity strength the ESP is fullfilled (cyan areas). For a strong connectivity, the ESP is not fulfilled anymore (green). C2+ represents symbolically the improvement of the sufficient condition according to [13,14]. For general matrices there can be a non-zero gap between C2+ and C1 which is drawn diagonally shaded. The transition from ESP to non-ESP happens somewhere within this gap. All three conditions C1, C2 and C2+ describe an open set. In the case of normal matrices and for one-neuron networks the gap is closed except for the separation line itself. The proof of Section 4 shows that there it depends on the transfer function if the network is an ESN or not.

One ultimate target of the present work is to find a way to organize reservoirs as recurrent filters with a memory compression feature. In order to bring concepts of data compression into the field of reservoir computing and in order to project as much as possible of the input history to the limit size reservoir, principles of memory compression have to be transferred into reservoir computing. However, the reservoir computing techniques that are analogous to classic memory compression have not been identified so far.

Another topic that needs further investigation is entropy in time series. Power law forgetting is only possible if the time series that relates to the criticality is either of a finite entropy, i.e., from a certain point in time all following entries of the time series can be predicted from the previous entries, or if the network simply ignores certain aspects of the incoming time series.

There also potential analogies in biology. Several measurements of memory decay in humans exist that reveal that there the forgetting follows a power law at least for a large fraction of the investigated examples [20,21].

**Acknowledgments:** The Nation Science Council and the Ministry of Science and Technology of Taiwan provided the budget for our laboratory. Also thanks go to AIM-HI for various ways of support.

**Conflicts of Interest:** The author declares no conflict of interest.

## Notations in Formulas

| | | |
|---|---|---|
| $s(M)$ | | vector of singular values of matrix $M$ |
| $\mathbf{S}$ | | diagonal matrix of singular values |
| $s_{max}$ | | largest singular value of a matrix |
| $\lambda(M)$ | | vector of eigenvalues |
| $\|\lambda\|_{max}$ | | the maximum of the eigenvalues in modulus |
| $\theta(.)$ | $\mathbb{R} \to \mathbb{R}$ | transcendental transfer function |
| $\theta', \dot{\theta}$ | | its derivative |
| $\mathbf{w}^{in}$ | $\mathbb{R}^{n \times k}$ | input matrix |
| $\mathbf{W}$ | $\mathbb{R}^{k \times k}$ | recurrent transfer matrix |
| $\mathbf{w}^{out}$ | $\mathbb{R}^{k \times m}$ | output matrix |
| $\mathbf{u}_t$ | $\mathbb{R}^n$ | input time series |
| $\bar{\mathbf{u}}^\infty$ | | complete infinite input time series |
| $\mathbf{x}_t$ | $\mathbb{R}^k$ | hidden layer states |
| $\mathbf{y}_t$ | $\mathbb{R}^k$ | alternative hidden layer states to check convergence |
| $\mathbf{o}_t$ | | training set |
| $\bar{\mathbf{o}}_t$ | | trained network output |
| $d(.,.)$ | $\|\|.\|\|_2$ | distance measure that is used to check convergence |
| $\mathbf{F}(.)$ | | vector of a set of linearly independent, non-linear functions |
| $\mathcal{F}_{\bar{\mathbf{u}}^\infty}$ | $\in \mathcal{F}$ | dynamics of the network with regard to the input |
| $\mathcal{F}_{crit}$ | | all dynamics that show power law forgetting |
| $q_{\bar{\mathbf{u}}^\infty, t}$ | | time series related to forgetting in ESNs |
| $q_t$ | | time series variable |
| $\phi_1()$ | | cover function to estimate contraction in a 1 neuron net |
| $\phi_k()$ | | dto. for a $k$ neuron network |
| $\eta, \kappa, \gamma$ | | const parameters of both $\phi_1$ and $\phi_k$ |
| $\Lambda_{\bar{\mathbf{u}}^\infty}$ | | Lyapunov exponent of a neural network wrt. an input time series $\bar{\mathbf{u}}^\infty$ |

## Appendix A. Analyze $q_{t+1} = q_t(1 - \eta q_t^\kappa)$

We can consider the sequence $q_t$:

$$q_{t+1} = q_t(1 - \eta q_t^\kappa). \tag{A1}$$

Convergence can be analyzed in the following way:

$$\Delta q_t = q_{t+1} - q_t = -\eta q_t^{\kappa+1}. \tag{A2}$$

Thus, the series $q_{t+1}$ can be written as

$$q_{t+1} = \sum_{t' < t} \Delta q_{t'}$$

Thus one comes up with the following discrete formula

$$\Delta q_{t'} = -\eta q_{t'}^{\kappa+1}. \tag{A3}$$

Since $\kappa \geq 1$ and $1 > \eta > 0$, convergence towards null is obvious here. In addition since the right side of Equation (A3) is decreasing continuously it is obvious that it is converging fast than the corresponding solution of the differential equation for a function $q_*$:

$$\frac{dq_*}{dt} = -\eta q_*^{\kappa+1},$$

which is easily solvable to:

$$-\frac{1}{\kappa\eta}q_*^{-\kappa} = t + C,$$

where $C$ is the integration constant. From this solution, by setting $q_0 = q_*(0)$ one can derive

$$q_*(t) = [\tfrac{\eta}{\kappa}t + q_0^{-\kappa}]^{-1/\kappa}.$$

Note that

$$\lim_{t\to\infty} q_* = 0.$$

Simple algebraic considerations show that $q_*(t)$ covers the sequence $q_t$. For a strict proof see Appendix B.

Thus,

$$q_{*,t} \geq q_t$$

for all $t > 0$ if

$$q_{*,0} = q_0$$

So, the sequences $q_{*,t}$ and $q_t$ converge to zero.

## Appendix B. Sequence $q_{*,t}$ Covers up Sequence $q_t$

In the following, identical definitions to Section 4 are used. One can start from the statement

$$-\frac{\eta}{\kappa} \geq -\eta. \tag{B1}$$

Since $\eta \geq 0$ and $\kappa \geq 1$, it can easily be seen that the inequality is fulfilled for any combination of $s$, $\eta$ and $\kappa$. One can now extend the numerator and denominator of the right side by $(\frac{\eta}{\kappa}t + C)$ and add also $(\frac{\eta}{\kappa}t + C)$ to both sides of the inequality. Here and in the following $C$ is defined as

$$C = q_0^{-\kappa} \geq 1. \tag{B2}$$

One obtains

$$\tfrac{\eta}{\kappa}t + C - \tfrac{\eta}{\kappa} \geq (\tfrac{\eta}{\kappa}t + C) - (\tfrac{\eta}{\kappa}t + C)\tfrac{\eta}{\frac{\eta}{\kappa}t+C}. \tag{B3}$$

A rearrangement of the right side results in

$$\tfrac{\eta}{\kappa}t + C - \tfrac{\eta}{\kappa} \geq (\tfrac{\eta}{\kappa}t + C) \times (1 - \tfrac{\eta}{\frac{\eta}{\kappa}t+C}). \tag{B4}$$

One can add $\frac{\eta}{\kappa}$ to both sides and obtain

$$\tfrac{\eta}{\kappa}t + C \geq \tfrac{s\eta}{\kappa} + (\tfrac{\eta}{\kappa}t + C) \times (1 - \tfrac{\eta}{\frac{s\eta}{\kappa}t+C}). \tag{B5}$$

Now, one can use the fact that

$$1 \geq 1 - \frac{\eta}{\frac{\eta}{\kappa}t + C} \geq \left(1 - \frac{\eta}{\frac{\eta}{\kappa}t + C}\right)^{\kappa}$$

and

$$\frac{\eta}{\kappa} \geq \frac{\eta}{\kappa} \times \left(1 - \frac{\eta}{\frac{\eta}{\kappa}t + C}\right) \geq \frac{\eta}{\kappa} \times \left(1 - \frac{\eta}{\frac{\eta}{\kappa}t + C}\right)^{\kappa}$$

and thus rewrite inequality Equation (B5) as

$$\frac{\eta}{\kappa}t + C \geq \left(\frac{\eta}{\kappa}t + C + \frac{\eta}{\kappa}\right) \times \left(1 - \frac{\eta}{\frac{\eta}{\kappa}t + C}\right)$$

and finally arrive at

$$\left(\frac{\eta}{\kappa}t + C\right) \geq \left(\frac{\eta}{\kappa}t + C + \frac{\eta}{\kappa}\right) \times \left(1 - \frac{\eta}{\frac{\eta}{\kappa}t + C}\right)^{\kappa}.$$

One can multiply both sides by $\left(\frac{\eta}{\kappa}t + C + \frac{\eta}{\kappa}\right)^{-1} \times \left(\frac{\eta}{\kappa}t + C\right)^{-1}$. So,

$$\left(\frac{\eta}{\kappa}t + C + \frac{\eta}{\kappa}\right)^{-1} \geq \left(\frac{\eta}{\kappa}t + C\right)^{-1} \times \left(1 - \frac{\eta}{\frac{\eta}{\kappa}t + C}\right)^{\kappa}.$$

Taking the $k$th root on both sides one gets

$$\left(\frac{\eta}{\kappa}t + C + \frac{\eta}{\kappa}\right)^{-1/\kappa} \geq \left(\frac{\eta}{\kappa}t + C\right)^{-1/\kappa} \times \left(1 - \frac{\eta}{\frac{\eta}{\kappa}t + C}\right).$$

One can rearrange the inequality to:

$$\left(\frac{\eta}{\kappa}t + C + \frac{\eta}{\kappa}\right)^{-1/\kappa} \geq \left(\frac{\eta}{\kappa}t + C\right)^{-1/\kappa} \times \left(1 - \eta\left(\frac{\eta}{\kappa}t + C\right)^{-1}\right)$$
$$\left(\frac{\eta}{\kappa}(t+1) + C\right)^{-1/\kappa} \geq \left(\frac{\eta}{\kappa}t + C\right)^{-1/\kappa} \times \left(1 - \eta\left(\left(\frac{\eta}{\kappa}t + C\right)^{-1/\kappa}\right)^{\kappa}\right) \tag{B6}$$

Using the definitions of $q_{*,t}$ and $q_{*,t+1}$, i.e.,

$$q^*(t) = \left[\frac{\eta}{\kappa}t + q_0^{-\kappa}\right]^{-1/\kappa},$$

one has finally

$$q_{*,t+1} \geq q_{*,t} \times \left(1 - \eta(q_{*,t})^{\kappa}\right). \tag{B7}$$

proven as a true statement. Thus,

$$q_{*,t} \geq q_t \geq d(\mathbf{y}_t, \mathbf{x}_t)$$

for all $t > 0$ if

$$q_{*,0} = q_0 = d(\mathbf{y}_0, \mathbf{x}_0).$$

Thus, the positive definite sequence $q_{*,t}$ covers $d(\mathbf{y}_t, \mathbf{x}_t)$.

## Appendix C. Weak Contraction with the Present Transfer Function

In this appendix a test function of the form of Equation (26), is verified for the function of Equation (16) and hyperbolic tangent. Within this section we test the following values

$$\eta = \frac{1}{48 \times n^2}, \gamma = \frac{1}{2} \text{ and } \kappa = 2, \tag{C1}$$

for both transfer functions and square norm ($||.||_2$), and $n$-neurons. In order to derive these values, one can start by considering linear responses $||\mathbf{y}_{lin,t} - \mathbf{x}_{lin,t}||$ and the final value $||\mathbf{y}_t - \mathbf{x}_t||$ within one single neuron as

$$\mathbf{y}_t = \theta(\mathbf{y}_{lin,t})$$
$$\text{and } \mathbf{x}_t = \theta(\mathbf{x}_{lin,t}). \tag{C2}$$

We can define

$$y_{lin,t,i} = \Delta_i + \zeta_i, \ x_{lin,t,i} = \zeta_i, \ \text{and} \ x_{t,i} - y_{t,i} = \delta_i \tag{C3}$$

where $i$ is the index of the particular hidden layer neuron. Note that

$$||\mathbf{y}_{lin,t} - \mathbf{x}_{lin,t}||^2 = \sum_i \Delta_i^2, \ ||\mathbf{y}_t - \mathbf{x}_t||^2 = \sum_i \delta_i^2.$$

*Appendix C.1. One-Neuron Transfer*

In this section considerations are restricted to the case in which one has only one neuron in the hidden layer. For the sake of simplicity, the subscript index $i$ is left out in the following considerations.

Setting in the definitions of Equation (C3) into the square of Equation (25) and for a single neuron and for any $\zeta$ we get

$$\Delta^2 \, \phi_1(\Delta^2) \geq (\theta(\Delta + \zeta) - \theta(\zeta))^2 \, .$$

Thus, it suffices to consider

$$\phi_1(\Delta^2) \geq \max_\zeta \omega(\Delta, \zeta), \tag{C4}$$

where

$$\omega(\Delta, \zeta) = \left( \frac{\theta(\Delta + \zeta) - \theta(\zeta)}{\Delta} \right)^2$$

The $\max_\zeta$ can be found by basic analysis. Extremal points can be found as solutions of

$$\frac{\partial}{\partial \zeta} \omega =$$
$$\frac{\partial}{\partial \zeta} \left[ \left( \frac{\theta(\Delta + \zeta) - \theta(\zeta)}{\Delta} \right)^2 \right] = \tag{C5}$$
$$\frac{2(\theta(\Delta + \zeta) - \theta(\zeta)) \times (\dot{\theta}(\Delta + \zeta) - \dot{\theta}(\zeta))}{\Delta^2} = 0.$$

This can only be fulfilled if

$$\dot{\theta}(\Delta + \zeta) - \dot{\theta}(\zeta) = 0$$

Since $\theta'$ is an even function for both suggested transfer functions, one gets $\zeta = -z/2$ as the extremal point. Fundamental analysis shows that this point in both cases is a maximum. Thus, requiring

$$\phi_1(\Delta^2) \geq \left[ \frac{2\theta(\Delta/2)}{\Delta} \right]^2 \tag{C6}$$

also would satisfy Equation (C4).

Numerically, one can find parameters for $\phi_1$ that are $\eta = 1/48$ and $\kappa = 2$ of as in Equation (C1). For $\Delta^2 > \gamma$, it suffices to check if

$$\frac{4\tanh^2(\Delta/2)}{\Delta^2} \leq 1 - \eta \gamma^\kappa$$

First the inequality is fulfilled for $\Delta^2 = \gamma$. Since the left side of the equation above is strictly decreasingfulfilled for all values $\Delta^2 > \gamma$.

Analogous considerations lead to the same parameters to cover up the transfer function from Equation (16).

*Appendix C.2. Multi-Neuron Parameters*

For several neurons one has to consider the variational problem of all possible combinations of values of $\Delta_i$. One can start from the proven relation from the previous section,

$$\delta_i^2 \leq \Delta_i^2 \phi_1 \left( \Delta_i^2 \right)$$

Thus,

$$\sum_i \delta_i^2 \leq \sum_i \left( \Delta_i^2 \phi_1 \left( \Delta_i^2 \right) \right),$$

implying that

$$\sum_i \left( \Delta_i^2 \phi_1 \left( \Delta_i^2 \right) \right) \leq \sum_i \Delta_i^2 - \Delta_{max}^2 + \Delta_{max}^2 \phi_1 (\Delta_{max}^2), \tag{C7}$$

where $\Delta_{max}^2 = \max_i \Delta_i^2$. The smallest possible $\Delta_{max}^2$ is

$$\frac{\sum_i \Delta_i^2}{n},$$

Substituting into Equation (C7), we get

$$\sum_i \Delta_i^2 - \Delta_{max}^2 + \Delta_{max}^2 \phi_1 (\Delta_{max}^2) \leq$$

$$\sum_i \Delta_i^2 - \frac{\sum_i \Delta_i^2}{n} + \frac{\sum_i \Delta_i^2}{n} \phi_1 (\frac{\sum_i \Delta_i^2}{n}) =$$

$$(\sum_i \Delta_i^2) \phi_1 (\frac{\sum_i \Delta_i^2}{n^2}).$$

Thus, the inequality (which is equivalent to Equation (25)),

$$\sum_i \delta_i^2 \leq \left( \sum_i \Delta_i^2 \right) \phi_k \left( \sum_i \Delta_i^2 \right),$$

is fulfilled if $\phi_k$ is defined as

$$\phi_k(x) = \phi_1 \left( \frac{x}{n^2} \right).$$

Thus, the parameters from Equation (C1) fulfil the inequality of Equation (25).

## References

1. Werbos, P.J. Backpropagation through time: What it does and how to do it. *Proc. IEEE* **1990**, *78*, 1550–1560.
2. Lukoševičius, M.; Jaeger, H. Reservoir computing approaches to recurrent neural network training. *Comput. Sci. Rev.* **2009**, *3*, 127–149.
3. Maass, W.; Natschläger, T.; Markram, H. Real-time computing without stable states: A new framework for neural computation based on perturbations. *Neural Comput.* **2002**, *14*, 2531–2560.
4. Jaeger, H. The "echo state" approach to analysing and training recurrent neural networks—With an erratum note. In *GMD German National Research Insitute for Computer Science 2001*; GMD Report 148. Available online: https://pdfs.semanticscholar.org/8430/c0b9afa478ae660398704b11dca1221ccf22.pdf (accessed on 22 December 2016).
5. Boedecker, J.; Obst, O.; Lizier, J.; Mayer, N.M.; Asada, M. Information processing in echo state networks at the edge of chaos. *Theory Biosci.* **2012**, *131*, 205–213.
6. Manjunath, G.; Jaeger, H. Echo state property linked to an input: Exploring a fundamental characteristic of recurrent neural networks. *Neural Comput.* **2013**, *25*, 671–696.
7. Mayer, N.M. Adaptive critical reservoirs with power law forgetting of unexpected input events. *Neural Comput.* **2015**, *27*, 1102–1119.

8. Ng, A. CS 229 Machine Learning, UC Stanford. URL. 2014. Available online: http://cs229.stanford.edu/notes/cs229-notes1.pdf (accessed on 22 December 2016).

9. Jaeger, H.; Maass, W.; Principe, J. Special issue on echo state networks and liquid state machines. *Neural Netw.* **2007**, *20*, 287–289.

10. Schrauwen, B.; Verstraeten, D.; Van Campenhout, J. An overview of reservoir computing: Theory, applications and implementations. In Proceedings of the 15th European Symposium on Artificial Neural Networks, Bruges, Belgium, 25–27 April 2007.

11. Jaeger, H. Adaptive nonlinear system identification with echo state networks. In Proceedings of the Neural Information Processing Systems: Natural and Synthetic, Vancouver, BC, Canada, 9–14 December 2002.

12. Wainrib, G.; Galtier, M.N. A local Echo State Property through the largest Lyapunov exponent. *Neural Netw.* **2016**, *76*, 39–45.

13. Buehner, M.R.; Young, P. A Tighter Bound for the Echo State Property. *IEEE Trans. Neural Netw.* **2006**, *17*, 820–824.

14. Yildiz, I.B.; Jaeger, H.; Kiebel, S.J. Re-visiting the echo state property. *Neural Netw.* **2012**, *35*, 1–20.

15. Appeltant, L.; Soriano, M.C.; Van der Sande, G.; Danckaert, J.; Massar, S.; Dambre, J.; Schrauwen, B.; Mirasso, C.R.; Fischer, I. Information processing using a single dynamical node as complex system. *Nat. Commun.* **2011**, *2*, 468.

16. Ortín, S.; Soriano, M.C.; Pesquera, L.; Brunner, D.; San-Martín, D.L.; Fischer, I.; Mirasso, C.R.; Gutiérrez, J.M. A unified framework for reservoir computing and extreme learning machines based on a single time-delayed neuron. *Sci. Rep.* **2015**, *5*, 14945.

17. Natschläger, T.; Bertschinger, N.; Legenstein, R.A. At the edge of chaos: Real-time computations and self-organized criticality in recurrent neural networks. In Proceedings of the Advances in Neural Information Processing Systems 17, Bonn, Germany, 7–11 August 2005.

18. Legenstein, R.A.; Maass, W. Edge of chaos and prediction of computational performance for neural circuit models. *Neural Netw.* **2007**, *20*, 323–334.

19. Mayer, N.M. Critical Echo State Networks that Anticipate Input Using Adaptive Transfer Functions. *arXiv* 2016, arXiv:1411.6757.

20. Rubin, D.C.; Wenzel, A.E. One hundred years of forgetting: A quantitative description of retention. *Psychol. Rev.* **1996**, *103*, 734.

21. Kahana, M.J.; Adler, M. *Note on the Power Law of Forgetting*; University of Pennsylvania: Philadelphia, PA, USA, 2002.

*entropy*

MDPI

Article
# Complexity and Vulnerability Analysis of the *C. Elegans* Gap Junction Connectome

James M. Kunert-Graf *, Nikita A. Sakhanenko and David J. Galas *

Pacific Northwest Diabetes Research Institute, Seattle, WA 98122, USA; nsakhanenko@pnri.org
* Correspondence: jkunert@pnri.org (J.M.K.-G.); dgalas@pnri.org (D.J.G.);
  Tel.: +1-800-745-1527 (J.M.K.-G. & D.J.G.)

Academic Editor: Mikhail Prokopenko
Received: 30 December 2016; Accepted: 3 March 2017; Published: 8 March 2017

**Abstract:** We apply a network complexity measure to the gap junction network of the somatic nervous system of *C. elegans* and find that it possesses a much higher complexity than we might expect from its degree distribution alone. This "excess" complexity is seen to be caused by a relatively small set of connections involving command interneurons. We describe a method which progressively deletes these "complexity-causing" connections, and find that when these are eliminated, the network becomes significantly less complex than a random network. Furthermore, this result implicates the previously-identified set of neurons from the synaptic network's "rich club" as the structural components encoding the network's excess complexity. This study and our method thus support a view of the gap junction Connectome as consisting of a rather low-complexity network component whose symmetry is broken by the unique connectivities of singularly important rich club neurons, sharply increasing the complexity of the network.

**Keywords:** complexity; computational neuroscience; *C. elegans*; neural connectome; rich club; vulnerability

## 1. Introduction

It is increasingly clear that the specific connectivities of biological networks are far from random. In neuroscience, the overall behavior and dynamics of a neuronal network is dictated both by the properties of individual neurons, and by the specific complex structure of the network through which they are connected [1,2]. Understanding the behavior of such a neuronal network must therefore involve inferring its underlying structural properties and design principles. However, it is unclear a priori which structural features of a network are most important in doing so. Distinguishing these important features is one of the goals in quantifying the complexity of a network, as a measure of network complexity could allow us to determine which features of a network cause it to be complex, and to infer then the purpose of these complex structures.

As of yet, there exists no broadly accepted definition of network complexity which would allow for its unambiguous quantification, though many quantitative measures have been proposed and applied [3,4]. A necessary property of any such measure, however, is that it must distinguish complex, structured networks from networks which are either completely random or completely ordered. One such measure was developed by Sakhanenko and Galas [5] who defined a general complexity measure, based in the Kolmogorov complexity, which correctly vanishes in these limiting cases. Indeed, even this simple approach appears capable of indicating certain complex structures existing within the network [5,6].

If a network is determined via some measure to be highly complex, the question yet remains: what is the source of that complexity? That is, what are the substructures or features of the network which make it "complex"? Can these measures be applied in such a way so as to identify complexity-causing structures within the network, and can we attribute biophysiological meaning to these structures?

The nematode *C. elegans* is an ideal system in which to attempt to address such questions. It remains the only organism for which the full connectivity of its neuronal network (its "Connectome") is known. The *C. elegans* Connectome consists of both directed synaptic connections and undirected gap junction connections among its 302 neurons [7]. These neurons can be broadly classified as sensory neurons (those receiving external sensory input), motorneurons (those synapsing onto muscle), or, otherwise, as interneurons (if lacking explicit sensory input or motor output) [7]. Despite not directly receiving external input or driving motor output, many interneurons play important computational roles in the network. For example, the "command" interneurons are individually crucial to the control of locomotion; for example, if the command interneuron pair AVAL/R is ablated, the ability of the worm to crawl backwards is severely diminished, whereas ablating the command interneuron pair AVBL/R diminishes the ability of the worm to crawl forwards [8].

Even with its relatively small nervous system, *C. elegans* is capable of a fairly broad range of behaviors. In addition to responding to a range of mechanical and chemical stimuli [8–10], the worm must navigate [11], mate [12,13], and lay eggs [14,15]. The network controls these various behaviors through shared pathways of overlapping subcircuits, all while the structure appears to approximately minimize the wiring cost between nodes [16]. Previous studies indicate that, as with other neuronal networks [17–23], the network partially accomplishes this trade-off through its "rich club" structure, having a highly-connected hub of "rich" neurons with high betweenness centrality [24]. The specifically-tuned complex structure of the network may ultimately encode behavioral responses to inputs. For example, simulations of Connectome dynamics which treat all neurons as identical units are capable, through their connectivity structure alone, of generating biophysiologically reasonable dynamical responses to specific inputs (e.g., in [25]).

It was recently demonstrated by Kim et al. [26] that vulnerability analysis is capable of identifying many important functional structures within the *C. elegans* connectome. The core idea of vulnerability analysis is that important links or nodes within the network can be identified by considering the amount by which different structural properties of the network change when said link/node is removed. Kim et al. computationally analyzed the effects of removing any one given node/edge from the Connectome on its clustering coefficient, global efficiency, and betweenness centrality, and found that such an approach identified biophysiologically relevant subcircuits.

In this paper, we study the complexity of the *C. elegans* gap junction Connectome based on our previously defined measure and find that it is vastly more complex than a random graph with the same degree distribution. Its complexity score is 16.5 standard deviations above the randomly-expected mean (see Methods Section 4.5 for specifics). We then extend the use of vulnerability analysis on the *C. elegans* connectome by considering how the complexity of the graph is altered if any given edge is deleted. A large fraction of the network's complexity is seen to be caused by a relatively small set of connections involving the known "command" interneurons. We then extend the idea of vulnerability analysis by using a greedy algorithm to iteratively delete these "complexity-causing" connections. When these links are eliminated, the network becomes significantly less complex than a random graph. Furthermore, the deleted structure is seen to have a clear biophysical interpretation: our algorithm implicates a set of edges involving neurons from the synaptic network's "rich club" as being the source of the network's excess complexity. Thus, this study supports a view of the Connectome as consisting of a low-complexity structure whose complexity is dramatically enhanced by the addition of unique connectivities of the network's rich club.

## 2. Results

### 2.1. Investigation of Small, Random Networks

We begin our exploration of complexity/vulnerability analysis by considering graphs small enough to be easily visualized. We generated 10,000 random Erdős–Rényi networks, all of which had exactly 12 nodes each of degree four. This node count and degree were chosen heuristically such that the network would be small, but still large enough to allow for a very large number of possible connection schemes and complex structures. We investigate the complexity of these graphs by calculating the measure $\Psi(G)$, the aforementioned complexity measure developed in [5]. As detailed further in the Methods section, this measure is calculated as:

$$\Psi(G) = \frac{1}{N(N-1)} \sum_i \sum_{j \neq i} \max(K_i, K_j) m_{ij}(1 - m_{ij}), \tag{1}$$

where the graph $G$ consists of $N$ nodes, $K_i$ is a measure of the complexity of the individual node $i$, and $m_{ij}$ is the mutual information between the connection patterns of nodes $i$ and $j$. By construction, the summand vanishes when either $m_{ij} = 0$ (as in random graphs) or $m_{ij} = 1$ (when $i$ and $j$ are completely redundant). Complex networks reside between these two limiting cases.

Calculating $\Psi(G)$ for each of the small random graphs gives the distribution shown in Figure 1a. The very lowest and very highest values found were $\Psi(G_{\min}) = 0.120$ and $\Psi(G_{\max}) = 0.553$, respectively. The graphs corresponding to these extreme values are shown in Figure 1a.

(a) Network Complexity Measure

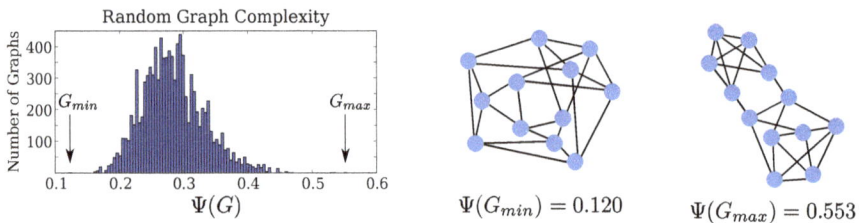

$\Psi(G_{min}) = 0.120$    $\Psi(G_{max}) = 0.553$

(b) Complexity Vulnerability Greedy Algorithm

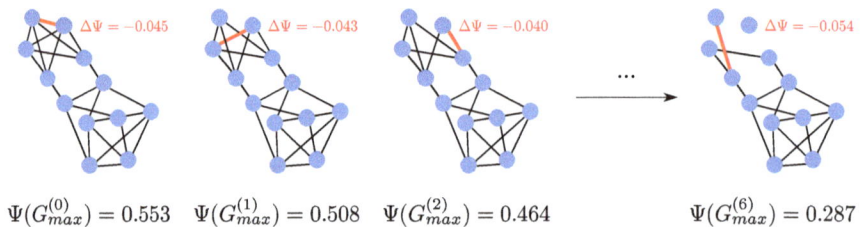

$\Psi(G_{max}^{(0)}) = 0.553$    $\Psi(G_{max}^{(1)}) = 0.508$    $\Psi(G_{max}^{(2)}) = 0.464$    $\Psi(G_{max}^{(6)}) = 0.287$

**Figure 1.** (**a**) To illustrate our method, we generated 10,000 random Erdős-Rényi networks, all with 12 nodes of four degrees. The distribution of our complexity measure $\Psi(G)$ is shown, along with the networks having the lowest and highest $\Psi(G)$ values; (**b**) $\Delta\Psi$ of an edge is the amount by which $\Psi(G)$ is reduced if that edge is removed. We progressively delete the edges with the highest $\Delta\Psi$ at each step. In this example, most of the complexity in the graph is contained within the upper pentagram-shaped connection structure; the elimination ordering reveals these "complexity-causing" structures.

In Figure 1b, we illustrate our greedy edge-elimination procedure, starting with the high-complexity graph $G_{\max}$. At each step, we individually delete every edge and calculate what the complexity of the graph would be without said edge. The edge causing the largest drop in $\Psi$ (i.e., the highest magnitude of $\Delta\Psi$) is then chosen for deletion, giving the partially-deleted graph $G_{\max}^{(1)}$. This procedure is then iterated, recalculating each edge's $\Delta\Psi$ within the partially deleted graph (i.e., $\Delta\Psi$ is re-calculated for all edges at each step).

It is interesting that the greedy procedure eliminates almost all of the edges from the pentagram-shaped structure before deleting any edges from the rest of the network. This suggests that the iterative elimination procedure may best reduce the network complexity by initially "attacking" a distinct structure within the network, rather than multiple, or more widely distributed structures. This suggests that we can usefully search for the property that unites the edges chosen for early elimination, as it may carry some structural significance. Specifically, we should consider any biophysiological significance associated with these edges; we suspect that the identified complex structures may be biophysically relevant, as complex network structures are capable of encoding biophysically important network dynamics [2,27].

### 2.2. Application to C. Elegans Data

Using the *C. elegans* connectome data from Varshney et al. [7], we then proceed to apply the same edge-eliminating procedure to the *C. elegans* gap junction network. We focus here on the subnetwork of 253 somatic neurons which have both synaptic and gap junction connections. For this analysis, we ignore the weighting of each connection (i.e., number of gap junctions and gap junction connection strengths), simply labeling each pair of neurons as connected or unconnected. Thus, we represent the network as a binary, undirected graph with 253 nodes, connected by individual 514 edges.

Calculating $\Psi$ for this network reveals that, by this measure, the *C. elegans* gap junction network is extraordinarily complex. We compare the actual gap junction network's complexity to the complexity of 10,000 randomly-rewired networks, all of which have the exact same degree distribution (see Methods Section 4.5 for more information). Given its degree distribution, a random graph would have a complexity within an approximately normal distribution with a mean and standard deviation of $\Psi(G_{rand}) = 0.001173 \pm 0.000015$. The neuron graph, however, has a complexity of $\Psi(G_{gap}) = 0.00143$, about 16.5 standard deviations above the random graph average. This distribution is plotted in comparison with the actual value in Figure 2.

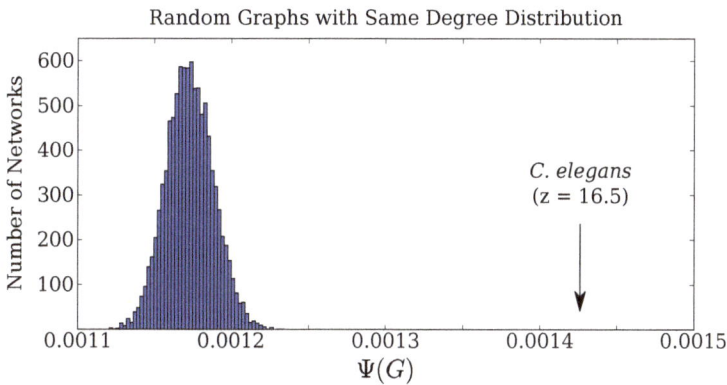

**Figure 2.** The *C. elegans* gap junction connectome has a complexity score of $\Psi(G_{gap}) = 0.00143$. For comparison, we generated 10,000 random networks with the same degree distribution and calculated $\Psi(G)$ for each. The distribution of $\Psi(G_{rand})$ is approximately normal, with an average score of $\Psi(G_{rand}) = 0.001173 \pm 0.000015$. Thus, the actual *C. elegans* gap junction network has a complexity 16.5 standard deviations above the mean value for its degree distribution.

Which edges are most responsible for this unusually high complexity? We use our edge deletion approach to explore this question. We begin by calculating $\Delta\Psi$ for every edge (i.e., the amount by which the network complexity would change were that edge deleted). The distribution of $\Delta\Psi$ values is shown in Figure 3. A few outliers are immediately apparent: the link between the command interneuron pair AVAL/AVAR, and the link between another command interneuron pair AVBL/AVBR. This suggests that the important biophysiological role of these neurons may be reflected in the relative complexity of their connection schemes. It is interesting that all edges to the left of the red dotted line implicate at least one interneuron.

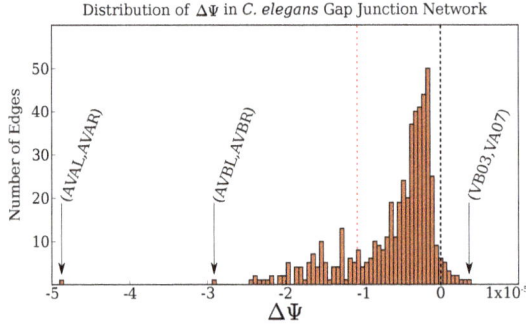

**Figure 3.** Distribution of $\Delta\Psi$ values for the intact *C. elegans* gap junction connectome. The link between the command interneuron pair AVAL/AVAR is a clear outlier, causing by far the largest drop in network complexity. It is notable that every edge below the red line at $\Delta\Psi = -1.083$ involves at least one interneuron. Another notable feature is that some deletions will actually increase the complexity: deleting the edge between the motor neurons VB03/VA07 causes $\Psi(G)$ to increase slightly.

We repeat the iterative, greedy edge-elimination procedure illustrated in Figure 1b, this time altering the *C. elegans* network. As it is intended to do, the procedure reduces the complexity of the network monotonically, eventually reducing the complexity to zero by deleting all edges. The complexity decay curve for the neuron graph is plotted in blue in Figure 4a. At each iteration, we generated 256 random graphs with the same modified degree distribution and plotted this distribution of random graph complexities in red.

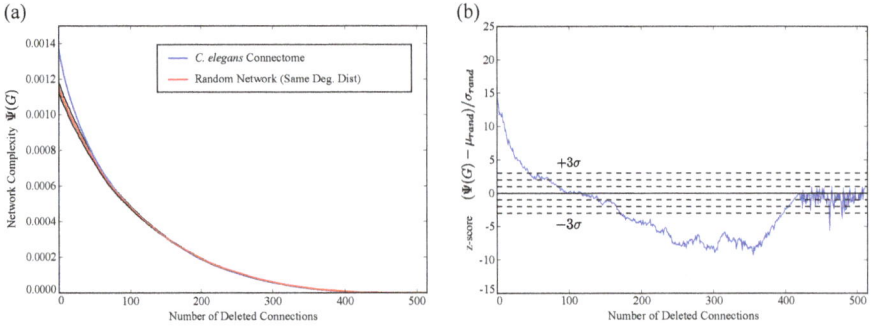

**Figure 4.** (**a**) We iteratively delete connections with the highest $\Delta\Psi$, causing the complexity to decay as shown by the blue curve. At each point, we calculate $\Psi(G_{rand})$ for 256 random graphs with the same degree distribution. The red line shows the mean $\Psi(G_{rand})$, with the red band showing the range $\pm 2\sigma$; (**b**) the same data converted to z-score (i.e., the number of standard deviations by which the actual network differs from random). The *C. elegans* gap junction network is initially much more complex than randomly expected, but as we successively delete edges, it reveals an underlying network that is much *less* complex than random.

This comparison is made clearer by converting the same data to z-scores (i.e., the number of standard deviations by which the *C. elegans* curve is away from the mean). This z-score curve is plotted in Figure 4b. As edges are progressively deleted, the complexity decays towards the randomly expected value, lying near the middle of the random range after about 100 deletions (about 25% of the edges). As edge deletion continues, however, the graph ultimately becomes much less complex than what one expects for a random graph. This suggests that the the graph consists of a low complexity component, connected with a smaller component connected to it that is responsible for the resulting excess complexity of the entire structure.

## 2.3. Complex Structure and the Rich Club

The discovery of a minority of edges that are particularly important to the complex structure of the network is reminiscent of the graph's "rich club" property. Rich clubs within a graph are sets of high-degree nodes (the so-called "rich" nodes) which are highly interconnected. A rich node is more likely to connect with other rich nodes than one would expect at random, forming a highly-interconnected hub within the network that is important to the network's function. This rich club structure appears in many complex networks [28] and helps to achieve a topologically desirable network (e.g., with high efficiency and centrality) for a relatively small wiring cost [16]. In neuroscience, a rich club architecture appears to be present in systems ranging from *C. elegans* to humans [17–24].

The *C. elegans* synaptic network was shown by Towlson et al. [24] to possess a rich club structure. We consider the set of neurons which they identified as the "rich club" due to their associated biological significance, though it should be noted that the gap junction network does not itself have the same rich club structure. Using the terminology of Towlson et al. [24], each edge in the graph can be classified as a "club" edge (if it connects two rich nodes), a "feeder" edge (if it connects a rich, high degree node to a poor node), or a "local" edge (if it connects two poor nodes). This is illustrated in Figure 5a. In Figure 5b, we label each edge in our deletion order according to these classes. This shows that the iterative procedure disproportionately targets club and feeder edges: out of the first 100 edge deletions, 19% are local, 74% are feeder, and 7% are club. By comparison, the intact network edges are about 69% local, 29% feeder, and 2% club. A $\chi^2$-test shows that the probability of randomly drawing as many feeder and club edges out of the network is less than $p < 10^{-25}$; thus, the edge selection process heavily targets the rich club. This disproportionate targeting can be seen by looking at the percentage of each class that has been deleted at each iteration, as shown in Figure 5c.

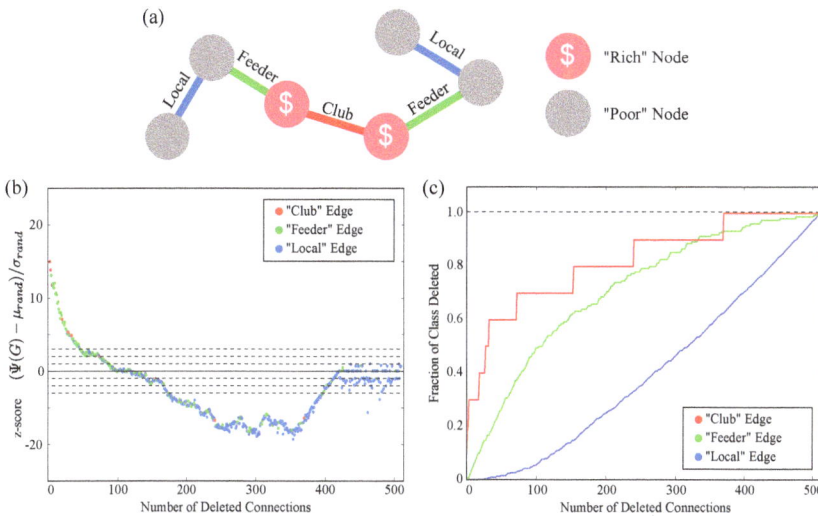

**Figure 5.** (**a**) As in [24], the *C. elegans* synaptic Connectome can be understood to have a "rich club" structure. Edges are classified as "Club" (if between two rich nodes), "Feeder" (if between a rich and poor node), or "Local" (if between two poor nodes); (**b**) the same curve as Figure 4b, with each deleted edge labeled by class. In the region where the graph is much more complex than average, the procedure disproportionately targets Club and Feeder edges; (**c**) the fraction of each class which has been deleted at each iteration.

## 2.4. Complex Structure beyond Degree

The rich club is important to several complex topological properties of the network, but a node's membership in the rich club is ultimately determined by its degree. The "rich club" nodes we consider are defined from their synaptic connectivity, and we are considering the gap junction network, but there is a positive correlation between a node's gap junction degree and synaptic degree. It is therefore possible that this complexity-reduction procedure could yield a similar result (i.e., apparent targeting of the rich club nodes) if it were simply reducing the degrees of the highest-degree nodes. Figure 6a plots the initial degree distribution of the graph and shows that the first implicated edge $(i_0, j_0)$ is indeed between the first and second mostly highly-connected nodes. Could such a simple criterion explain these results, or is the procedure indeed targeting a more sophisticated structure?

Referring to Figure 1b, we suspect that the edge elimination order is not so trivial. After all, the procedure continued to target the pentagram-shaped structure even after those nodes had been trimmed to a very low degree. Indeed, we see similar results in the *C. elegans* network: Figure 6b shows how highly ranking the degree of the targeted nodes are at each step of the elimination procedure (i.e., whether or not the edge includes the 1st most highly connected node, the 2nd most highly connected node, and so on). Note that, at any given step, there are no more than 18 unique degrees (as there are many nodes sharing the same number of connections). For example, the second edge to be deleted is the edge between AVBR and AVBL, which have degrees 29 and 24, respectively. These are the 3rd and 4th highest degrees in the network at this point of the procedure, and so a blue marker is plotted at a node degree rank of 3. As can be seen in Figure 6b, many of the subsequent edge deletions do not implicate the most highly connected nodes. The highest-complexity edges are not simply determined by the degree of the implicated nodes.

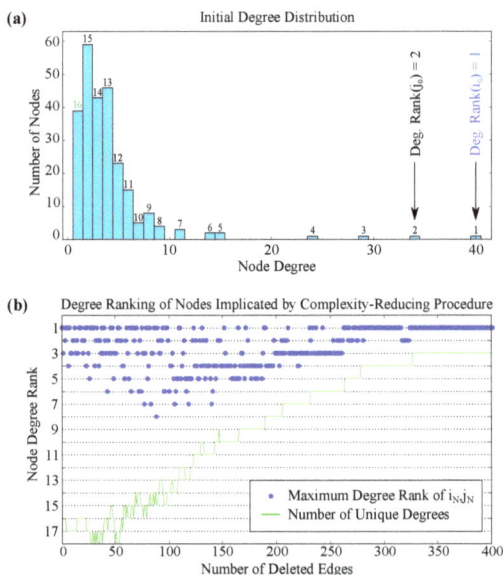

**Figure 6.** (**a**) the initial degree distribution of the graph. Each node has one of 16 unique degree values, which we label by their relative rank from highest to lowest. The first implicated edge $(i_0, j_0)$ connects nodes with degree ranks 1 and 2, such that its "Maximum Degree Rank" is 1; (**b**) the maximum degree rank of the subsequently targeted edges, plotted in blue. The green line indicates the number of unique degrees, which changes as the degree distribution is altered. The procedure *does* tend to trim edges to relatively highly connected nodes, but this relationship is not the driving criterion, and it does not simply choose edges based upon connectivity level alone.

## 2.5. Robustness to Specific Elimination Order

How robust are these results to the specific order in which edges are eliminated? To address this, we conduct 50 trials in which we calculate the decay of $\Psi(G)$ when we delete all club edges in a random order, then delete all feeder edges in a random order, and then delete all local edges in a random order. The resulting distribution of complexity curves is shown in Figure 7. The random Club/Feeder/Local edge deletion orderings result in a smaller decrease in complexity than the algorithmically-determined ordering, which is unsurprising given the "greedy" nature of our edge-elimination procedure. However, the result is much more drastic when we conduct 50 trials in which we first delete the local edges, then the feeder edges, and then the club edges. This suggests that it is indeed appropriate to think of the network's high complexity as being disproportionately caused by the club and feeder edges.

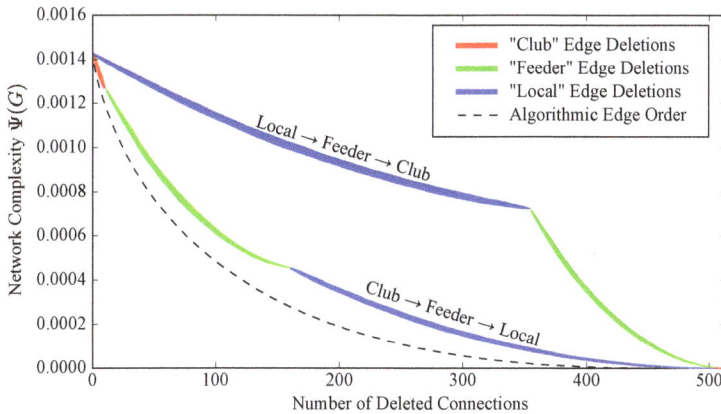

**Figure 7.** In 50 trials, we deleted all club edges in a random order, then all feeder edges, and then all local edges. The red/green/blue bands show the average resulting complexity curve within one standard deviation. This was repeated for 50 trials in which we instead deleted local edges, then feeder edges, and then club edges. The resulting distribution of complexity curves is indicated by the blue/green/red bands. The edge deletion order prescribed by our algorithm (i.e., the blue curve in Figure 4a) is shown by the black dotted line. A random Club/Feeder/Local deletion order results in a significantly slower complexity decay than our algorithm prescribes, but leads to a much larger decrease in complexity than the Local/Feeder/Club ordering. Thus, the results implicating the rich club are robust to the specific edge deletion order.

## 2.6. Robustness to Initial Edge Choice

We may similarly investigate the effect of choosing a different edge at the first deletion step. Consider again the example in Figure 1b, in which the iterative procedure initially targets nearly all of the edges within the upper pentagram-shaped structure. We interpreted this to mean that the pentagram-shaped structure was primarily responsible for the high complexity of the network. However, one could posit that the lower portion of the graph was similarly complex and that initially deleting an edge in the upper portion simply biased the iterative procedure towards targeting that portion of the graph. Does the initial edge deletion bias the procedure in this manner?

We directly investigated this possibility by repeating the analysis of Figure 5, but with an arbitrary edge chosen at the first deletion step (followed by the greedy iterative deletion procedure). This was repeated for all possible choices of initial edge deletions (i.e., for all 514 edges). When AVAL/AVAR is chosen initially (as in the full greedy procedure), the results are, of course, identical to the previous results. For all of the other 513 initializations, however, the greedy procedure chooses AVAL/AVAR as the very next edge to be deleted. Furthermore, the procedure continues to target club/feeder edges in largely the same manner. Figure 8 shows the order in which edges of each class are deleted for all possible initial edge choices. Despite minor variations in edge ordering, the order in which edges are targeted by the greedy procedure appears to be largely robust to the initial edge choice, and the disproportionate targeting of club/feeder edges does not depend upon this initialization.

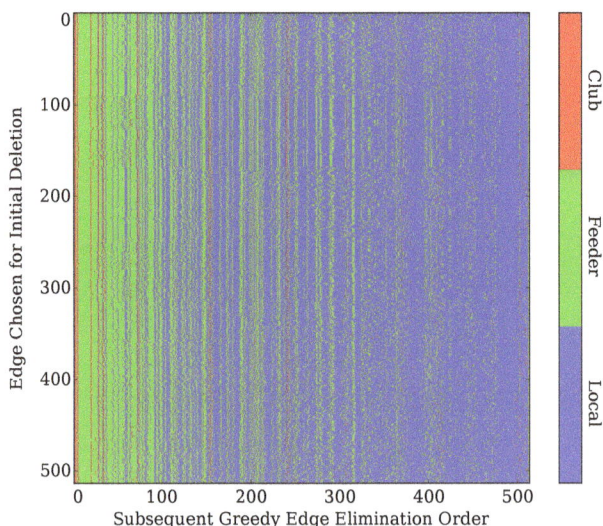

**Figure 8.** The greedy edge-elimination procedure was repeated for all possible choices of initial edge deletions (i.e., the edge chosen for deletion at the first step). Each row corresponds to a different choice of initial edge, with each column showing the class of the edges subsequently deleted by the iterative procedure. Club and Feeder edges are targeted disproportionately regardless of the initial edge choice.

## 3. Discussion

This study shows that the *C. elegans* neuron gap–junction network is vastly more complex than one would expect at random given its degree distribution. Expanding the concept of vulnerability analysis, we analyzed the complexity vulnerability of the graph and used an iterative procedure to successively eliminate edges based upon how much each deletion reduced graph complexity. After many edge deletions, we were left with a graph much *less* complex than an equivalent random graph. This suggests that the majority of the graph consists of a low complexity component that can be described fairly simply, and a subset of edges, which is responsible for the excess complexity of the network. This complexity-causing set of edges belongs disproportionately to the neurons which are members of the associated synaptic network's previously-identified rich club.

The fact that the procedure reduces the graph to one with a remarkably *low* complexity raises interesting questions concerning the topology of graphs. Many graphs, particularly in biology and neuroscience and including the *C. elegans* neuronal network, are rather well-described by a collection of over-represented motifs [7,29–31]. A common view of neuronal networks, as consisting of a highly structured graph of repeated motifs with the repetitive structure violated by important hub neurons, is interesting to compare with the results in Figure 4. Is the iterative procedure simply stripping away these hub structures to leave a simple, highly-repetitive core? Given that this complexity measure disappears when the mutual information between nodes approaches 1, it is plausible that particularly motif-heavy graphs, consisting of sets of nodes having similar connectivity patterns/high mutual information, may be classified as having a particularly low complexity by this measure. The exact relationship between motif structures, symmetry and symmetry-breaking and their effects upon network complexity measures deserves a deeper theoretical investigation, which will be the subject of future work.

These results also suggest that future work should investigate the broader range of structural properties to which this complexity measure is sensitive. It is important to note that this procedure does *not* simply identify rich-club structure; recall that we consider the "rich club" set of neurons due to its previously-identified importance, but this set is defined from the *synaptic* network, not the gap junction network, which lacks such structure. What functionally relevant structures does this approach reveal, and what are the associated topological properties? For example, one could consider the change in average betweenness centrality as nodes are iteratively deleted. Initially, the network has an average betweenness which is a factor of about 1.29 higher than that of a random graph (with the same degree distribution). After 40 deletions, this factor is reduced to 1.22. This small reduction does not appear to explain the large reduction in relative complexity over the same range (as seen in Figure 4). Other preliminary work has been similarly inconclusive, suggesting that the relationship between complexity and other topological measures is nontrivial. The more general relationship between network complexity and various other global topological properties of the network is beyond the scope of this paper, but the subject of future work.

How would the identified structures differ if one uses a different quantification of network complexity? There is no universally recognized quantification of network complexity, and many such measures exist [3,4,32,33]. Furthermore, the $\Psi$ measure is inherently dependent only on pairwise measures of connectivity in the sense that it is a sum over pairs of mutual information between nodes based on their connectivity. Measures that include higher numbers of nodes and their informational interdependence might reveal other features [23,32,33]. It should be possible to perform a similar deconstruction procedure using other measures, which may distinguish different features and could prove useful both as a tool for the exploration of the structure of networks and for illuminating the differences between the measures themselves.

Further development of network complexity quantification and exploration of different measures could incorporate information which we have ignored thus far in our analysis. Since we used an undirected network complexity measure, we focused upon the undirected gap junction network. However, the directed synaptic network is also vitally important, and the method can be modified to incorporate directed connections. Similarly, we consider the unweighted binary graph, ignoring edge-weighting data such as the number of gap junctions between each node. Future analysis could make use of such information, perhaps by using previously-developed extensions of this complexity measure to include multiple edge types [6].

In spite of these current limitations, this study strongly suggests that quantifying a network's global complexity, performing vulnerability analysis using this complexity measure, and then iteratively eliminating edges based upon their vulnerability at each step is a useful direction for identifying physiologically important structures within complex graphs. Such methods are increasingly important as increasingly large and difficult-to-comprehend neuronal networks are measured by the burgeoning field of connectomics [34–36]. In a real network, the interaction between the structural features, the behavior and flexibility of the network function, the vulnerability to damage and the costs of wiring the networks are fundamental. The trade-offs can only be fully understood by careful quantitative analysis of the kind begun here. Given the potentially scale-invariant nature of neuroscience networks [24], the architectural principles we infer from these complex networks may yield broader design insights into the function and structure of nervous system networks.

## 4. Materials and Methods

### 4.1. Network Complexity $\Psi(G)$

The graph complexity measure $\Psi(G)$ is an information theoretic measure with a basis in the Kolmogorov complexity and was introduced and explored for undirected binary graphs by Sakhanenko and Galas [5]. Consider an undirected binary graph with $N$ nodes, characterized by a symmetric adjacency matrix $A = \{a_{ij}\}$, where $a_{ij} = 1$ if nodes $i$ and $j$ are connected, and $a_{ij} = 0$ if they are not. The measure is based on a Shannon-like description of information, such that the fundamental properties of a node are its connection probabilities. We denote the connection probability of node $i$ by:

$$p_i(1) = \frac{1}{N-1} \sum_{j \neq i} a_{ij}. \tag{2}$$

Similarly, the disconnection probability is given by:

$$p_i(0) = \frac{1}{N-1} \sum_{j \neq i} (1 - a_{ij}). \tag{3}$$

The complexity of an individual node is then given by:

$$K_i = \sum_{a=0}^{1} p_i(a) \log_2(p_i(a)). \tag{4}$$

We can then calculate the mutual information between two nodes $i$ an $j$:

$$m_{ij} = \sum_{a=0}^{1} \sum_{b=0}^{1} p_{ij}(a,b) \log_2 \left( \frac{p_{ij}(a,b)}{p_i(a) p_j(b)} \right). \tag{5}$$

The graph complexity $\Psi$ can then be calculated from the mutual informations and individual node complexities:

$$\Psi = \frac{1}{N(N-1)} \sum_i \sum_{j \neq i} \max(K_i, K_j) m_{ij} (1 - m_{ij}). \tag{6}$$

Note that the summands will disappear when either $m_{ij} = 0$ or $m_{ij} = 1$. The former case is when the connectivity patterns of the two nodes contain no information about the other, as is the case of infinite random graphs. The latter case occurs when two nodes carry perfect information about the other, as is the case with completely connected or completely disconnected graphs. Thus, the graph complexity will disappear in both of these limiting cases.

### 4.2. $\Delta\Psi(G)$ and Iterative Edge Removal

Our greedy algorithm for complexity vulnerability analysis consists of iteratively deleting the graph edges whose deletions cause the largest reduction in graph complexity. That is, for an initial adjacency matrix $A^{(0)}$, we consider all one-edge deletions and recalculate the resulting $\Psi$, finding the edge $(i_1, j_1)$ which causes the largest reduction in $\Psi$ when deleted. We refer to the new adjacency matrix, with the single edge deletion, as $A^{(1)}$. We then repeat this process with $A^{(1)}$ to identify $(i_2, j_2)$. This is repeated until all $N$ edges are deleted, yielding an edge deletion order $e = \{(i_1, j_1), (i_2, j_2), (i_3, j_3), ..., (i_N, j_N)\}$.

For notational purposes, we define a matrix $\Delta(i, j)$ which is equal to one at entry $(i, j)$ and is zero for all other entries. That is,

$$\Delta(i', j')_{ij} = \begin{cases} 1, & \text{if } (i, j) = (i', j'), \\ 0, & \text{otherwise.} \end{cases} \tag{7}$$

This allows us to write the adjacency matrix with a single edge deletion at $(i', j')$ as $A - \Delta(i', j')$. We select the first edge to delete by calculating:

$$(i_0, j_0) = \underset{i,j}{\operatorname{argmin}} [\Psi(A^{(0)} - \Delta(i, j))]. \tag{8}$$

That is, we choose the indices of the edge deletion resulting in the graph with the least complexity. We can write the resulting adjacency matrix as:

$$A^{(1)} = A^{(0)} - \Delta(i_0, j_0). \tag{9}$$

We write the corresponding change in complexity as:

$$\Delta\Psi^{(1)} = \Psi(A^{(1)}) - \Psi(A^{(0)}). \tag{10}$$

This process is repeated iteratively until all edges are deleted. At the $n$th edge deletion we may write:

$$(i_n, j_n) = \operatorname*{argmin}_{i,j}[\Psi(A^{(n)} - \Delta(i, j))], \tag{11}$$

$$A^{(n+1)} = A^{(n)} - \Delta(i_n, j_n). \tag{12}$$

### 4.3. Illustrative Example: 12 Nodes of Four Degrees

Using the Python package *graph-tool* [37], we generated 10,000 random Erdős–Rényi graphs, all consisting of 12 nodes which all have degree four. We calculated $\Psi$ for all of these 10,000 graphs, giving the distribution seen in Figure 1a. We then applied the iterative edge-removal procedure from Section 4.2 to the random graph which had the highest value of $\Psi$. This gave the edge removal ordering as partially visualized in Figure 1b.

### 4.4. The C. Elegans Connectome

The *C. elegans* neuronal network consists of 302 neurons connected via both a directed synaptic network (with 6393 synapses) and undirected gap junction network (with 890 gap junctions) [7]. The bulk of these neurons, 282 out of 302, belong to the somatic nervous system. We use the connectivity data for the giant component of the somatic nervous system, consisting of 279 neurons, as provided by Varshney et al. [7], who consolidated and updated earlier connectome measurements [38–40].

Since this work focuses on the structure of undirected graphs, our analysis is of the gap junction network. All of the 279 neurons within the network are connected synaptically, but many lack gap junctions entirely. We eliminate the neurons that have no gap junction connections, leaving the 253 somatic neurons with both synaptic and gap junction connections. Many nodes share multiple connections, but we simplify our analysis by considering the unweighted network: $a_{ij} = 1$ if nodes $i$ and $j$ have one or more gap junction connections, or $a_{ij} = 0$ if they have no gap junction connections. Thus, our adjacency matrix $A^{(0)}$ consists of 253 nodes connected by 514 binary edges.

### 4.5. Comparison of C. elegans to Random Connectivity

To compare the complexity of the actual Connectome against what we might expect at random, we use the *random_rewire* function included within the Python package *graph-tool* [37], selecting the "uncorrelated" rewiring model. This procedure is described in the *graph-tool* documentation and can be summarized as follows: for each edge $(i, j)$, the algorithm randomly selects a second edge $(i', j')$. It then attempts to swap the target of each edge, such that the edges would then become $(i, j')$ and $(i', j)$. This swap is rejected if it would result in parallel edges or self-loops. This swapping procedure is repeated for all edges $(i, j)$ within the graph. This rewires a graph to randomize connections while preserving the exact degree sequence.

For the distribution of Figure 2, we generated 10,000 randomly-rewired graphs with the same degree distribution as the *C. elegans* gap junction network, yielding the displayed distribution with a mean and standard deviation of $\Psi(G_{rand}) = 0.001173 \pm 0.000015$. As we iteratively eliminated edges, we wished to continue comparing the reduced complexity against what we would expect at random.

*Entropy* **2017**, *19*, 104

At each step $n$, we therefore randomly rewired the partially deleted graph $A^{(n)}$ to generate 256 graphs with the same (partially deleted) degree distribution.

*4.6. C. elegans Rich Club*

The "rich club" is defined by the rich club coefficient $\Phi(k)$, which is defined by the connection probability between nodes of degree greater than $k$. For some degree $k$, define $N_{>k}$ as the number of nodes with degree greater than $k$, and $M_{>k}$ as the number of connections between said nodes. The rich club coefficient is then just the ratio between the actual number of connections $M_{>k}$ and the number of possible connections[24,28,41]:

$$\Phi(k) = \frac{2M_{>k}}{N_{>k}(N_{>k} - 1)}. \tag{13}$$

Towlson et al. [24] found that the *C. elegans* Connectome has a significant rich club coefficient for degrees $35 \leq k \leq 73$, which implicates the following 11 nodes as belonging to the rich club: AVAL/R, AVBL/R, AVDL/R, AVEL/R, PVCL/R, and DVA. It is important to note that this is the set of neurons which we use as the "rich club" of the network, and we do *not* do any re-calculation of the rich club based on our particular graph; Towlson calculates the rich club set from a binary form of the synaptic network, whereas we consider a binary form of the gap junction network. We refer to this set of neurons due to its known biological significance.

**Acknowledgments:** This work was supported in part by the NIH Common Fund, the Extracellular RNA Communication Consortium (ERCC) 1U01HL126496-01, the Bill and Melinda Gates Foundation, and the Pacific Northwest Research Institute.

**Author Contributions:** All authors conceived of and planned the research. James M. Kunert-Graf carried out the analysis with advice and discussions with Nikita A. Sakhanenko and David J. Galas. James M. Kunert-Graf wrote the initial draft, and Nikita A. Sakhanenko and David J. Galas edited and modified the manuscript with James M. Kunert-Graf.

**Conflicts of Interest:** The authors declare no conflict of interest.

## References

1. Sporns, O. The Non-Random Brain: Efficiency, Economy, and Complex Dynamics. *Front. Comput. Neurosci.* **2011**, *5*, doi:10.3389/fncom.2011.00005.
2. Gollo, L.L.; Zalesky, A.; Hutchison, R.M.; van den Heuvel, M.; Breakspear, M. Dwelling quietly in the rich club: Brain network determinants of slow cortical fluctuations. *Philos. Trans. R. Soc. Lond. B Biol. Sci.* **2015**, *370*, doi:10.1098/rstb.2014.0165.
3. Dehmer, M.; Barbarini, N.; Varmuza, K.; Graber, A. A Large Scale Analysis of Information-Theoretic Network Complexity Measures Using Chemical Structures. *PLoS ONE* **2009**, *4*, e8057.
4. Emmert-Streib, F.; Dehmer, M. Exploring Statistical and Population Aspects of Network Complexity. *PLoS ONE* **2012**, *7*, e34523.
5. Sakhanenko, N.A.; Galas, D.J. Complexity of Networks I: The SetComplexity of Binary Graphs. *Complexity* **2011**, *17*, 51–64.
6. Ignac, T.M.; Sakhanenko, N.A.; Galas, D.J. Complexity of Networks II: The Set Complexity of Edge-colored Graphs. *Complexity* **2012**, *17*, 23–36.
7. Varshney, L.R.; Chen, B.L.; Paniagua, E.; Hall, D.H.; Chklovskii, D.B. Structural Properties of the Caenorhabditis elegans Neuronal Network. *PLoS Comput. Biol.* **2011**, *7*, e1001066.
8. Chalfie, M.; Sulston, J.E.; White, J.G.; Southgate, E.; Thomson, J.N.; Brenner, S. The neural circuit for touch sensitivity in Caenorhabditis elegans. *J. Neurosci.* **1985**, *5*, 956–964.
9. Sawin, E. Genetic and Cellular Analysis of Modulated Behaviors in *Caenorhabditis elegans*. Ph.D. Thesis, Massachusetts Institute of Technology, Cambridge, MA , USA, 1996.
10. Sawin, E.R.; Ranganathan, R.; Horvitz, H.R. *C. elegans* locomotory rate is modulated by the environment through a dopaminergic pathway and by experience through a serotonergic pathway. *Neuron* **2000**, *26*, 619–631.
11. Gray, J.M.; Hill, J.J.; Bargmann, C.I. A circuit for navigation in *Caenorhabditis elegans*. *Proc. Natl. Acad. Sci. USA* **2005**, *102*, 3184–3191.

12. Liu, K.S.; Sternberg, P.W. Sensory regulation of male mating behavior in *Caenorhabditis elegans*. *Neuron* **1995**, *14*, 79–89.

13. Macosko, E.Z.; Pokala, N.; Feinberg, E.H.; Chalasani, S.H.; Butcher, R.A.; Clardy, J.; Bargmann, C.I. A hub-and-spoke circuit drives pheromone attraction and social behaviour in *C. elegans*. *Nature* **2009**, *458*, 1171–1175.

14. Bany, I.A.; Dong, M.Q.; Koelle, M.R. Genetic and cellular basis for acetylcholine inhibition of *Caenorhabditis elegans* egg-laying behavior. *J. Neurosci.* **2003**, *23*, 8060–8069.

15. Hardaker, L.A.; Singer, E.; Kerr, R.; Zhou, G.; Schafer, W.R. Serotonin modulates locomotory behavior and coordinates egg-laying and movement in *Caenorhabditis elegans*. *J. Neurobiol.* **2001**, *49*, 303–313.

16. Bassett, D.S.; Greenfield, D.L.; Meyer-Lindenberg, A.; Weinberger, D.R.; Moore, S.W.; Bullmore, E.T. Efficient physical embedding of topologically complex information processing networks in brains and computer circuits. *PLoS Comput. Biol.* **2010**, *6*, e1000748.

17. Van den Heuvel, M.P.; Sporns, O. Rich-Club Organization of the Human Connectome. *J. Neurosci.* **2011**, *31*, 15775–15786.

18. Bullmore, E.; Sporns, O. The economy of brain network organization. *Nat. Rev. Neurosci.* **2012**, *13*, 336–349.

19. Harriger, L.; van den Heuvel, M.P.; Sporns, O. Rich club organization of macaque cerebral cortex and its role in network communication. *PLoS ONE* **2012**, *7*, e46497.

20. Ball, G.; Aljabar, P.; Zebari, S.; Tusor, N.; Arichi, T.; Merchant, N.; Robinson, E.C.; Ogundipe, E.; Rueckert, D.; Edwards, A.D.; et al. Rich-club organization of the newborn human brain. *Proc. Natl. Acad. Sci. USA* **2014**, *111*, 7456–7461.

21. Schroeter, M.S.; Charlesworth, P.; Kitzbichler, M.G.; Paulsen, O.; Bullmore, E.T. Emergence of rich-club topology and coordinated dynamics in development of hippocampal functional networks in vitro. *J. Neurosci.* **2015**, *35*, 5459–5470.

22. Nigam, S.; Shimono, M.; Ito, S.; Yeh, F.C.; Timme, N.; Myroshnychenko, M.; Lapish, C.C.; Tosi, Z.; Hottowy, P.; Smith, W.C.; et al. Rich-club organization in effective connectivity among cortical neurons. *J. Neurosci.* **2016**, *36*, 670–684.

23. Pedersen, M.; Omidvarnia, A. Further Insight into the Brain's Rich-Club Architecture. *J. Neurosci.* **2016**, *36*, 5675–5676.

24. Towlson, E.K.; Vértes, P.E.; Ahnert, S.E.; Schafer, W.R.; Bullmore, E.T. The Rich Club of the *C. elegans* Neuronal Connectome. *J. Neurosci.* **2013**, *33*, 6380–6387.

25. Kunert, J.; Shlizerman, E.; Kutz, J.N. Low-dimensional functionality of complex network dynamics: Neurosensory integration in the *Caenorhabditis elegans* connectome. *Phys. Rev. E* **2014**, *89*, 052805.

26. Kim, S.; Kim, H.; Kralik, J.D.; Jeong, J. Vulnerability-Based Critical Neurons, Synapses, and Pathways in the *Caenorhabditis elegans* Connectome. *PLoS Comput. Biol.* **2016**, *12*, e1005084.

27. Hu, Y.; Brunton, S.L.; Cain, N.; Mihalas, S.; Kutz, J.N.; Shea-Brown, E. Feedback through graph motifs relates structure and function in complex networks. *arXiv* **2016**, arXiv:1605.09073.

28. Colizza, F.; Flammini, A.; Serrano, M.; Vespignani, A. Detecting rich-club ordering in complex networks. *Nat. Phys.* **2006**, *2*, 110–115.

29. Milo, R.; Shen-Orr, S.; Itzkovitz, S.; Kashtan, N.; Chklovskii, D.; Alon, U. Network Motifs: Simple Building Blocks of Complex Networks. *Science* **2002**, *298*, 824–827.

30. Sporns, O.; Kötter, R. Motifs in Brain Networks. *PLoS Biol.* **2004**, *2*, e369.

31. Qian, J.; Hintze, A.; Adami, C. Colored Motifs Reveal Computational Building Blocks in the *C. elegans* Brain. *PLoS ONE* **2011**, *6*, e17013.

32. Galas, D.J.; Sakhanenko, N.A.; Skupin, A.; Ignac, T. Describing the Complexity of Systems: Multivariable "Set Complexity" and the Information Basis of Systems Biology. *J. Comput. Biol.* **2014**, *21*, 118–140.

33. Sakhanenko, N.; Galas, D. Biological Data Analysis as an Information Theory Problem: Multivariable Dependence Measures and the Shadows Algorithm. *J. Comput. Biol.* **2015**, *22*, 1005–1024.

34. Bohland, J.W.; Wu, C.; Barbas, H.; Bokil, H.; Bota, M.; Breiter, H.C.; Cline, H.T.; Doyle, J.C.; Freed, P.J.; Greenspan, R.J.; et al. A Proposal for a Coordinated Effort for the Determination of Brainwide Neuroanatomical Connectivity in Model Organisms at a Mesoscopic Scale. *PLoS Comput. Biol.* **2009**, *5*, e1000334.

35. Chiang, A.S.; Lin, C.Y.; Chuang, C.C.; Chang, H.M.; Hsieh, C.H.; Yeh, C.W.; Shih, C.T.; Wu, J.J.; Wang, G.T.; Chen, Y.C.; et al. Three-Dimensional Reconstruction of Brain-wide Wiring Networks in *Drosophila* at Single-Cell Resolution. *Curr. Biol.* **2011**, *21*, 1–11.

36. Oh, S.; Harris, J.; Ng, L.; Winslow, B.; Cain, N.; Mihalas, S.; Wang, Q.; Lau, C.; Kuan, L.; Henry, A.; et al. A mesoscale connectome of the mouse brain. *Nature* **2014**, *508*, 207–214.

37. Peixoto, T.P. The graph-tool python library. *Figshare* **2014**, doi:10.6084/m9.figshare.1164194.

38. White, J.G.; Southgate, E.; Thomson, J.N.; Brenner, S. The structure of the nervous system of the nematode Caenorhabditis elegans. *Philos. Trans. R. Soc. Lond. B Biol. Sci.* **1986**, *314*, doi:10.1098/rstb.1986.0056.

39. Hall, D.H.; Russell, R.L. The posterior nervous system of the nematode Caenorhabditis elegans: Serial reconstruction of identified neurons and complete pattern of synaptic interactions. *J. Neurosci.* **1991**, *11*, 1–22.

40. Durbin, R. Studies on the Development and Organisation of the Nervous System of *Caenorhabditis elegans*. Ph.D. Thesis, University of Cambridge, Cambridge, UK, 1987.

41. Zhou, S.; Mondragon, R. The rich-club phenomenon in the internet topology. *IEEE Commun. Lett.* **2004**, *8*, 180–182.

*entropy*

MDPI

*Article*

# Cockroach Swarm Optimization Algorithm for Travel Planning

**Joanna Kwiecień [1],\*** and **Marek Pasieka [2]**

[1]   Faculty of Electrical Engineering, Automatics, Computer Science and Biomedical Engineering,
     AGH University of Science and Technology, Al. Mickiewicza 30, 30-059 Kraków, Poland
[2]   Swisscom (Schweiz) AG, Waldeggstrasse 51, 3097 Liebefeld, Switzerland; marek.pasieka@gmail.com
\*   Correspondence: kwiecien@agh.edu.pl; Tel.: +48-12-617-4320

Academic Editor: Mikhail Prokopenko
Received: 27 February 2017; Accepted: 3 May 2017; Published: 6 May 2017

**Abstract:** In transport planning, one should allow passengers to travel through the complicated transportation scheme with efficient use of different modes of transport. In this paper, we propose the use of a cockroach swarm optimization algorithm for determining paths with the shortest travel time. In our approach, this algorithm has been modified to work with the time-expanded model. Therefore, we present how the algorithm has to be adapted to this model, including correctly creating solutions and defining steps and movement in the search space. By introducing the proposed modifications, we are able to solve journey planning. The results have shown that the performance of our approach, in terms of converging to the best solutions, is satisfactory. Moreover, we have compared our results with Dijkstra's algorithm and a particle swarm optimization algorithm.

**Keywords:** cockroach swarm optimization; swarm intelligence; journey planning; public transport; optimization of travel time

## 1. Introduction

Intensive studies on journey planning problems produced several models and many algorithms over the last few decades. The popularity of automated planning systems have motivated researchers to search for methods that are sufficient for practical applications and meet travellers' expectations. There was considerable progress in the performance methods for journey planning in public transit networks in recent years. Upon consideration of public transport timetable models in respect of how they provide the best possible routes, we can divide them into graph-based models, representing the timetable as a graph, and array-based models, using an array for the given timetable. Among the models belonging to the first type, well-known examples include the time-expanded model [1–3] and the time-dependent model [3,4]. In this paper, we focus on the time-expanded model, which is based on the concept of the shortest path problem and is still used in many practical applications.

It should be mentioned that the performance of the methods for solving the journey planning problems is receiving attention in various papers and depends on the complexity of the problems. Various nature-inspired metaheuristics based on the existing mechanisms of a biological phenomenon have been widely used to solve many optimization problems with success. The behavior of social insects and animals, including foraging, nest building, hunting, and cooperative transport has become a fascinating topic in the various problem-solving tasks in the last few years. Some of the mechanisms underlying the collective activities show that complex group behavior may emerge from many relatively simple interacting individuals. The growing interest of many researchers in the emergent collective intelligence of insects, birds, and mammals led to the design of a special group of algorithms, known as swarm intelligence, belonging to computational intelligence. Although the algorithms do not ensure obtaining optimum solutions, they achieve good results in a reasonable computation time. For a survey on numerous examples of these algorithms and their applications, one can refer to [5,6].

The cockroach swarm optimization (CSO) algorithm is one of the new methods belonging to the aforementioned algorithms. The algorithm is modeled after the habits of cockroaches looking for food [7–9]. It can be used to solve various problems, for example the flow shop scheduling [10] or the traveling salesman problem [9,11,12]. That algorithm is fairly simple to adapt, although it requires adjustment to the problems being solved by introduction of certain movement modifications. According to our knowledge, the CSO algorithm has never been used in solving travel planning, involving several modes of transport. However, other applications of that algorithm to generally known combinatorial problems, for example solving scheduling problems, indicate the algorithm's potential for solving real problems.

The purpose of this paper is not to provide a new model, but rather to demonstrate that the swarm-based approach for solving a journey planning problem is possible through the design of case studies, the characteristic properties of solutions and the type of movement mechanisms implemented. Therefore, we assumed that the CSO algorithm can be applied to the travel planning problem after its modification, concerning generation of an initial population, movement performance in particular procedures, and proper definition of specific parameters. For that reason, the algorithm had to be implemented and tested in several generated test instances. We should emphasize that we concentrated on the practical application of the CSO algorithm in the travel planning problem, with the use of the time-expanded model. We show that appropriate modifications are needed to ensure the admissibility of solutions and the convergence of the CSO algorithm. In addition, we implemented the particle swarm optimization (PSO) and Dijkstra's algorithms to be able to compare our results and evaluate the quality of our CSO approach.

The rest of the paper is organized as follows: Section 2 gives more insight into journey planning in a public transit network and briefly describes the time-expanded model. In order to cope with the application of the CSO algorithm to solve such journey planning, we present some adaptations of the algorithm in Section 3. These include: (a) an appropriate definition of the cockroach position which carries a solution, (b) the involvement of rules for movement through the time-expanded graph, because any replacement of vertices in solutions carried by cockroaches can lead to obtaining incorrect solutions, and (c) the assumption that, in the case of chase-swarming procedure, the nodes shared between two cockroaches are found, and then a part of the nodes of one cockroach is conveyed to the second individual. In Section 4, we provide a description of test instances, results of conducted experiments and comparison of the CSO, the PSO, and Dijkstra's algorithms with respect to their performance on selected instances. We also present the influence of the selected parameters of the CSO approach on the quality of the obtained solutions, using the variance analysis (ANOVA). Finally, Section 5 presents a discussion of the results and summarizes the conclusions.

## 2. The Journey Planning Problem

### 2.1. Related Work

Many papers focus on road networks and public transportation networks. These important and challenging problems are extensively investigated by a lot of researchers and solved by different methods. The problem of journey planning was considered by many researchers. As reported in [13,14], several techniques and algorithms have been proposed for solving that problem. We can characterize this problem by using graph theory as a shortest path problem. Mohemmed et al. [15] used the particle swarm optimization algorithm with a modified priority-based encoding for path finding problems. Zhang et al. [16] integrated the artificial immune system and chaos operator in structure of the particle swarm optimization for a realistic freeway network. Effati and Jafarzadeh [17] included neural networks for solving the shortest path problem. In [18] the improved matrix multiplication method for solving the all pairs shortest path problem was presented. Moreover, the pulse-coupled neural networks have been applied to realize parallel computation. Rajabi-Bahaabadi et al. [19] proposed a new model to find optimum paths in road networks, with time-varying stochastic travel times, and solved it by genetic algorithms. Wang et al. [20] studied a biogeography-based optimization method for solving multi-objective path finding. Many researchers dealt with genetic algorithms to solve route planning problems. In [21], the proposed approach uses a priority-based encoding method to represent all paths. In [22], the genetic-algorithm-based strategy was used to find the shortest path in a dynamic network. Lozano and Storchi [23] solved the shortest viable path problem in a multi-modal network using label correcting methods. In turn, in [24], an A∗ label-setting algorithm was presented to solve a constrained shortest path problem. Zhang et al. [25] investigated the multi-modal shortest path problem, in which travel time and travel costs were uncertain variables.

A number of papers discussed various ways of finding the shortest path in a multi-modal network, but most articles often refer to the use of the label-setting algorithm [26], label-correcting algorithm [27], and genetic algorithms [28]. It should be noted that, in the basic effective solutions for journey planning in public transit networks, the timetable is formulated as a graph, hence, travel corresponds to a path in the graph. Therefore, we can solve the problem for example by Dijkstra's algorithm [1,29]. For a comprehensive study on heuristic approaches in transportation applications, see [30].

As mentioned in the previous section, among the most studied approaches that model timetable information as the shortest path problem, one can find the time-expanded [1,2] and time-dependent [4] models that construct the digraphs. Pyrga et al. [3] discussed and examined both models in respect of their theoretical considerations and practical use. They proposed those models along with some speed-up techniques. Concerning CPU time, the time-dependent model was faster, but it was experimentally proved that the time-expanded approach was more robust than the second one in the case of realistic problems. It is worth mentioning that various studies focusing on design and optimization of public transportation networks incorporate approaches based on the time-expanded model, which is much more flexible and easily extendable, as concluded in [14]. For example Dib et al. [31] formulated route planning in multimodal transportation networks as the time-expanded model and proposed a combination of genetic algorithms (AG) and variable neighborhood search (VNS) to compute multimodal shortest paths. Another way to tackle journey planning, instead of using one of the graph-based models mentioned above, consists in developing approaches that directly operate on the timetable. These methods involve CSA that assembles connections into one single array (*connection scan algorithm*) [32], RAPTOR (*round-based public transit routing*) that operates on the timetable using a dynamic programming approach [33], MCR (*multimodal multicriteria RAPTOR*) [34], and FBS (*frequency-based search*) [35].

## 2.2. Time-Expanded Model

Due to the great importance of the time-expanded model and its common use in journey planning, we will briefly describe this approach.

As we know, the basic transport networks consist of nodes that represent stops and edges that represent links connecting nodes. Itinerary of transit line is formulated as the sequence of traversed nodes. It should be noted that in the time-expanded model, every time event (e.g., departure or arrival) at a stop (or a station) is presented as a node and connections between the two events or waiting within a stop are represented by weighted edges. In other words, in this approach, we have three types of nodes belonging to a station: arrival and departure nodes that are used to represent connections in the timetable, and transfer nodes representing modeling transfers. In the simplified version of the time-expanded model, transfer time between vehicles at a station is negligible. The weight of each edge represents the time difference between the departure ($t_d$) and arrival ($t_a$) times, where $t_d$ and $t_a$ represent times in minutes after midnight [3]. For each elementary connection from station $X$ to the next one $Y$, there is an edge connecting a departure node belonging to the first station with associated time $t_d$, to an arrival node of station $Y$ with associated time $t_a$. In turn, for the realistic version of this model, one should ensure a minimum transfer time at a station [36]. Therefore, for every arrival node, two additional edges are assumed: one edge to ensure the possibility of departure by the same vehicle, and a second edge for the first transfer node to allow transfers. For a detailed description, see [3]. Unfortunately, such an approach yields a high number of edges. Taking into account the earliest arrival problem, in the case of removing some nodes (nodes having an outgoing degree of one), the original size of the graph in the time-expanded model can be reduced [3,33].

## 3. Cockroach Swarm Optimization Algorithm for Transport Planning

Transport planning, which has been defined in the previous section, can be solved with various algorithms. One way of solving the described problem would be to use the cockroach swarm optimization algorithm with some of the proposed modifications described here. The issues presented in the subsequent parts of this paper concern designing of the cockroach swarm optimization algorithm to solve a specific travel planning problem, taking into account the proper representation of the solution, the determination of the neighborhood, the distance of individuals, procedures of the movement in the space of solutions, and the selection of the parameters that control the algorithm operation. We assumed the TE model [36] and restricted ourselves to travel time (arrival at the target) as a single optimization criterion.

Formally, we considered a set of stations $\Omega$, a set of stop events $\Psi_S$ per station $S \in \Omega$, and a set of elementary connections $\Delta$, whose elements were tuples of the form $\delta = \{Z_d, Z_a, S_d, S_a, t_d, t_a\}$, where [37]:

- $Z_d$—stop event of the departing vehicle,
- $Z_a$—stop event of the arriving vehicle,
- $S_d$—station from which the vehicle departs,
- $S_a$—station at which the vehicle arrives, and
- $t_d, t_a$—the departure and arrival times, respectively.

Given the start station ($A$) and the end station ($B$), the task was to find the sequence $P \in \Delta$, $P = (\delta_1, \delta_2, \dots, \delta_k)$ so that $S_d(\delta_1) = A$ and $S_a(\delta_k) = B$ and minimize the travel time, taking into account the limitation of maintaining a minimum time buffer ($b$) for safe transfer between public transportation vehicles. The departure station of $\delta_{i+1}$ is the arrival station of $\delta_i$ [3,14].

In this paper, the objective function we want to minimize is defined as a sum of times taken between departing from the previous node and arriving to the next node (including the transfer time) until the final station ($B$) is reached. Therefore, the objective function $f$ is defined as follows:

$$f = \sum_{i=1}^{k} (t_a(\delta_i) - t_d(\delta_i)) + \sum_{i=1}^{k-1} (t_d(\delta_{i+1}) - t_a(\delta_i)), \tag{1}$$

with the constraint of:

$$t_d(\delta_{i+1}) - t_a(\delta_i) \geq b. \tag{2}$$

### 3.1. Cockroach Swarm Optimization Algorithm—Basic Approach

The cockroach swarm optimization (CSO) algorithm is inspired by the behavior of cockroaches looking for food, such as moving in swarms, scattering or escaping from light [7–9]. Hence, a set of rules that models the collective cockroach behavior is employed in the CSO algorithm. In its initial step, the algorithm focused on creating a set of possible solutions. In general, the initial solutions are randomly generated in the search space. Furthermore, at each iteration, the CSO algorithm involves three procedures for solving different optimization problems such as chase-swarming, dispersing, and ruthless behavior.

In the chase-swarming procedure, in the new cycle, the strongest cockroaches carry the local best solutions ($P_i$), form small swarms, and move forward to the global optimum ($P_g$). Within this procedure, each individual ($X_i$) moves to its local optimum in the range of its visibility. There can occur a situation when a cockroach moving in a small group becomes the strongest by finding a better solution, because individuals follow in other ways that their local optimums. A lonely cockroach, within its own scope of visibility, is its local optimum and it moves forward to the best global solution.

Another procedure concerns the dispersion of individuals. It is performed from time to time to preserve the diversity of cockroaches. The procedure involves each cockroach performing a random step in the search space. We can also deal with ruthless behavior when a random individual is replaced by the currently best individual. That process corresponds to the phenomenon of eating weaker cockroaches in the case of inadequate food availability.

The main steps of the basic CSO algorithm can be described as below:

- STEP 1: generate a population of $n$ individuals and initialize algorithm's parameters (*step*, *visual*—visual scope, $D$—space dimension, stopping criteria).
- STEP 2: Search $P_i$ (within the visual scope of the $i$th individual) and $P_g$.
- STEP 3: Implement behavior of chase-swarming and update $P_g$ at the end; if a cockroach $X_i$ is local optimum, then it goes to $P_g$ according to $X_i = X_i + step \cdot rand \cdot (P_g - X_i)$, where *rand* is a random number within [0,1]; otherwise, the cockroach $X_i$ goes to $P_i$ (within its visibility) through formula $X_i = X_i + step\ rand \cdot (P_i - X_i)$.
- STEP 4: Implement dispersing procedure and update $P_g$.
- STEP 5: Implement ruthless procedure ($X_k = P_g$ or $X_k = 0$, where $k = 1, \dots, n$).
- STEP 6: Repeat steps 2–5 until a termination criterion is satisfied and output the final results.

The stopping criterion can include the maximum number of iterations, number of iterations without improvement, computation time, obtaining an acceptable error of a solution, and so on.

### 3.2. Proposed Adaptation of Cockroach Swarm Optimization Algorithm to Time-Expanded Model

An adaptation of the cockroach swarm algorithm to work with the time-expanded model requires additional operations. Therefore, our approach is an extension of the basic CSO algorithm with some modifications.

As we know, each cockroach generates one solution at the beginning of the CSO algorithm. Possible solutions encoded in the form of real variables, concerning at least the position of individuals in the cockroach swarm optimization algorithm, do not reflect the nature of the problem. Therefore, the solution represented by the cockroach is a set of successive vertices in the graph leading from the start to the final destination. It must be correct and consist only of the permitted moves. In order to generate the initial population of solutions, we used specific rules for movement through the graph shown in the activity diagram (Figure 1), because a random choice of the next node did not result in achievement of the destination node. When generating the initial solution, beginning with the start node, subsequent nodes are searched in the neighborhood (belonging to the same line), until the final node is reached. If another node is not found in the neighborhood of the current node within the same line, either the solution generation is interrupted (after the limit of steps has been reached), or random selection of a new line is effected from among those available in the current node, followed by the search of a new node in the neighborhood, within a new line. If a new node is not found or another line is not selected, the procedure is ended, without returning a correct solution.

Upon selection of initial solutions, the solution quality (determined by the travel time in our case) is estimated. The purpose of the subsequent steps of the CSO algorithm is to improve solutions and select the best one, with the fastest time of travel to the destination stop. In the chase-swarming procedure, a weaker cockroach tends toward the better solution representing a shorter destination time. It should be mentioned that appropriate interpretation of cockroach movement is necessary to effectively solve various optimization problems. Therefore, in order to increase the efficiency of the CSO algorithm, we assume that a step in the search space consists in taking over several nodes from a better one and the number of said nodes is determined by the step size. In addition, the visibility parameter (*visual*) denotes the minimum number of common nodes that two cockroaches should have in order to be visible to each other. Thus, *visual* = 2 means that the intersection of routes carried by both cockroaches at two points would be sufficient.

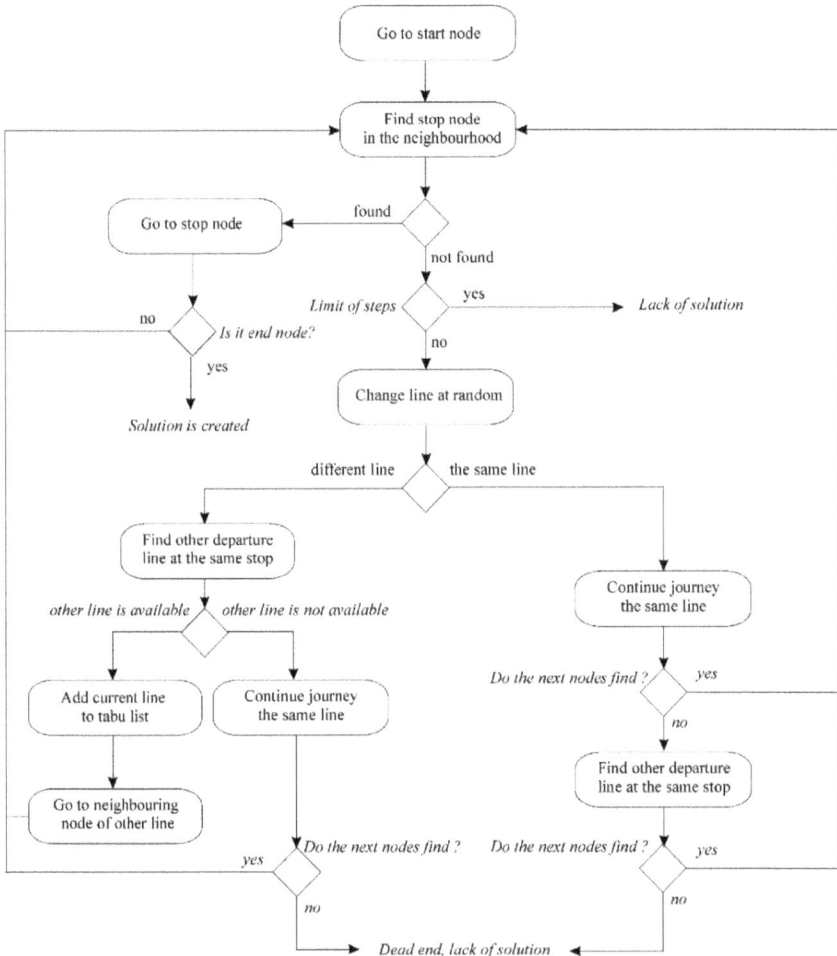

**Figure 1.** Rules to create the initial population.

If two cockroaches have no common nodes, they cannot move with respect to each other. In detail, the process of creating a new solution in this procedure starts with a random selection of one of the common edges ($e_n$) for a weaker and a stronger cockroaches. Next, the edges are copied to the new solution, from the first edge to the $e_n$ from the weaker cockroach and the $r$ ($r$ is a randomly chosen number) of the edges following the $e_n$ from the stronger cockroach. If, as a result of this operation, the end node has not been reached, the missing edges are created in the same way as in the process of initial solution generation.

Taking into account the described modifications, the steps of implementing our chase swarming procedure is outlined as follows:

---

*procedure* chase-swarming

---

**initialize**:
  parameter *graph* from TE model;
  parameter list of $k$ cockroaches;
  parameters *visual, maxAttempt*—maximum attempts to improve chase-swarming;
  $n$–target node
**for each** cockroach $k$ **do**:
    *oldSolution, newSolution* := *k.currentSolution*
    *visibleRoaches* := empty list
    **for each** cockroach $c$ **do**:
      **if** $k == c$ continue;
      **if** $c$ contains at least *visual* common edges with $k$ and $c$ solution is better than $k$ solution:
        add $c$ to *visibleRoaches*
      **end if**
    **end for**
    **if** *visibleRoaches* is empty **break** loop
    $v$ := select the best cockroach from *visibleRoaches*
    $e_n$ := select random common edge for $v$ and $k$
    *newSolution* := remove all edges after $e_n$ from *newSolution*
    *newSolution* := append random number of edges r from $v$ after $e_n$ ($e_{n+1}, \ldots , e_{n+r}$) // *(following the better*
*cockroach v)*
    **for** $i = 0$ to *maxAttempt*:
      *newSolution* := randomly generate remaining path for *newSolution*
      **if** n*ewSolution* reached $n$ **break** loop
    **end for**
    **if** n*ewSolution* is better than *oldSolution*:
      *k.currentSolution* := *newSolution*
    **else**
      *k.currentSolution* := *oldSolution*
    **end if**
**end for**

---

In the case of a dispersion procedure, we decided to select a random number of the *n*th node from the end of the path, and then generate a new route from this node to the destination. The pseudo-code of this procedure can be stated as follows:

---

*procedure* dispersing

---

**initialize**:
  parameter *graph* from TE model;
  parameter list of $k$ cockroaches;
  parameters *maxStep*—maximum dispersion step, *maxAttempt*—maximum attempts to improve dispersing;
  $n$–target node
**for each** cockroach $k$ **do**:
    *oldSolution, newSolution* := *k.currentSolution*
    *step* := random number from range <1; *maxStep*>
    *newSolution* := remove last *step* nodes from *newSolution*
    **for** $i = 0$ to *maxAttempt*:
      *newSolution* := randomly generate remaining path for *newSolution*
      **if** *newSolution* reached $n$ **break** loop
    **end for**
    **if** *newSolution* is better than *oldSolution*:
      *k.currentSolution* := *newSolution*
    **else**
      *k.currentSolution* := *oldSolution*
    **end if**
**end for**

---

The searching procedure ceases if the stopping criterion, defined as the maximum number of iterations or the number of unimproved iterations, is met.

## 4. Experiments and Results

We developed many experiments to assess the performance of the presented CSO algorithm. Many runs of the proposed approach were executed and the solution quality was taken into account. We have prepared seven benchmarks for the construction of problems of varying complexity (see Table 1). The time-expanded model was simulated. We assumed that all lines had regular departures (12 or 48 daily) at equal intervals, regardless of the time of day. In order to test the effectiveness of the CSO algorithm, experiments were performed 10 times for each test instance, with the same setting of parameters. In all experiments we assumed fixed parameters of the CSO algorithm during all iterations. The performance of our CSO adaptation was evaluated in comparison with Dijkstra's [1,29] and the particle swarm optimization [15,38,39] algorithms. All algorithms were implemented in the Java programming language, using a Linux operating system. We ran all experiments using an Intel Core i5-5200U 2.20 GHz processor with 16 GB RAM.

**Table 1.** Characteristics of the time-expanded model.

| Number of Lines | Number of Stops | Daily Number of Departures | Number of Arrival Nodes | Number of Departure Nodes | Number of Transfer Nodes | Total Number of Nodes |
|---|---|---|---|---|---|---|
| 1 | 21 | 12 | 217 | 229 | 228 | 674 |
| 1 | 21 | 48 | 865 | 913 | 912 | 2690 |
| 2 | 33 | 12 | 361 | 385 | 384 | 1130 |
| 2 | 33 | 48 | 1441 | 1537 | 1536 | 4514 |
| 6 | 52 | 12 | 673 | 745 | 744 | 2162 |
| 6 | 52 | 48 | 2689 | 2977 | 2976 | 8642 |
| 7 | 52 | 12 | 674 | 747 | 746 | 2167 |

### 4.1. Description of the Considered Test Instances

In order to verify the correctness and the quality of the implemented algorithm, we designed several test instances with simple timetables. Below, we present a short description of those.

In the first case, we constructed only one transportation line, containing 21 stops with 12 departures per day. It is worth noting that, for this line, a graph with 674 vertices will be chosen. Consequently, the increase of the number of departures to 48 per day required a more complex graph to be adopted, as shown in Table 1. On this basis, the implemented solutions and processes of graph construction were validated.

In the second experiment, we decided to investigate the correct detection of transfers between the transportation lines, also for the two versions of timetables (12 and 48 departures). For this purpose, the second transportation line intersecting with the first one at a single stop was added. It should be emphasized that the only possible route between the desired points required transfer operations.

In another experiment, the area of travel was slightly expanded. Therefore, we increased the number of public transportation lines to six. In addition, we assumed that at least two transfers were required to reach the target point.

In the last experiment, we set the number of transportation lines at seven. In that case, one night line was also taken into account, which was necessary to ensure diversity among instances.

All of the test instances conducted in the context of the time-expanded model are summarized in Table 1. Therefore, we have a summary describing the dependence of the number of created nodes in the graph on the complexity of the transportation plan. Note that the number of transfer nodes is always one less than the number of departure nodes.

*4.2. CSO Performance*

By implementing the CSO algorithm described in the previous section, its performance was obtained and then compared with Dijkstra's and PSO algorithms. In addition, the variance analysis (ANOVA) was applied in the statistical evaluation of the CSO results. We checked how the variable containing the travel time obtained was influenced by the population size and the *visual* coefficient. We tested the settings of the parameter: *visual* = 1, 2, 3, 4 and the population size: 5, 15, and 50 individuals. Depending on the test instance, our analysis indicated either an essential influence or no essential influence of the *visual* parameter on the travel time obtained. For each test instance, there existed, however, an essential dependence of that variable that contained the travel time obtained on the population size. However, no interactions between the *visual* and population size factors occurred. Selected results of the significance levels *p* for two analyses (instances with six lines, marked as 6/12 and 6/48) and the measures of the effect magnitude are presented in Table 2.

**Table 2.** Results of the ANOVA test for instances 6/12 and 6/48.

| Instance | Effect | *p* | $\eta^2$ |
|----------|--------|-----|----------|
| 6/12 | visual | 0.79 | 0.01 |
| | population size | 0 | 0.43 |
| | visual × population size | 0.17 | 0.07 |
| 6/48 | visual | 0.01 | 0.1 |
| | population size | 0 | 0.44 |
| | visual × population size | 0.74 | 0.03 |

As we can see, the population size presents a strong effect (the *travel time* variable is explained by the population size variable in the proportion of more than 40%) in both cases; however, the *visual* either fails to indicate any influence (test instance 6/12) or shows a medium-size effect (test instance 6/48). The influence of the *visual* and population size parameters on the travel time obtained (vertical bars refer to 0.95 confidence intervals) are presented for both test instances in Figures 2 and 3.

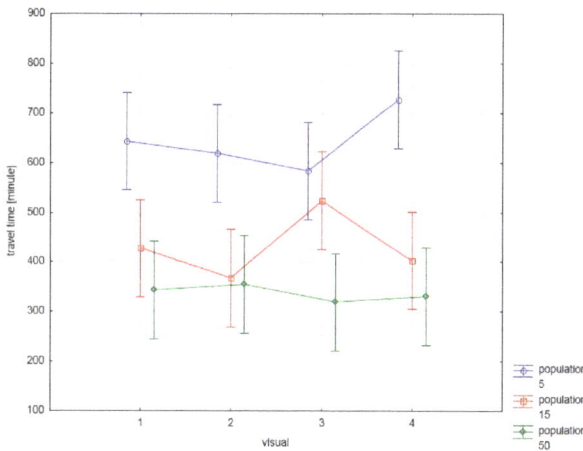

**Figure 2.** The influence of the selected parameters (test instance 6/12).

To investigate which of the parameters being tested, *visual* or population size, are different from each other, post hoc tests (Tukey tests) were conducted. Selected test results for test instance 6/48 are presented in Table 3. For the *visual* parameter, Tukey test showed essential differences of the average values between 1 and 2 and 4. In turn, for the population size parameter, tests showed essential differences between all the groups. For test instance 6/12, essential differences also occurred in all of the values of the population size parameter. Therefore, in our following tests, we assumed various sizes of population (5, 15, and 50) and *visual* = 3.

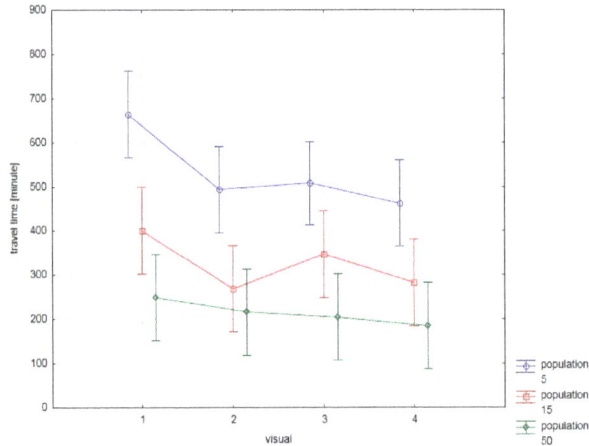

**Figure 3.** The influence of the selected parameters (test instance 6/48).

**Table 3.** Results of Tukey test (test instance 6/48).

| Visual | {1} | {2} | {3} | {4} |
|--------|---------|---------|---------|---------|
| 1 | | 0.032497 | 0.196031 | 0.010639 |
| 2 | 0.032497 | | 0.853130 | 0.979342 |
| 3 | 0.196031 | 0.853130 | | 0.628842 |
| 4 | 0.010639 | 0.979342 | 0.628842 | |

Each run of the CSO approach was terminated after 1000 iterations or if there was no improvement of the best solution through 25 iterations. In all tests: *maxStep* = 15, *maxAttempt* = 100. Table 4 shows the selected results of the CSO algorithm for the test instance with only one line, relating, however, to various sizes of population. For one line with 12 departures per day (marked as 1/12), the population consisting of 5 individuals is sufficient to find the best solution of travel time. Through 10 independent runs of the CSO approach, that solution was found in four cases. It turned out, however, that such a small number of solutions was not enough to solve the problem with 48 departures. The shortest travel time amounted to 1 h and 9 min. That value was obtained in three runs of the CSO algorithm. In the same test instance, the longest travel time amounted to 5 h and 21 min. Upon the increase of the population of individuals up to 15, we obtained the shortest travel time of 41 min in three runs. It is worth noting that the CSO algorithm, with the population of 50 solutions, generated the shortest travel time (41 min) in eight runs. What is interesting is that the worst solution for said population size was the same as the best solution for the population equal to 5.

Table 4. Results for one line.

| Population | Daily Number of Departures | Travel Time [h:min] | Computational Time [ms] |
|---|---|---|---|
| 5 | 12 | 00:45, 00:45, 02:45, 00:45, 02:45, 02:45, 02:45, 02:45, 06:45, 00:45 | 38, 23, 9, 10, 6, 10, 5, 4, 8, 3 |
| 15 | 12 | 00:45, 00:45, 00:45, 00:45, 02:45, 00:45, 00:45, 02:45, 00:45, 00:45 | 108, 56, 13, 31, 16, 9, 18, 8, 7, 9 |
| 50 | 12 | 00:45, 00:45, 00:45, 00:45, 00:45, 00:45, 00:45, 00:45, 00:45, 00:45, | 110, 82, 52, 56, 31, 50, 43, 64, 50, 35 |
| 5 | 48 | 03:29, 02:05, 05:21, 03:01, 05:21, 01:09, 03:01, 03:57, 01:09, 01:09 | 17, 11, 6, 4, 2, 4, 2, 3, 3, 2 |
| 15 | 48 | 00:41, 00:41, 01:37, 03:01, 01:37, 00:41, 04:25, 01:09, 01:09, 01:37 | 58, 15, 10, 12, 12, 11, 14, 11, 6, 12 |
| 50 | 48 | 00:41, 00:41, 00:41, 00:41, 00:41, 01:09, 00:41, 00:41, 01:09, 00:41 | 28, 29, 33, 94, 29, 30, 46, 43, 26, 47 |

The number of correct solutions obtained during 10 runs, depending on the population size, is shown in Figure 4. As expected, the increased population improved the chance of finding the best solution. Note that in the test instance 6/48 (six lines and 48 departures per day), the best result could not be attained anyway.

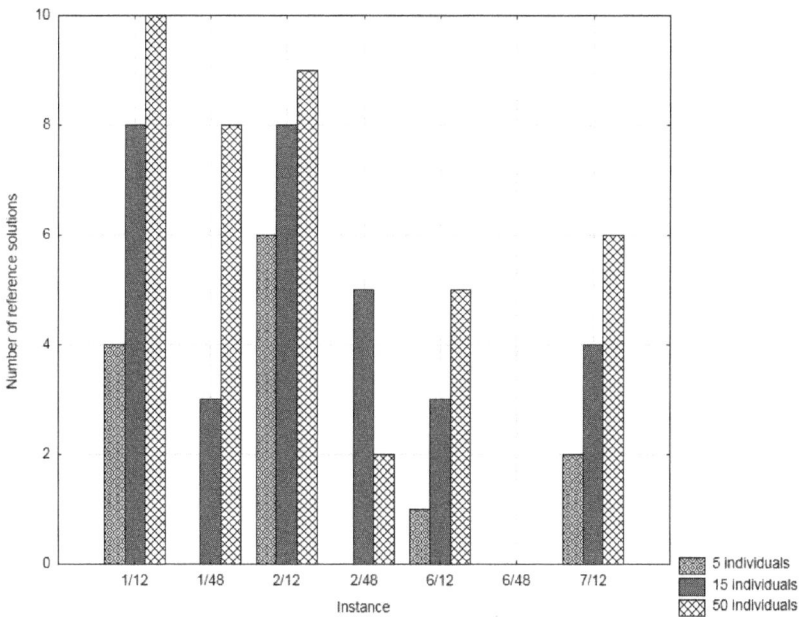

Figure 4. Number of the best solutions depending on the population size.

For simpler TE models, relating to 12 departures a day, we obtained the best solution, although only in several runs.

However, with the population composed of five solutions, it was not possible to obtain the best known solution for any of the test instances in which 48 departures a day were taken into account (marked as 1/48, 2/48, and 6/48, respectively).

In addition, we implemented the particle swarm optimization algorithm. We assume that following the neighboring particles consist of attempts at "taking over" part of the graph edge from the solution of a better particle; that is, the particle that "follows" another one is trying to add to its route a node from the particle being followed. The number of nodes that is tried to be added is determined by the particle's velocity. Moreover, we assumed that the difference of one hour in reaching the target node caused the increase of one node in the particle's velocity.

Table 5 shows the experimental results of the CSO algorithm performance in comparison with the results obtained by Dijkstra's and PSO algorithms for each test instance. Among the results obtained from the swarm algorithms, the best and the worst values of travel time were gathered. The first column shows the population size, the second one presents specific test instances, described by the numbers of lines and departures. The columns "Best travel" and "Worst travel" represent the best and worst travel times found by the CSO and the PSO algorithms, respectively. The fifth and the eighth columns give the average computational times of 10 independent runs of swarm algorithms. The last column shows reference solutions obtained by Dijkstra's algorithm.

**Table 5.** Selected results of seven test instances.

| Population | Instance | CSO | | | PSO | | | Dijkstra [h:min] |
|---|---|---|---|---|---|---|---|---|
| | | Best Travel [h:min] | Worst Travel [h:min] | Mean Time [ms] | Best Travel [h:min] | Worst Travel [h:min] | Mean Time [ms] | |
| 5 | 1/12 | 00:45 | 06:45 | 11.6 | 00:45 | 08:45 | 11:5 | |
| 15 | 1/12 | 00:45 | 02:45 | 27.5 | 00:45 | 02:45 | 30.8 | 00:45 |
| 50 | 1/12 | 00:45 | 00:45 | 57.3 | 00:45 | 00:45 | 172.8 | |
| 5 | 1/48 | 01:09 | 05:21 | 5.4 | 00:41 | 11:53 | 2.7 | |
| 15 | 1/48 | 00:41 | 04:25 | 16.1 | 00:41 | 05:21 | 16.3 | 00:41 |
| 50 | 1/48 | 00:41 | 01:09 | 40.5 | 00:41 | 01:37 | 90.0 | |
| 5 | 2/12 | 02:38 | 08:38 | 4.2 | 02:38 | 16:38 | 5.6 | |
| 15 | 2/12 | 02:38 | 04:38 | 33.8 | 02:38 | 12:38 | 25.2 | 02:38 |
| 50 | 2/12 | 02:38 | 04:38 | 84.1 | 02:38 | 04:38 | 215.3 | |
| 5 | 2/48 | 02:31 | 09:43 | 16.3 | 02:31 | 13:43 | 3.5 | |
| 15 | 2/48 | 01:07 | 03:55 | 28.3 | 01:07 | 11:23 | 16.5 | 01:07 |
| 50 | 2/48 | 01:07 | 02:03 | 131.6 | 01:07 | 08:35 | 187 | |
| 5 | 6/12 | 04:07 | 18:07 | 14.8 | 14:07 | 22:07 | 9.3 | |
| 15 | 6/12 | 04:07 | 12:07 | 31.7 | 08:07 | 14:07 | 74.5 | 04:07 |
| 50 | 6/12 | 04:07 | 06:07 | 122.3 | 06:07 | 10:07 | 272.3 | |
| 5 | 6/48 | 03:22 | 11:46 | 11.6 | 11:18 | 17:22 | 3.8 | |
| 15 | 6/48 | 01:58 | 08:30 | 42.1 | 05:42 | 13.38 | 36 | 01:02 |
| 50 | 6/48 | 01:30 | 05:42 | 187.5 | 02:54 | 13:38 | 301.1 | |
| 5 | 7/12 | 04:07 | 12:07 | 15.1 | 08:07 | 16:07 | 12 | |
| 15 | 7/12 | 04:07 | 8:07 | 38.5 | 04:07 | 16:07 | 56.3 | 04:07 |
| 50 | 7/12 | 04:07 | 6:07 | 153.1 | 04:07 | 08:07 | 284 | |

Depending on particular tests, and upon comparison with Dijkstra's algorithm, the best results obtained with the CSO algorithm were similar to those of the second algorithm. In all of the test instances the results show that application of a larger population (50 individuals) reduced travel time, and produced the same solutions as Dijkstra's algorithm in most cases. It needs to be pointed out that such an improvement was reached at the expense of computational time. Note that it does not imply that the optimum result was achieved. In the cases presented here, the result was not satisfying in one test instance with six lines and 48 departures per day.

When analyzing the results obtained with the use of the PSO algorithm, one can conclude that the CSO algorithm performs better than the PSO does. We should notice that the best travel times obtained with the PSO algorithm for two instances (6/12 and 6/48) are worse than those obtained with the use of the CSO algorithm. We should further consider the worst solutions, in all of the analyzed cases, there was a distinct domination of the CSO algorithm over the PSO one. Additionally, the average time of calculation performance, with the population increased to 50 individuals, indicates that the CSO algorithm provides a better solution when such specific parameters have been selected.

## 5. Conclusions

As we mentioned at the beginning of this work, we are merely presenting our research about the cockroach swarm optimization algorithm used to solve travel planning. Hence, we have shown that it is possible to use the CSO algorithm with the time-expanded model to solve travel planning, with some additional assumptions. To apply the CSO approach, we had to define movement in the search space. Therefore, we introduced a set of rules for determining paths in the time-expanded graph and some modifications of the searching capability during the whole algorithm. The proposed CSO approach was tested on seven instances of varying complexity. Solving travel planning was a complex process, but the results obtained were satisfactory. We compared the performance of the CSO, the PSO, and Dijkstra's algorithms applied to the time-expanded model. Many experiments were conducted for different test problems and various sizes of population (5, 15, and 50). We also conducted some tests relating to the influence of *visual* and population size parameter settings on the results obtained, using the variance analysis (ANOVA). Our research proved that, in the majority of cases, the value of *visual* parameter did not have any significant influence on the travel time obtained. We observed that, for all the test instances, the increase of the population to up to 50 individuals improved the performance of the CSO algorithm, in respect of the travel time. However, it took more computational time, and it could be too slow to be used when solving larger problems. It should be mentioned, that for the test instance with six lines and 48 departures per day, the obtained solutions were not as good as those obtained with Dijkstra's algorithm, but the improvement of travel time for the population of 50 individuals was clearly visible. However, our experiments indicated that the CSO method outperforms the PSO in terms of the best travel time.

Analyzing results, we observe that by introducing some modifications into the framework of the CSO approach, the approach can produce good solutions, but it requires the use of speed-up techniques or parallel computing. Therefore, one possibility for future research could examine the performance of a GPU implementation of our approach. Furthermore, the use of other models, instead of the time-expanded model, could lead to better results.

**Acknowledgments:** This paper was supported by the statutory research of AGH University of Science and Technology (no. 11.11.120.396).

**Author Contributions:** The concept of the CSO algorithm with the proposed modifications was designed by Joanna Kwiecień. The CSO approach, the PSO, and Dijkstra's algorithms were implemented by Marek Pasieka. Material of Sections 1–3 was written by Joanna Kwiecień. The rest of the paper was written by both authors. All experiments were conceived, performed, and described by both authors. Authors have read and approved the final manuscript.

**Conflicts of Interest:** The authors declare no conflict of interest.

## References

1. Schulz, F.; Wagner, D.; Weihe, K. Dijkstra's algorithm on-line: An empirical case study from public railroad transport. *ACM J. Exp. Algorithmics* **2000**, *5*, 1–23. [CrossRef]
2. Schulz, F.; Wagner, D.; Zaroliagis, C. Using multi-level graphs for timetable information in railway systems. In *Algorithm Engineering and Experiments*; Mount, D.M., Stein, C., Eds.; Lecture Notes in Computer Science (LNCS); Springer: Heidelberg, Germany, 2002; Volume 2409, pp. 43–59.
3. Pyrga, E.; Schulz, F.; Wagner, D.; Zaroliagis, C. Efficient Models for Timetable Information in Public Transportation Systems. *ACM J. Exp. Algorithmics* **2008**, *12*, 1–39. [CrossRef]
4. Brodal, G.S.; Jacob, R. Time dependent networks as models to achieve fast exact time-table queries. *Electron. Notes Theor. Comput. Sci.* **2004**, *92*, 3–15. [CrossRef]
5. Xing, B.; Gao, W. *Innovative Computational Intelligence: A Rough Guide to 134 Clever Algorithms*; Springer: Cham, Switzerland, 2014.
6. Yang, X.-S. *Nature-Inspired Metaheuristic Algorithms*, 2nd ed.; Luniver Press: Frome, UK, 2010.
7. Chen, Z. A modified cockroach swarm optimization. *Energy Proced.* **2011**, *11*, 4–9.

8. Chen, Z.; Tang, H. Cockroach swarm optimization for vehicle routing problems. *Energy Procedia* **2011**, *13*, 30–35. [CrossRef]

9. Cheng, L.; Wang, Z.; Yanhong, S.; Guo, A. Cockroach swarm optimization algorithm for TSP. *Adv. Eng. Forum* **2011**, *1*, 226–229. [CrossRef]

10. Kwiecień, J.; Filipowicz, B. Comparison of firefly and cockroach algorithms in selected discrete and combinatorial problems. *Bull. Pol. Acad. Sci. Tech. Sci.* **2014**, *62*, 797–804. [CrossRef]

11. Kwiecień, J. Use of different movement mechanisms in cockroach swarm optimization algorithm for traveling salesman problem. In *Artificial Intelligence and Soft Computing*; Rutkowski, L., Korytkowski, M., Scherer, R., Tadeusiewicz, R., Zadeh, L.A., Zurada, J.M., Eds.; Lecture Notes in Computer Science (LNCS); Springer: Cham, Switzerland, 2016; Volume 9693, pp. 484–493.

12. Obagbuwa, I.C.; Abidoye, A.P. Binary cockroach swarm optimization for combinatorial optimization problem. *Algorithms* **2016**, *9*, 59. [CrossRef]

13. Bast, H.; Delling, D.; Goldberg, A.; Müller-Hannemann, M.; Pajor, T.; Sanders, P.; Wagner, D.; Werneck, R.F. Route Planning in Transportation Networks. In *Algorithm Engineering*; Kliemann, L., Sanders, P., Eds.; Lecture Notes in Computer Science (LNCS); Springer: Cham, Switzerland, 2016; Volume 9220, pp. 19–80.

14. Müller-Hannemann, M.; Schulz, F.; Wagner, D.; Zaroliagis, C. Timetable information: Models and algorithms. In *Algorithmic Methods for Railway Optimization*; Geraets, F., Kroon, L.G., Schoebel, A., Wagner, D., Zaroliagis, C.D., Eds.; Lecture Notes in Computer Science (LNCS); Springer: Heidelberg, Germany, 2007; Volume 4359, pp. 67–90.

15. Mohemmed, A.W.; Sahoo, N.C.; Geok, T.K. Solving shortest path problem using particle swarm optimization. *Appl. Soft Comput.* **2008**, *8*, 1643–1653. [CrossRef]

16. Zhang, Y.; Jun, Y.; Wei, G.; Wu, L. Find multi-objective paths in stochastic networks via chaotic immune PSO. *Expert Syst. Appl.* **2010**, *37*, 1911–1919. [CrossRef]

17. Effati, S.; Jafarzadeh, M. Nonlinear neural networks for solving the shortest path problem. *Appl. Math. Comput.* **2007**, *189*, 567–574. [CrossRef]

18. Zhang, Y.; Wu, L.; Wei, G.; Wang, S. A novel algorithm for all pairs shortest path problem based on matrix multiplication and pulse coupled neural network. *Digit. Signal Process.* **2011**, *21*, 517–521. [CrossRef]

19. Rajabi-Bahaabadi, M.; Shariat-Mohaymany, A.; Babaei, M.; Ahn, C.W. Multi-objective path finding in stochastic time-dependent road networks using non-dominated sorting genetic algorithm. *Expert Syst. Appl.* **2015**, *42*, 5056–5064. [CrossRef]

20. Wang, S.; Yang, J.; Liu, G.; Du, S.; Yan, J. Multi-objective path finding in stochastic networks using a biogeography-based optimization method. *Simulation* **2016**, *92*, 637–647. [CrossRef]

21. Gen, M.; Cheng, R.; Wang, D. Genetic algorithms for solving shortest path problems. In Proceedings of the 1997 IEEE International Conference on Evolutionary Computing, Indianapolis, IN, USA, 13–16 April 1997; pp. 401–406.

22. Davies, C.; Lingras, P. Genetic algorithms for rerouting shortest paths in dynamic and stochastic networks. *Eur. J. Oper. Res.* **2003**, *144*, 27–38. [CrossRef]

23. Lozano, A.; Storchi, G. Shortest viable path algorithm in multimodal networks. *Transp. Res. Part A Policy Pract.* **2001**, *35*, 225–241. [CrossRef]

24. Ma, T.Y. An A* label-setting algorithm for multimodal resource constrained shortest path problem. *Procedia Soc. Behav. Sci.* **2014**, *111*, 330–339. [CrossRef]

25. Zhang, Y.; Liu, P.; Yang, L.; Gao, Y. A bi-objective model for uncertain multi-modal shortest path problems. *J. Uncertain. Anal. Appl.* **2015**, *3*, 8. [CrossRef]

26. Horn, M.E.T. An extended model and procedural framework for planning multi-modal passenger journeys. *Transp. Res. Part B Methodol.* **2003**, *37*, 641–660. [CrossRef]

27. Liu, L.; Yang, J.; Mu, H.; Li, X.; Wu, F. Exact algorithm for multi-criteria multi-modal shortest path with transfer delaying and arriving time-window in urban transit network. *Appl. Math. Model.* **2014**, *38*, 2613–2629. [CrossRef]

28. Yu, H.; Lu, F. A multi-modal route planning approach with an improved genetic algorithm. *Adv. Geo-Spat. Inform. Sci.* **2012**, *38*, 193–202.

29. Dijkstra, E.W. A note on two problems in connexion with graphs. *Numer. Math.* **1959**, *1*, 269–271. [CrossRef]

30. Fu, L.; Sun, D.; Rilett, L.R. Heuristic shortest path algorithms for transportation applications: State of the art. *Comput. Oper. Res.* **2006**, *33*, 3324–3343. [CrossRef]

31.  Dib, O.; Manier, M.-A.; Caminada, A. Memetic algorithm for computing shortest paths in multimodal transportation networks. *Transp. Res. Procedia* **2015**, *10*, 745–755. [CrossRef]

32.  Dibbelt, J.; Pajor, T.; Strasser, B.; Wagner, D. Intriguingly simple and fast transit routing. In *Experimental Algorithms*; Bonifaci, V., Demetrescu, C., Marchetti-Spaccamela, A., Eds.; Lecture Notes in Computer Science (LNCS); Springer: Heidelberg, Germany, 2013; Volume 7933, pp. 43–54.

33.  Delling, D.; Pajor, T.; Werneck, R.F. Round-based public transit routing. *Transp. Sci.* **2015**, *49*, 591–604. [CrossRef]

34.  Delling, D.; Dibbelt, J.; Pajor, T.; Wagner, D.; Werneck, R.F. Computing Multimodal Journeys in Practice. In *Experimental Algorithms*; Bonifaci, V., Demetrescu, C., Marchetti-Spaccamela, A., Eds.; Lecture Notes in Computer Science (LNCS); Springer: Heidelberg, Germany, 2013; Volume 7933, pp. 260–271.

35.  Bast, H.; Storandt, S. Frequency-based search for public transit. In Proceedings of the 22nd ACM SIGSPATIAL International Conference on Advances in Geographic Information Systems, Dallas, TX, USA, 4–7 November 2014; pp. 13–22.

36.  Delling, D.; Pajor, T.; Wagner, D. Engineering time-expanded graphs for faster timetable information. In *Robust and Online Large-Scale Optimization*; Ahuja, R.K., Möhring, R.H., Zaroliagis, C.D., Eds.; Lecture Notes in Computer Science (LNCS); Springer: Heidelberg, Germany, 2009; Volume 5868, pp. 182–206.

37.  Geisberger, R. Advanced Route Planning in Transportation Networks. Ph.D. Thesis, Karlsruhe Institute of Technology, Baden-Württemberg, Germany, February 2011.

38.  Poli, R.; Kennedy, J.; Blackwell, T. Particle swarm optimization. An overview. *Swarm Intell.* **2007**, *1*, 33–57. [CrossRef]

39.  Toofani, A. Solving routing problem using particle swarm optimization. *Int. J. Comput. Appl.* **2012**, *52*, 16–18. [CrossRef]

*entropy*

MDPI

*Article*

# Can a Robot Have Free Will?

**Keith Douglas Farnsworth**

School of Biological Sciences, Queen's University Belfast, Belfast BT97BL, UK; k.farnsworth@qub.ac.uk;
Tel.: +44-2890-97-2352

Academic Editor: Mikhail Prokopenko
Received: 27 February 2017; Accepted: 15 May 2017; Published: 20 May 2017

**Abstract:** Using insights from cybernetics and an information-based understanding of biological systems, a precise, scientifically inspired, definition of free-will is offered and the essential requirements for an agent to possess it in principle are set out. These are: (a) there must be a self to self-determine; (b) there must be a non-zero probability of more than one option being enacted; (c) there must be an internal means of choosing among options (which is not merely random, since randomness is not a choice). For (a) to be fulfilled, the agent of self-determination must be organisationally closed (a "Kantian whole"). For (c) to be fulfilled: (d) options must be generated from an internal model of the self which can calculate future states contingent on possible responses; (e) choosing among these options requires their evaluation using an internally generated goal defined on an objective function representing the overall "master function" of the agent and (f) for "deep free-will", at least two nested levels of choice and goal (d–e) must be enacted by the agent. The agent must also be able to enact its choice in physical reality. The only systems known to meet all these criteria are living organisms, not just humans, but a wide range of organisms. The main impediment to free-will in present-day artificial robots, is their lack of being a Kantian whole. Consciousness does not seem to be a requirement and the minimum complexity for a free-will system may be quite low and include relatively simple life-forms that are at least able to learn.

**Keywords:** self-organization; downward causation; autocatalytic set; goal-oriented behaviour; autopoiesis; biological computing

## 1. Introduction

Why do things do what they do? We have a hierarchy of explanations that roughly reflects a gradient in complexity, matched by the epistemic hierarchy which starts with the physics of Hamiltonian mechanics (and Shrödinger's equation), extends through statistical mechanics and complex systems theory, but then declines in power as we try to account for the behaviour of living systems and finally of the human condition, for which we have no satisfactory scientific explanation. One of the most persistent open questions, at the far end of the complexity gradient, is whether we as humans have free will. Here, I attempt to address this question with respect to a broader category of active agent (sensu Sharov [1]): anything that can make decisions and act in the physical world. The premise of this paper is that an understanding of the interaction and dynamics among patterns, of the distribution of matter and energy in space and time, may bring such high-level phenomena into resolution. That means a focus on information and its interactions using cybernetics and computation theory, but it also requires a broad concept of information addressing the relationship among patterns (as data) in general, rather than just the statistics of data transmission. If the emergence of material reality (as we experience it) is the assembly of stable configurations (of matter), undergoing transformations and combining as stable composites (e.g., molecules forming materials, forming structures [2]), then a deep understanding of it requires a mathematically precise account of the physics of patterns that are simultaneously the product of material structure and the cause of it [3]. For this investigation, we must focus on the kind of "information" that is embodied by, and processed by natural systems. This concept of intrinsic structural "information" (surely it ought to have a word of its own), is not necessarily ontological (as in the theories of Weizsäcker [4] and Stonier [5]), but refers to at least observable patterns having observable effects [6], is objectively quantifiable [7] and useful in understanding biological processes in terms of cybernetic systems [1], and functions [8,9]. To avoid ambiguity (and conflict) I will refer to this kind of "information" as pattern and the "information" which reduces "uncertainty" in a receiver as Shannon information.

My aim is to discover whether free will can be understood in terms that make no reference to specifically human qualities and whether free will could be attributed to a broader class of active agent (sensu Sharov [1]). I aim to do this by identifying the criteria that must be met for free will, and from that, to identify the kinds of active agent which might meet those criteria.

I will start by defining what I mean by "free-will" and then give a very brief overview of the philosophical debate about free-will, identifying some of the problems. Then I will introduce ways of thinking about these problems based on cybernetic/computation theory and use these to identify the necessary and sufficient conditions for free-will according to my definition. The class of systems for which these conditions are fulfilled will then be identified and with this, the minimum complexity compatible with autonomous action will be implied. The term "robot" is specifically used here to mean any cybernetic system coupled to a physical system that allows it to independently act in the physical world; this includes living systems (see e.g., the requirements specified for a molecular robot in Hagiya et al. [10]) and artificial systems, including the subject of cognitive robotics which concerns embodied artificial intelligence [11].

## 1.1. A Definition of Free-Will

There is no generally agreed definition for free will. To find the conditions allowing active agents (sensu Sharov [1]) to have free-will, a working definition is first provided. It is defined here as the condition in which all of the following are jointly true:

- FW1: there exists a definite entity to which free-will may (or may not) be attributed;
- FW2: there are viable alternative actions for the entity to select from;
- FW3: it is not constrained in the exercising of two or more of the alternatives;
- FW4: its "will" is generated by non-random process internal to it;
- FW5: in similar circumstances, it may act otherwise according to a different internally generated "will".

In this definition the term "will" means an intentional plan which is jointly determined by a "goal" and information about the state (including its history) of the entity (internal) and (usually, but not necessarily) its environment. The term "goal" here means a definite objective that is set and maintained internally. The terms used here will be explained and justified in what follows.

This list of criteria is chosen to address the main features that most philosophers have thought important (though they do not necessarily agree with one another about what is important and some philosophers would leave out some items of the list). FW2 and FW5 are intended to examine the effect of determinism and FW4 represents the "source arguments" for and against free will, whist FW3 ensures freedom in the most obvious (superficial) sense. Only one of the list (FW1) is not usually included in any philosophical discussion of free will, perhaps because it is usually considered to be self evident, but it will play an important role here.

## 1.2. The Philosophical Background

In philosophy, freedom "to act as one wills" is often referred to as "superficial freedom" (or first order volition, [12]) and the freedom "to determine what one wills" as "deep freedom" (or second order volition, [12]). Kane [13], McKenna [14] and Coeckelbergh [15] provide a broad introduction to the subject, Westen [16] gives a deeper criticism of it. As an example, one may be free to drink a bottle of whisky in one sitting and as an alcoholic, or in an irresponsible mood, or emotional turmoil one may will it, but knowing the consequences, one may master the desire and will otherwise: rejecting the opportunity of this poisonous pleasure. If the alcoholism had taken over, or one was under duress, the "freedom to choose otherwise" might be denied and deep freedom would be lost with it, although the superficial freedom to drink would remain. The classical philosophical argument on free-will consists of a) whether it is compatible or not with determinism and b) whether at least some agents (usually people) are at least in part the ultimate cause of their actions.

Determinism is the idea that there is, at any instant, exactly one physically possible future [16,17], summarised in the slogan "same past: same future" (see List [18] for a more rigorous analysis). Cybernetics captures determinism in the definition of a Determinate Machine (DM) as a series of closed, single valued transformations (for example describing a Finite State Automaton (FSA)). Superficial freedom is often seen as the absence of constraint, leading to the (relatively trivial) conclusion that it is compatible with determinism. However, deep freedom needs more than an absence of constraints, it requires alternative paths into the future to provide the "freedom to do otherwise" [17]. The cybernetic model of a system with this capacity is of course the non-determinate machine (NDM). However, the NDM is usually conceived as a probabilistic process in which a set of possible states $S : \{s_1, ...., s_n\}$ of the system, given the present conditions $C$ (in general including the previous history), may occur at random, with probability set $P : \{p_1, ..., p_n\}$ where $\sum p_i = 1$ and each $p_i$ is the probability of each possible state $s_i$. Most philosophers agree that randomness is not compatible with self-determination, indeed it seems to be the opposite, so they reject random spontaneity as a means of achieving deep freedom (and so do I). They reason that if an agent's action were ultimately caused by e.g., a quantum fluctuation or thermal noise, then we could not reasonably hold the agent responsible for it. This indicates that philosophers supporting the existence of deep free-will are searching for a non-random ultimate cause of actions within the agent of those actions.

Unfortunately for them, a paradox arises: since any agent is the product of its composition and of its previous experiences (which may play a role in its formation) and these are beyond its control. If it did not make itself and select its own experiences, then its behaviour must be determined by things other than itself. An agent is not free in the deep sense unless it has control over all the events that led to it's choice of action. Recognising that all events in the universe belong to a chain of cause and effect that extends back before the existence of the agent, some philosophers conclude that either (a) this deep freedom cannot exist and is considered an illusion (e.g., Van Inwagen [17], reiterated in [19]); or (b) the agent is indeterminate so that we get "same past: different futures". If they also rule out randomness, then (b) suggests that an agent which could act in more than one possible way from exactly the same state and history (i.e., it is indeterminate) must act without cause. They conclude that this is self-contradictory, hence deep freedom cannot exist. This line of thinking is closely related to Strawson's [20] "Basic Argument" against free-will, which starts from the premise that for an agent to have free-will it must be the cause of itself and shows, via infinite regress, that this is not possible. It is an axiom of these positions that the motivation for an agent's action can only be either (a) random; (b) exogenous or (c) self-generated, with only (c) being compatible with free-will.

## 2. Systems with Identity: The Closure Condition

The first criterion (FW1) implies that we need to determine what parts of a system must be included in identifying an agent (i.e., the extent of its identity) before being able to determine if it has free-will or not. Free-will requires a definite boundary between the internal and external, not necessarily a physical boundary (as supplied by e.g., a casing or skin), but more profoundly one of organisation and control. For example a computer controlled robot must have all the necessary provisions for physical independence (as in the extraterrestrial exploration robots), but this still leaves it organisationally linked to humanity because its existence is entirely dependent on our gathering and processing the materials for its "body" and assembling these, implicitly embodying it with functional information [1] and programming its control computer (including with the goal for operation). For these reasons, such robots remain extensions of ourselves: tools just as sophisticated hammers would be. In general, for free-will, the control information of an agent must be independent of anything beyond a cybernetically meaningful boundary. Put the other way round, for the identification of free-will, we must first identify the boundary of the agent, which is defined by independence of control. The existence of such a cybernetic boundary enclosing the agent is here termed the "closure condition". Given this, the Mars Rover coupled with its human design team seems to meet the closure condition, but the Mars Rover alone does not.

This idea of a boundary surrounding a system, such that whatever is within the boundary has the property of organisational independence from what lies without, was encapsulated by the concept of the "Kantian whole" by Kauffman [21] and can be formally described in cybernetic terms as organisational closure. This idea is used in a particular definition of autonomy which I shall next argue to be a pre-requisite for free will.

## 2.1. Systems with Causal Autonomy

Froese et al. [22] distinguishes behavioural (based on external behaviour) from constitutive (based on internal organisation) autonomy. They state that for the former, the identity of the system may be imposed by an external observer (it could even be no more than an thermostatic system) and that it is sufficient for the system to demonstrate a capacity for stable and/or flexible interaction with its environment, "without human intervention, supervision, or instruction". This, they argued, left behavioural autonomy so "ambiguous and inclusive ... it threatens to make the concept of autonomy meaningless". Behavioural definitions of autonomy are inadequate in relation to questions of free will because they do not address the source of will (more precisely, the origin of the goal/s) which motivate the observed actions. To attribute free will to an agent, we need to identify the source of will as a part of the agent (under FW4) and this requires us to consider the composition of the agent. To exclude external (generalised from "human") "intervention, supervision, or instruction", we must have an agent that is separated from external causation of its actions. According to most authors seeking to explain the apparent independence of action found among living systems, this requirement leads directly to constitutive autonomy, e.g., "every autonomous system is organizationally closed" (Varela [23], p. 58). This idea has a relatively long history in a multi-disciplinary literature (Froese et al. [22], Zeleny [24], Rosen [25], Vernon et al. [26], Bich [27] and references therein), but it is not clear if it is restricted to living systems, or may be broader. Therefore, rather than taking this literature as sufficient justification for a constitutive autonomy requirement, let us examine the options for matching with the following tasks:

- to answer Strawson's [20] "Basic Argument" of ultimate responsibility;
- to separate internal from external (i.e., to give formal meaning to internal and external);
- to unambiguously break the physical causal link between the agent and its environment, allowing "leeway" from determinism.

This will start with task (2) because from its conclusion, task (3) may follow, given a specification that relations are strictly causal and if the answer to task (2) does specify constitutive autonomy, then task (1) may be implied from that, though it leaves unproven that a causally autonomous system must be self-made. This gap may not be serious, as later shown.

Therefore, we seek a structure for which "internal" is causally distinct from "external", giving a clear definition to both. For this we need to define an object $A$, properly composed of parts (e.g., $x$, $y$), none of which is a part of any other object $B$. More formally: $\exists\ A$ composed of parts $a_i \in a : \{a_1...a_N\}$ in which no part of $A$ is also a part of any object $B$ unless $B$ contains or is $A$: $(B \supseteq A)$. Assume the mereology in which the reflexivity and transitivity principles are true and also the antisymmetry postulate is true. That is: two distinct things cannot be part of each other. This is expressed by the axioms [28]:

- $x\mathrm{P}x$
- $(x\mathrm{P}y \wedge y\mathrm{P}z) \rightarrow x\mathrm{P}z$

- $(x\mathrm{P}y \wedge y\mathrm{P}x) \to x = y$,

in which P is the "part of" relation. For these, the auxiliary relations are defined:

- Overlap: $x \circ y := \exists z(z\mathrm{P}x \wedge z\mathrm{P}y)$,
- Exterior: $x \perp y := x\neg \circ y$.

Therefore, we can define an isolated object $y$ : no part of $y$ can overlap with anything other than $y$. Hence, Isolation: $\forall z \ z\neg \circ y$, hence $\forall z \ z \perp y$ (this definition is not a part of standard mereology). The mereological sum (noting that alternative definitions exist) is defined as: an object $y$ is a mereological sum of all elements of a set X iff every element of X is a part of $y$ and every part of $y$ overlaps some element $x \in$ X. (Effingham [29], p. 153) puts it this way: "the $x$s compose $y$ by definition if (i) each $x$ is a part of $y$; (ii) no two of the $x$s overlap; (iii) every part of y overlaps at least one of the $x$s". This definition of sum allows overlap with objects that are not parts of the mereological sum, so $y$ is not necessarily isolated, therefore also let $|y(X)|$ denote that $y$ is an object that is both isolated and composed of a set X (of $x$s).

So far, partness (the $x\mathrm{P}y$ relation) has not been defined. The definition of partness depends on what condition must be met for objects to be associated as parts. This is the Van Inwagen [30] special composition question: "what constitutes being a part; what connection or relationship qualifies as 'partness'?" Specifically, we need a criterion for "restricted composition" that is relevant to the question of separating internal from external in terms of causation. For this, define causation as a binary relation: $x\mathrm{C}y$, specifying that the state of object $y$ is strictly determined by the state of object $x$. With this, we can define a transitive closure for causation (transitive closure is the minimal transitive binary relation R on a set X).

First, note that any relation R on a set X is transitive iff $\forall a, b, c \in$ X, whenever $a\mathrm{R}b$ and $b\mathrm{R}c$ then $a\mathrm{R}c$. The conditions for a transitive closure can then be written as follows [31] (using the notation $\mathrm{R}^+$ to represent a transitive relation): (i) if $\{a,b\} \in \mathrm{R} \ \to \{a,b\} \in \mathrm{R}^+$ (ii) if $\{a,b\} \in \mathrm{R}^+ \wedge \{b,c\} \in \mathrm{R} \to \{a,c\} \in \mathrm{R}^+$ (iii) nothing is in $\mathrm{R}^+$ unless by (i) and (ii).

Next, if R is specified as the C relation (from above), then (iii) specifies all the causal relations among members of a set $\mathrm{C}^+$, so that no relation $\{x,a\} \notin \mathrm{C}^+$ can be causal. This has the effect of causally isolating the elements of $\mathrm{C}^+$ as well as ensuring that they are causally related to one another.

Finally, let Y be the set of elements that are included in a transitive causal closure $\mathrm{C}^+$, where $x\mathrm{C}y$ is the condition for association as parts: $x$ is a part of $y$. Under these restrictions, mereologically, the elements of Y are the sum of an object $y$ and $|y(Y)|$. This means that the transitive closure for causation meets the requirement for formally separating internal from external (task 2). Since the relations in the closure are defined to be causal, this achieves task 3 as well. An object (agent) with the property of transitive causal closure among its parts is a causally autonomous system. All that is missing is a way to ensure that the agent is a cause of itself from the beginning of its existence, which requires it to be self-constructing. That will next be addressed with a more concrete example.

## 2.2. The Kantian Whole as a Material System

A system composed of parts, each of whose existence depends on that of the whole system is here termed a "Kantian whole", the archetypal example being a bacterial cell [32]. The origin of this terminology lies in Immanuel Kant's definition of an organised whole [33]. To make the closure condition concrete and include an answer to Strawson's [20] "Basic Argument" it will now be narrowed to a requirement for self-construction, since this implies the embodiment of self with the pattern-information that will then produce the agents behaviour (i.e., we require strictly constitutive autonomy as defined by Froese et al. [22]). In other words, we are to consider a cybernetic system that, by constructing itself materially, determines its transition rules, by and for itself (material self-construction may not be essential to ensuring self-determination, but assuming that the cybernetic relations embodied in it are essential, we may proceed without loss of generality). An autocatalytic chemical reaction network with organisational closure (and this is also what Kauffman [21] considered a Kantian whole) is an anabolic system able to construct itself [34].

Hordijk and Steel [34] and Hordijk et al. [35] define their chemical reaction system by a tuple $Q = \{X, \mathcal{R}, C\}$, (their symbols) in which $X$ is a set of molecular types, $\mathcal{R}$ a set of reactions and $C$ a set of catalytic relations specifying which molecular types catalyse each member of $\mathcal{R}$. The system is also provided with a set of resource molecules $F \subseteq X$, freely available in the environment, to serve as raw materials for anabolism (noting that whilst we are defining an organisational closure, we may (and indeed must) permit the system to be materially and thermodynamically open). The autocatalytic set is that subset of reactions $\mathcal{R}' \subseteq \mathcal{R}$, strictly involving the subset $X' \subseteq X$, which is:

- reflexively autocatalytic: every reaction $r \in \mathcal{R}'$ is catalysed by at least one of molecular type $x' \in X'$ and
- composed of $F$ by $\mathcal{R}'$: all members of $X'$ are created by the actions of $\mathcal{R}'$ on $F \cup X'$.

This definition of an autocatalytic set is an application of the broader mathematical concept of closure and more specifically of transitive closure of a set, since when the autocatalytic set is represented as a network (of reactions), this network has the properties of transitive closure. The concept of autocatalytic set has been implemented in experiments for exploring aspects of the origin of life (e.g., the GARD system simulating "lipid world" [36]). Clearly with the two conditions for an autocatalytic set met, everything in the system is made by the system, but there is a more important consequence. The system is made from the parts (only) and can only exist if they do. Organisational closure of this kind has been identified as a general property of individual organisms [37], many biochemical sub-systems of life [21] and embryonic development [38].

As it is defined above, living systems fulfil the closure condition, but can we conceive of a non-living system also reaching this milestone? Von Neumann's [39] self-replicating automata show that some purely informational (algorithm) systems have the capacity to reproduce within their non-material domain, but they cannot yet assemble the material parts necessary, nor can they build themselves from basic algorithmic components (they rely on a human programmer to make the first copy). What is needed for the physical implementation of a Kantian whole is the ability to "boot-strap" from the assembly of simple physical components to reach the point of autonomous replication (i.e., the system must be autopoietic [37,40]). This is necessary to answer Strawson's [20] "Basic Argument": that for deep free-will an entity must be responsible for shaping its own form and it provides a motivation for rejecting dualism (the idea that the "mind" is not created from the material universe).

### 2.3. Emergence and Downward Causation

Considering the forgoing, we might ask what is responsible for making an autopoietic system (e.g., an organism); is it the components themselves, or is it the organisational system. We might further ask: in either case, what really is the "system". Cybernetics provides an answer to the second question, in that the system is the organisational pattern-information embodied in a particular configuration of interactions among the component parts. Because it is abstract of its material embodiment, it is "multiply realisable", i.e., composed of members of functionally equivalent parts (see Auletta et al. [41]; and Jaeger and Calkins [42] for biological examples). It is not the identity of the components that matters, rather it is the functions they perform (e.g., a digital computer may be embodied by semiconductor junctions, or water pipes and mechanical valves, without changing its identity). Crucially, "function" is defined by a relationship between a component and the system of which it is a part. According to Cummins [43], "function" is an objective account of the contribution made by a system's component to the "capacity" of the whole system. At least one process performed by the component/s is necessary for a process performed by the whole system. This implies that the function of a component is predicated on the function of the whole. This definition was recently modified to more precisely specify the meaning of "capacity" and of whole system, thus: "A function is a process enacted by a system A at organisational level $L$ which influences one or more processes of a system B at level $L + 1$, of which A is a component part" [44].

In this context, organisational level means a structure of organisation that is categorically different from those above and below in the hierarchy because it embodies novel functional information (levels may be ontological or merely epistemic in meaning: that is an open debate in philosophy). The self-organisation of modular hierarchy has been described as a form of symmetry-breaking phase transition [45], so the categories either side are quantitatively and qualitatively different. Organisational levels were defined precisely in terms of meshing between macro and micro dynamics (from partitioning the state-space of a dynamic system) by Butterfield [46] and also using category theory to specify supervenience relations and multiple-realisability among levels by List [47]. Neither definition, though, deals specifically with the phenomenon of new pattern-information "emerging" from the organisation of level $L$ components at level $L + 1$, which is responsible for the emergence of new phenomena.

Ellis [48] shows that a multiply realisable network of functions, self-organised into a functional whole, emerges to (apparently) exercise "downward causation" upon its component parts [48,49]. The organisational structure is selecting components from which to construct itself, even though it is materially composed of only the selected components. Since it is purely cybernetic (informational) in nature, the downward control is by pattern-information [42,48] which transcends the components from which it is composed. The pattern-information arises from, and is embodied by, the interactions among the components, and for these reasons it was termed a "transcendent complex" by Farnsworth et al. [50]. Examples are to be found in embryonic development, where a growing cluster of cells self-organises using environmental signals created by the cells taking part [51] and the collective decision making of self-organising swarms (e.g., honey bees in which the hive acts as a unity [52]).

There is something significant here for those who conflate determinism with causation. All causal paths traced back would be expected to lead to the early universe. Despite the appearance of near maximum entropy from the uniformity of background microwave radiation, there is broad agreement that the entropy of the early universe was low and its embodied pattern-information (complexity) could not account for the present complexity, including living systems [53]. Novel pattern-information has been introduced by selection processes, especially in living systems, for which Adami et al. [54] draw the analogy with Maxwell's demon. Selection is equivalent to pattern matching, i.e., correlation, and is accompanied by an increase of information. Since its beginning, the entropy of the universe has been increasing [53] and some of this has been used as a raw material for transformation into pattern-information. This is achieved by creating the "order" of spatial correlation through physical self-assembly (atoms into molecules into molecular networks into living systems). This self-assembly embodies new information in the pattern of a higher level structure through the mutual provision of context among the component parts [3]. The process of self-assembly is autonomous and follows a boot-strap dynamic, so it provides a basis for answering Strawson's [20] "Basic Argument" in which the putative agent of free-will is an informational (pattern) structure of self-assembly.

### 2.4. Purpose and Will

Much of the literature on downward causation uses the idea of "purpose", though many are uncomfortable with its teleological implication. The aim of this section is to form a non-teleological account of purpose and its connection with will in non-human agents.

Cause creates correlation (usually, but not necessarily, in a time-series): the pattern of any action having a cause is correlated with its cause. An action without cause is uncorrelated with anything in the universe and accordingly considered random. If an action is fully constrained, then its cause is the constraint. Thus, freedom from at least one constraint allows the cause to be one of either: random, or exogenous control or agent control (in which "control" means non-random cause). By definition the cause is only taken to be the agent's will if it originated in agent control. Correlation alone, between some outcome variable $x$ and some attribute $a$ of the agent, is not sufficient to establish will: (a) because correlation has no direction (but metrics such as "integrated information" [55] can resolve direction) and (b) because $a$ may itself be random in origin and thereby not of the agent's making. Marshall et al. [56]

showed that cause can be established at the "macro" level of agent (as opposed to the "micro" level of its components) using an elaboration of integrated information, so the pattern in $x$ can be attributed to agent-cause. Because the agent-based cause could be random ((b) above), we must form and test a hypothesis about the effect of $x$ on the agent before we can attribute the cause to the alternative of agent-will. The hypothesis is that the effect of $a$ on $x$ is to increase the overall functioning $F$ of the agent. If this were true, then to act wilfully is to reduce the entropy of $x$, by increasing the probability of an outcome $x'$ where $x' \Rightarrow F' > \bar{F}$ (and $\bar{F}$ is the average $F$). That means that the mutual information between a wilful action $a(t)$ at time $t$ and the resulting function $F(t + \tau)$, $\tau \geq 0$ is greater than zero. This mutual information between action and future functioning is taken to imply a "purpose" for the action, so purpose is identified by the observations that:

$$H(a) + H(F) > H(a|F) \text{ and } F|a > F|r, \tag{1}$$

where $r$ is a random (comparator) variable. This is clearly an observational definition and is in some way analogous to the Turing test, but it is a test for purpose rather than "intelligence". It represents our intuition that if a behaviour repeatedly produces an objectively beneficial outcome for its actor, then it is probably deliberate (repeatedly harmful behaviour is also possibly deliberate, but all such actions are regarded as pathological and thereby a subject beyond the present scope).

To recap, for attributing the action to the agent's will, we must at least identify a purpose so that Equation (1) is true. The purpose is a pattern embodied in the agent, which acts as a template for actions of the agent that cause a change in future states (of the agent, its environment, or both). We may call this pattern a "plan" to attain an objective that has been previously set, where the objective is some future state to which the plan directs action. Specifically, let the objective be a state X (of the system or the world, etc.), which can be arrived at through a process P from the current state Y, then the purpose is a "plan" to transform $Y \rightarrow X$ by the effect of at least one P and at least one function $F$ is necessary for the process P to complete.

The homoeostatic response to a perturbation, for example, has maintenance at the set-point as its purpose. Y is the perturbed state, $F$ is some function of the internal system having the effect of causing a process P, i.e., some transition $Y \rightarrow X$. In general there is more than one P and more than one $F$ for achieving each. This results in a choice of which to use: it is a choice for the agent described by the system. To make a choice requires a criterion for choosing (else the outcome is random and therefore not a choice). The criterion for choosing is a "goal" G, consisting of one or more rules, which identify a location in a function describing the outcome (which we may call the objective function). In general, this location could be any and it is essential for freedom of will that it be determined by the agent of action alone. However, in practice it is most likely to be an optimisation point (in living systems, this is implied by Darwinian evolution and in designed systems, it is the basis of rational design). Therefore, narrowing the scope, but with justification, let us take the criterion for choosing to be a goal G, consisting of one or more optimisation rules. For example, of all the possible systems performing homoeostasis, the purposeful one is defined as enacting P' such that $Y \rightarrow X$, with $P' \in P| \max(\mathcal{G})$, where $\mathcal{G}$ is an objective function for which the optimisation goal $G(\mathcal{G})$ is satisfied, contingent upon the options (e.g., P proceeds as quickly as possible, or with minimum energy expenditure, etc.). Accordingly, "will" is defined by a purpose which is a plan to enact a process causing a transition in state, "as well as possible" (according to $G(\mathcal{G})$), notwithstanding the earlier comment about pathological purposes. A free-will agent has a choice of transition and a goal which identifies the most desirable transition and the best way to enact it, from those available. These two choices can be united (by intersection), without loss of generality, to one choice of best transition.

One of the reasons for objecting to teleological terms such as "purpose", "plan" and "goal" in relation to natural systems has been the belief that a plan implies a "designer", the concept at the centre of the most famous battles between science and religion. This implication is not necessary and is rejected here (following the argument of Mayr [57]). A plan is merely a pre-set program of steps taking the system from Y to X; it is the concept for which computation theory was developed. It may be designed (the work of an engineer), but also may have evolved by natural selection (which also supplies the goal, in which case it is a teleonomic system (sensu Mayr [57]).

A plan, as an ordered sequence of transformations, is an abstraction of information from the physical system, which for free-will must be embodied within the system. A more subtle implication of "plan" is that as a path leading from Y to X, it is one among several possible paths: different plans may be possible, perhaps leading to different outcomes. There is a fundamental difference between this and the inevitability of a dynamic system which follows the only path it may, other than by the introduction of randomness. The reason is that for a dynamical system all the information defining its trajectory is pre-determined in the initial (including boundary) conditions and the laws of physics. The initial conditions constitute its one and only "plan". If a system embodies pattern-information (by its structure) which constitutes a developed plan, then this pattern-information may direct the dynamics of the system along a path other than that set by the exogenous initial conditions (though we may consider the structure of the system to be a kind of initial condition). The point is that the embodied plan gives freedom to the system, since it "might be otherwise'; there could be a different plan and a different outcome. We see this in the variety of life-forms: each follows its own algorithm of development, life-history and behaviour at the level of the individual organism. The existence of a plan as abstract pattern-information is a pre-requisite for options and therefore freedom of action.

*2.5. Goal, Master Function and Will Nestedness*

Now let us complete the connections between will, goal and function. The previous argument reveals an important difference between downward and any other kind of causation (considered important by Walker [58]): the former must always be directed by a purpose, for which we need to identify a goal (upward and same level causation are satisfactorily explained by initial conditions [48]). Viewing entities and actions both as the consequence of information constraining (filtering) entropic systems, then the role that is taken by initial conditions in upward causation, is taken by system-level pattern-information (the transcendent complex [50]) in downward causation. Since the goal G is a fixed point in an objective function $\mathcal{G}$, it constitutes information (e.g., a homoeostatic set-point) that must be embodied in the agent's internal organisation. Since the objective function $\mathcal{G}$ represents the overall functioning of the system (at its highest level), it matches the definition given by Cummins [43] and Farnsworth et al. [44]. The highest level function from which we identify the purpose of a system was termed the "master function" by Jaeger and Calkins [42], so the will of an agent is instantiated in the master function. This then identifies $\mathcal{G}$ with "master function" and "will" with G($\mathcal{G}$).

Ellis [48] identifies five types of downward causation, the second being "non-adaptive information control", where he says "higher level entities influence lower level entities so as to attain specific fixed goals through the existence of feedback control loops..." in which "the outcome is not determined by the boundary or initial conditions; rather it is determined by the goals". Butterfield [46] gives a more mathematically precise account of this, but without elaborating on the meaning or origin of "goals". Indeed, as both Ellis [48] and Butterfield [46] proceed with the third type: downward causation "via adaptive selection" they refer to fitness criteria as "meta-goals" and it is clear that these originate before and beyond the existence of the agent in question. Ellis [48] describes meta-goal as "the higher level "purpose" that guides the dynamics" and explains that "the goals are established through the process of natural selection and genetically embodied, in the case of biological systems, or are embodied via the engineering design and subsequent user choice, in the case of manufactured systems".

This suggests a nested hierarchy of goal-driven systems and for each, the goal is the source of causal power and as such may be identified as the "will" (free or otherwise). We may interpret the definition of "deep freedom" [13] as meaning that an agent has at least two nested levels of causal power, the higher of which, at least, is embodied within the agent (as causal pattern-information). This concept may be formalised after introducing the discrete variable "will-nestedness" $\mathcal{N}$ which counts the number of levels of causal power exercised over a system, from within the agent as a whole (i.e., at the level of master function), the $\mathcal{N}$-th level being the highest-level internal cause of its actions.

Among organisms in general, the master function specifies the criteria by which the organism is to assess its possible future reactions to the environment. It is so much an integral part of the organism that without it, the organism would not exist. However, it was not chosen by the organism (in the sense of deep free-will) because it was created by evolutionary filtering and inherited from its parent(s); as all known life has been created by the previous generation copying itself. For single celled organisms the biological master function is to maximise their cell count by survival and reproduction, but in multicellular organisms, this master function exists, by definition, at the level of the whole organism (the unconstrained drive to proliferate a single cell line leads to cancer). Organisms with a central nervous system, regulated by neuro-hormone systems, with their corresponding emotions, can implement more complex (information rich) and adaptable (internally branched) algorithms for the master function, which may include will-nestedness $\mathcal{N} > 1$. In humans, this is taken to such an extent that the biological master function may seem to have been superseded (but the weight of socio-biological evidence may suggest otherwise [59,60]).

### 2.6. The Possibility of Choice and Alternative Futures

So far I have identified organisational-closure and the internal generation of a goal-based plan as prerequisites for free-will, but have not yet addressed the "alternative futures" problem relating to an agent constructed from elemental components that necessarily obey physical determinism. List [18] provides a philosophical argument for meeting this requirement, constructed from supervenience and multiple realisation of an agent in relation to its underlying (micro) physical level: "an agent-state is consistent with every sequence of events that is supported by at least one of its physical realizations" [18]. He shows that this may apply not only to multiple micro-histories up to $t$, but in principle includes subsequent $(t + \tau)$ sequences at the micro-level, which may map to different agent states and therefore permit different courses of action at the (macro) agent-level. To explain: for any given time $t$, the macro-state $Q_i(t)$ is consistent with a set of micro-states $\mathbf{s}$, at least one of which $s_i \in \mathbf{s}$ may lead (deterministically) to a new state $s_j \in \mathbf{s}$ at $t+1$, with which a different macro-state $Q_j$ is consistent, thus giving the agent a choice of which micro-state history to "ride" into $t+1$ (this idea is developed with rigour by List [18], and illustrated with "real-world" examples; it is the basis on which he concludes that agents may be "free to do otherwise", despite supervening on deterministic physical processes).

Alternative futures may be produced at multiple levels of system organisation within a hierarchical structure, by re-applying the principles identified by List [18] for each level of macro-micro relations. For any system level $L$ to have the potential for alternative futures, it must have the attributes of an "agent-level": supervenience and multiple realisation such that pattern-information with causal power emerges at level $L$ from $L - 1$: i.e., a transcendent complex exists at level $L$. However, this does not necessarily give free will to a system of that level, since for that, it must be organisationally closed. If it were not so, we would not be able to identify the system at level $L$ as an entity to which free-will could be ascribed. Thus will-nestedness cannot be attributed to levels of organisation below that of the Kantian whole. Since the Kantian whole is, by definition, the highest level of organisation to which free-will may be ascribed (any causal power beyond it rules out its free will), then will-nestedness can only apply at the level of the Kantian whole. Given this, the will-nestedness must be constructed from purely organisational, i.e., pattern-informational and therefore be purely computational in nature. This is an important deduction: free-will can only be an attribute of a Kantian whole and it can only result from the cybernetic structure at the level of the Kantian whole.

## 3. Choosing Possible Futures: The Computational Condition

We see that for free-will, an agent must have an independent and internally generated purpose for action and that this requires it to be organisationally closed. Free-will further requires the agent to use this purpose to choose among options. To do so, it needs an internal representation of possible futures from which to choose and an internally generated means of choosing. We now turn to the conditions which enable these essentially computational facilities.

*3.1. Information Abstraction*

The organisational boundary is where internal is distinguished from external. If the agent in question had no links of any kind between internal and external, then it would be unable to respond to, or use, its environment and in that case it could not be behaviourally autonomous as Froese et al. [22] defined it. Therefore to be both behaviourally and constitutively autonomous an agent must be disconnected from effective cause (also termed efficient cause), but able to perceive at least some aspect of its environment and act upon what it perceives (this being the foundation of cognition). The way this is achieved is via one or more transducer on the boundary, which acts as an intermediary between external and internal, allowing causal power without permitting effective cause to pass through the boundary. The transducer is a system which transforms information from one medium of embodiment to another (in analogy with an engine which transforms energy from one form to another). As information crosses the interface between one medium and another, it looses its effective causality. The reason is that information causality is mediated by physical forces. What is meant here by information causality, is that changes in an effective force (within a medium) constitute a pattern in force which as a pattern, constitutes information in the sense defined in Section 1. Forces are effective only within the medium to which they belong. The transducer has two sides, each belonging to a different medium, so each is causally linked to a different medium. Therefore fluctuations in force on one side exert no effective cause on the other side (because it is a different medium). However, the transducer allows a correlation between the fluctuation on one side and the other. What happens between the sides is a transformation of the information from one medium to the other and during that transformation the information looses its physical causal effect, but the transducer passes through its causal power via correlation, which itself may be modulated by the transducer. The disconnection of information from its medium is termed information abstraction because information without a medium of embodiment cannot really exist, it is an abstraction.

Information abstraction at the organisational boundary is crucial to achieving autonomy because it strips off the physical effect of the external environment to take only the abstract information which is then used as a signal in cognition. Causes are transformed into signals, their effects being rendered responses (which thereby may become optional). It is this separation of information (as signals) from the physical force of cause and effect that releases the agent from attachment to the cause-effect determinism of its environment. The material apparatus for performing this task is a transducer: the tegumental membranes of bacteria and other cells contain a wide variety of transducers (receptors) and we expect an artificial robot to be well equipped with them too. This is not a merely technical point. The closure condition gives the system a degree of causal independence and the boundary transducers give it a sensitivity to its environment whilst preserving this independence. The boundary is the place where the inevitability of cause and effect of the environment meets that of the internal processes of the system and the transducers are the interface between these causal chains. According to one interpretation, internally to the system, the environment is "reduced" to abstract, representational, information by the transducers [61]. Now the question is, what must the agent do with this information in order to exercise free-will?

Of course the answer is to compute: more specifically, to perform transformations on the data as a result of a sequence of physical changes in the physical structure of the agent. Such changes are described by automata theory, for which the most basic automaton has two states and can potentially change state on receiving a signal to which it responds: it is a switch (e.g., a protein molecule with two conformations is an "acceptor" of all strings from the alphabet $\{0, 1\}$, where these symbols may represent the presence/absence of e.g., another molecule or a level above a threshold (e.g., temperature)). Obviously, the switch is the elemental component for generating discrete options. A less obvious, but crucial property of the switch for the physical embodiment of computation is "thermodynamic indifference". Walker and Davies [62] focus on computation in explaining the origin of life, referring to genetics-first theories as "digital-first", emphasising the need for "programmability" and its provision by informational polymers (the genetic oligomers RNA, DNA etc.). By programmability they meant that components of a system are configured so that the system state can change reversibly, approximately independent of energy flow: i.e., changes of state are not accompanied by substantial changes in potential (stored) energy. If they were, then switching would always be biased by the difference in energetic cost between e.g., switching on and off. Energetically unbiased switching is the physical underlying mechanism of "information abstraction" referred to by Walker and Davies [62]. In reality, switching (and state-changes in general) always have energetic consequences (more deeply, there is always an exchange of entropy between the system and an external energy source), which is one of the reasons an autonomous agent must complete work cycles as Kauffman [63] specifies. What makes the informational polymers (e.g., DNA) of life special is the fact that they are reversible in a way that is thermodynamically indifferent (or very nearly so; see Ptashne [64]). Any ordered set of $n$ switch positions (e.g., 1,1,0,0) has very nearly the same potential energy as any other ordered set of $n$ (e.g., 0,1,0,1).

For free-will, both the self-assembly of autocatalytic systems and the computational requirements (switches and memory) are jointly necessary. Walker and Davies [62] and Walker [58] proposed that the autonomy of living systems arises from the combination of "analogue" chemical networks and "digital information processing". Accordingly, a hybrid automaton is a good model for the construction of a free-will agent. Figure 1 (from Hagiya et al. [10]) shows an example of a hybrid automaton which combines discrete-state with continuous dynamical systems, such that the discrete states (as modes of functioning) determine the system's responses to dynamic variables and these responses potentially influence the trajectory of the dynamics. In this pedagogical example from Hagiya et al. [10], there are two state variables $\alpha$ and $\tau$ which determine the set points for autonomous chemotaxis behaviour in the bacterium represented. This system can freely maximise its (experienced) environmental concentration of $x$ (e.g., glucose), so its goal G($\mathcal{G}$) is defined, but it cannot choose how (hence $\mathcal{N} = 1$), so it cannot express free-will in the deep sense. However, if it had independent control over $\alpha$ and $\tau$, with the ability to adjust these values according to a plan of its own making, then it would fulfil the condition of having willed its own behaviour ($\mathcal{N} = 2$), at least in the sense defined in the previous section. As part of a living bacterium, the system depicted would be a component of a Kantian whole (the free-living organism), so all that is missing for free-will is a plan for determining $\alpha$ and $\tau$ according to an internally generated goal (a master function) and some means of computing this. The computational requirements for free-will are identified as follows.

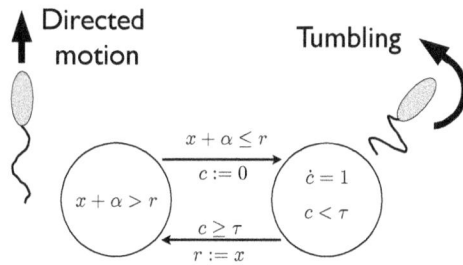

**Figure 1.** A bacterial chemotaxis controller described as a hybrid automaton, realised in practice by e.g., Escherichia coli species, but also as an engineering design for a molecular robot, using DNA-based components by Hagiya et al. [10], from which the figure, slightly modified, is taken. Note that $\alpha$ and $\tau$ are internal set-points, which constitute pattern-information embodied in the molecular robot's structure and are ultimately determined by Darwinian evolution (natural robots) or intentional design (human artefacts). The bacterium responds to the environmental concentration (e.g., of glucose); its objective function $x$, which is detected by a membrane transducer which generates the internal variable $r$. It is searching for a higher concentration when $x$ is below a set threshold (directed motion) and tumbles randomly for a time set internally by $\tau$ using an internal "clock" signal $c$, whenever the concentration is at least equal to the threshold. The swimming and tumbling are modes of action of its flagella. The objective function of this system is the experienced concentration $x$ and the goal is max($x$). The chemotaxis controller exhibits will-nestedness of one.

### 3.2. The Representation of Self

Firstly, a free-will agent must maintain an internal representation of itself, and also the effect of its environment on its internal state, to enable it to assess each of its options for action.

At the root of automata theory lies an attempt to fully describe (and therefore predict) the behaviour of a system without a detailed mechanistic account of its internal workings (the black box approach). The system is captured in the mapping between environmental stimuli and responses:

$$R(t+1) = W(\mathbf{S}(t), S(t)) \tag{2}$$

where $R(t)$ and $S(t)$ are the response and stimulus, and $\mathbf{S}(t)$ is the history of stimuli experienced by the system, from the beginning of its formation up to $t$ and $W(\cdot)$ is the mapping function. This presents an immediate problem, since in general $\mathbf{S}(t)$ is arbitrary and infinite in range (it is instructive to think of Strawson's [20] "Basic Argument" in terms of Equation (2): the response of a system, in general, depends on its environment from before the system came into existence). The solution to this indefinite $\mathbf{S}(t)$ problem (provided by Moore [65]) is to assume that the infinite set of $\mathbf{S}(t)$ may be partitioned into a finite number of disjoint equivalence sets, each containing the histories that are equivalent in their effect on $R(t+\tau)$, $\forall \tau$, where $\tau$ is in the interval $[0, \infty]$. These equivalence sets are represented as states $Q(t)$ of the system, so that:

$$R(t+1) = W(Q(t), S(t)) \text{ and } Q(t+1) = U(Q(t), S(t)), \tag{3}$$

where $U(Q(t), S(t))$ is the transformation function dictating the transition of system state given its present state and that of the stimulus (see Minsky [66], pp. 16–17). Thus $Q(t)$ represents how the system is now, given its previous history of experiences. $Q(t)$ corresponds to the agent-state of List [18], which is multiply-realisable and which is, at least in principle, free to take more than one value at some point in the future $t + \tau$, despite supervening on a a wholly determined set of micro-histories. List [18] showed the possibility of choice at the agent-level, but this does not necessarily mean that $Q(t)$ is indeterminate. Specifically, for free-choice (of $Q(t+1)$ and implied $R(t+1)$), the direction taken at the branching point $t$ must be determined by a process internal to the agent that represents its "purpose" (as defined earlier). For the choice to be purposeful, it must be based on an assessment of the outcomes that would arise from choosing each of the options. This entails a prediction of possible futures, for which a free-will system must have a model of itself in its environmental context.

The question now is, how can a system create such a representation by and for itself, not "programmed" by some exogenous source of information? The answer seems to be as it is with material self-assembly: a boot-strap, step by step, gathering of pattern-information embodied in form, such that as the form grows, it increases in complexity. In the particular case of building a model of self and environment, this process is one of learning, for which the field of "machine learning" provides our understanding. Well known advances in this field have already led to sophisticated learning among pre-existing (i.e., not self-assembled) computation systems such as deep neural networks etc. The difference here is that the learning is not merely a statistical problem, but one of simultaneous self-construction, which must begin with simple systems, so in the remainder of this section, only basic and simple systems capable of unsupervised learning are discussed.

#### 3.2.1. Learning in a Constant Environment

The most basic form of learning is operant conditioning (reinforcement learning), described mathematically by Zhang [67] (cited in Krakauer [68]) as follows from the description provided by [68]. Let $R_i$ be one of a set of $N$ possible responses ($R_i \in \mathbf{R}, i \in [1, N]$) in a constant environment, occurring with probability $r_i$. For each response there is a "reward" $\rho$ (which coincides with the objective function $\mathcal{G}$ that defines the goal of the system: $\rho \to f(\mathcal{G})$), so that the incremental change in probability of the $i$-th response is:

$$\Delta r_i = a\rho_k(\kappa_{k,i} - r_i), \tag{4}$$

where $\kappa_{k,i}$ is the Kronecker delta function (equal to 1 if $i = k$, else equal to zero) and $a$ is a learning rate constant. Since the average change in response over the ensemble of possible responses is the frequency-weighted sum: $\Delta \hat{r} = \sum_k^N r_k \Delta r_k$, the result is that the frequency of the $i$-th response incrementally increases in proportion to the difference between its reward and the average over all rewards:

$$\Delta r_i = a r_i (\rho_i - \hat{\rho}), \text{ where } \hat{\rho} = \sum_k^N r_k \rho_k. \tag{5}$$

The dynamic quantified in Equation (5) describes learning by maximising the reward experienced. Such learning is equivalent to making an increasingly accurate model of the (static) relationship between the agent's internal state and the environment, via (Bayesian) "trial and error" sampling of responses. Given a constant environment, the solution to Equation (5) yields a single, reward maximising response: $R^* = R_i$ such that $\rho_i = \hat{\rho}$ and $r_i = \kappa_{k,i}$.

To achieve this in practice the system must at least keep a record of the reward for the last response made and the average reward, for which an automaton with an external memory is required (e.g., a push-down automaton, though this is still essentially a determinate finite state automaton DFA). Quantifying the complexity of such a system is probably best achieved through a programme (algorithmic) complexity measure since the information instantiated by such an automaton is almost all in its transition mapping and there are robust methods for reducing this to the minimum description, leading directly to the Kolmogorov complexity. The process of learning can be interpreted in information terms: if the starting probability distribution of $\mathbf{R}$ is $\tilde{r}$, then the initial Shannon entropy of the system is $H = -\sum_i^N \tilde{r}_i \log(\tilde{r}_i)$ and the final entropy is zero: having completed its learning, the system has no uncertainty about the best way to respond to this environment. In this state, the automaton is a complete representation of its interaction with its environment (i.e., the distribution of rewards over its repertoire of responses) and it embodies exactly $H$ units of information: a quantity which should match the algorithmic complexity measure (though not tested here).

### 3.2.2. Extension to a Variable Environment

Generalising to a variable environment, for which a set of finite states $\mathbf{S}$ is an adequate representation, there would exist a reward maximising response for each state: $R_i^* \to (S_i)$, such that $R_i^*$ solves Equation (5). The agent may choose to maximise its reward over all $\mathbf{S}$, and to enable this, it must learn the best response for every $S_i \in \mathbf{S}$. In information terms, first let $H(\mathbf{s})$ be the entropy of the environment having probability distribution $\mathbf{s}$ and $H_t(\mathbf{r_t})$ be that of the responses, given their probability distribution $\mathbf{r_t}$ at time $t$. The Shannon information the agent has about its environment (in terms of its rewards) is:

$$I_t(\mathbf{s} : \mathbf{r_t}) = H(\mathbf{s}) + H(\mathbf{r_t}) - H(\mathbf{s}, \mathbf{r_t}) \tag{6}$$
$$= H(\mathbf{s}) + H(\mathbf{r_t}) - H(\mathbf{s}|\mathbf{r_t}),$$

meaning that the agent is learning both the distribution $\mathbf{s}$ and the reward associated with each $S_i \in \mathbf{S}$. This mutual information is embodied in the structure of the agent and can be used as a measure of its complexity. The structure of the agent may be too simple to embody as much as the maximum mutual information, in which case its learning will be limited and it will not make optimal responses, so Equation (6) is a measure of the minimum complexity required for optimal behaviour from the agent. It would be possible for an agent to implement this learning system by "growing" multiple copies of the DFA with memory (one for each $S_i \in \mathbf{S}$) that is used for a constant environment. The output of each of these would then be the input to a further DFA which it uses to maximise the reward across all of them. This "growth" would be enacting meta-learning: the agent would increase its complexity in response to rewards. The number of states in **S** is not known by the agent a-priori, so the number of DFAs needing to be "grown" is indeterminate. Further, account should be taken of the extended time needed to perform such laborious learning and the consequences of the agent being wrong in so many trials. It seems that for practical reasons there comes a point when a more powerful kind of computation becomes necessary. In computation terms, such a learning problem requires at least a finite and non-volatile memory, effectively to store multiple instances of the single learning problem encountered in a constant environment. For this reason a Turing Machine would be a more realistic option.

### 3.3. The Free-Will Machine

These computational requirements for free-will to become possible are brought together in the hypothetical "free-will machine" of Figure 2. This information processing must be implemented by the agent to which free-will is ascribed and that agent must, further, be a Kantian whole for the requirements of free-will to met. The current state of the environment (external) and of the agent (internal) are derived by information abstraction from the physical world: the array of receptor molecules in the cell membrane, the nervous senses of an animal, or the transducers of a human artefact all perform this task. The first Turing machine implementation TM1 constitutes a representation of the agent in the present (relevant aspects of its environment are included) and is informed (updated) by the state information $Q_t$ and $S_t$. The function of this representation is to identify the set of possible responses $\underline{R}_t$ that the agent can make, given $Q_t, S_t$ (underline notation now denotes a set).

TM2 uses these hypothetical responses to compute the set of possible futures at a time $t + n$ ($n$ may take any positive value) $\underline{F}_{t+n}$ for which in a simple case TM2 : $\underline{R}_t \rightarrow \underline{F}_{t+n}$ is a map of responses onto possible futures (simple, because there is not necessarily a 1 : 1 relation between $\underline{R}_t$ and $\underline{F}_{t+n}$, but we need not be concerned with that at present). There is no limit on the number of possible futures that TM2 may compute, but it must be at least 2 for a choice to exist. These possible futures are each represented by a set of states $\underline{f}$, each member $f_n$ being equivalent to a prediction of a possible $Q_{t+n}$. The finite state automaton FSA chooses from among these, using a selection criterion based on the objective function defined by a goal G, which is generated by the agent (not exogenously). This goal is the maximisation of the master function, (e.g., for a living agent this is life-time reproductive success). The goal enables the optimal possible future state to be recognised (it is the one which maximises the master function) and this future state $f'$ implies an optimal response $r'$ (in general there could be more than one, in which case the agent will be indifferent among them). Having selected an optimal response, the agent then must implement it in the physical realm. Since the computation of $r'$ has been conducted in the realm of information, this step appears to involve the control of material by information. In practice all the computation and indeed all the information is instantiated by material and energy acting in the physical realm, so our cybernetic model is merely an abstract representation of the organisation of the physical processes which lead to the implementation in the physical system and this return to physical reality is represent by the action (IPS). Such implementation inevitably results in a transformation of the agent into a new state $Q_{t+1}$, together with $S_{t+1}$ and this restarts the cycle. It may be noted that Von Neumann's self-replicating automata are proven universal Turing machines [39] and Turing machines are thought to be common among living systems, so this computational arrangement is not beyond the bounds of possibility.

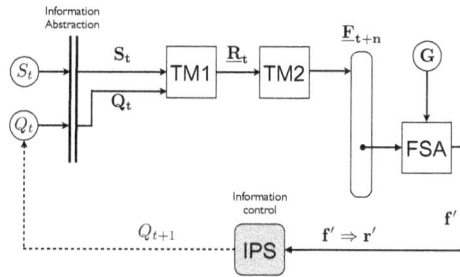

**Figure 2.** A conceptual "free-will" machine, which generates predictions of its state in alternative futures $\underline{F}_{t+n}$ using an internal representation of itself interacting with its environment, selecting the optimal from among these, using a goal-bases criterion G, in which the goal is internally determined (further explanation in the text).

## 4. Discussion and Synthesis

To summarise, the essential requirements for free will are:

R1    There must be a self to self-determine.
R2    There must be a non-zero probability of more than one option being enacted.
R3    There must be an internal means of choosing among options (which is not merely random, since randomness is not a choice).

For R3 to be fulfilled:

R4    Options must be generated from an internal model of the self which can calculate future states contingent on possible responses.
R5    Choosing among these options requires their evaluation using an internally generated goal.
R6    For "deep free-will", at least two nested levels of choice and goal (R4–R5) must be enacted by the agent.

R1 and references to "internally generated" are fulfilled by organisational closure. For R2, the possibility of options, which implies "multiple futures" has been established for the level of the agent by List [18]. R3 and its predicates R4–R5 imply a minimum level of computational power, which in principle can be met by a small set of Turing machines, which may in principle be implemented by a von Neumann architecture computer, a network or cellular automaton-based or any other sort of computer, including a biomolecular system such as found in higher animal life, but it seems to be beyond the power of a single living cell (though that last point is not yet established). R4 in particular seems to require a finite memory (the size depends on the complexity of the agent and its environment) and R6, the qualifier for deep free-will, adds a little more to the computing power necessary, but it is important to note that this extra is not a step-change: it is not qualitatively more demanding than the automated decision making required by R3.

The question of free-will is not one of whether an agent's actions are caused, since all actions ultimately have a cause. The ultimate cause of any action can be understood as resulting from selection over random actions by a pattern, which leaves a correlation with the pattern that caused the selection (instantiating pattern-information). All living organisms, including people, were produced by information-pattern filtering, proximally by molecular replication (creating inheritance) and ultimately by Darwinian selection. All human artefacts were created by following a design pattern (though it may not have been completed before artefact construction), so they correlate with their design. Even inanimate objects, such as stars, lakes and sand grains, owe their form to the information-pattern of underlying physical laws, Pauli's exclusion principle and the distribution of matter and energy in space following the big-bang. To this, we must add randomness which has been entering as "informational raw material" into the universe, disrupting the original patterns and opening opportunities for novelty (evolutionary for life) and more widely directing the course of the universe in unexpected ways as its history tracks a course in the highly ergodic space of possibilities.

Taking Strawson's [20] Basic Argument seriously, this pattern-correlation and the injection of randomness both deny free will. From them, we obtain a model in which the identity of all things, including human beings, is an illusion: as if the universe was all one complex manifestation which only appears to include separate agents. Closer inspection shows how the nested-hierarchical construction of this complexity entails the creation of genuinely new pattern-information, caused by and embodied in the interactions among component parts of putative agents. This novel pattern-information transcends its component parts and can exert downward causation upon them. Some structures (such as autocatalytic sets) created this way are organisationally closed (though materially and thermodynamically remain open systems). Because of this, their internal dynamics are, at least partially, separated from the external dynamics of their environment and this gives them an organisational boundary, enabling internal to be defined against external. At this boundary, external and internal chains of cause and effect interact through transducers which transform physical determinism into stimulus-response relations. Systems with these properties are essentially cybernetic and although their low-level processes are continuous with the rest of the universe, List [18] has shown that in principle they may have options for their next state and response: they are freed from physical determinism. To translate this freedom into free-will, requires that the (partially) independent agent chooses from among its options and this entails an internal computation of possible futures and their evaluation against a goal representing the fulfilment of the agent's "master function" (i.e., its purpose). This goal is a fixed point in an objective function which may be simple (as in a homoeostatic system), but also arbitrarily complex and multi-layered, taking account of multiple time-scales and interactions with other agents. If the objective function is at least two-layered (will-nestedness $\mathcal{N} \geq 2$), then it effectively has a choice of what to choose and thereby could fulfil the established definition of (deep) free-will [13,16]. This calls into question the idea that free-will is an all or nothing capacity, instead, it suggests free-will to be a discrete quantity and even something we could in principle measure as a trait of a system.

The reason is that deep free-will has so far been defined as the freedom to choose ones will, but the analysis presented here shows that to be wilful, a choice must be purposeful, which means optimising an objective function. Freedom is in the choice of objective function. Since therefore, the core of will is the objective function, deep freedom is the freedom to choose this, but to be wilful, this choice in turn must optimise a hierarchically superior objective function, which must have been determined by something. We can conceive of a large but finite nested set of such objective functions, but ultimately the highest of them all must be provided either arbitrarily (e.g., at random) or by natural selection (or its unnatural equivalent), or by design: in all cases, not the free choice of the agent. This applies even to human beings, who are still subject to "design" by inheritance, selected by evolution. The depth of free-will is therefore a discrete, finite variable (will-nestedness $\mathcal{N}$) and we may speak of one kind of agent being more deeply free than another, but no kind of agent can be ultimately free-willed in the sense meant by deep free-will (presumably, humans attain the highest will-nestedness of known agents).

Seeking an ultimate cause of will leads to an infinite regress because for (second order) will-setting to be "rational" (meaning a reason underlies it), it must be based on the setting of a (higher order) will. This introduces a third order rational basis of the will-setting and so on ad infinitum. Here the problem is explained in the more concrete and formal terms of fixed points (goals) in objective functions. The concept is made sufficiently specific to quantify (as nestedness) and used to conclude that no agent can be ultimately free willed in the strong source-theory sense of being ultimately responsible. Philosophers (such as Strawson) conclude that if the ultimate source is not of the agent, then the agent cannot have free will. The more quantitatively oriented analysis presented here adds a nuance to this, so we are not forced into an "all or nothing" conclusion about free will. It suggests that agents may be better characterised by the degree of free-will, expressed in terms of the will-nestedness, which can in principle be derived for any kind of agent under scrutiny. Therefore, whilst Frankfurt [12] asserts that only humans can have second order volition, but animals can have first order, I contend that any autonomous agent (including synthetic) can be characterised by the hierarchical order of their goal setting (be it zero, one, two or higher) and further that this order can be identified as the number of hierarchically arranged objective functions in their decision making (computation) that are embodied (as information) within their structure (what I call will nestedness).

If interpreting the philosopher's definition of deep free-will as $\mathcal{N} \geq 2$ is correct, then it is not hard to achieve in principle. The impediment to a non-life robot acquiring that sort of free-will is not computational, it is the closure constraint with its requirement for bootstrapping self-assembly (especially since this includes the "growth" of the sort of computational apparatus indicated in Figure 2). That is a problem already solved by life. Quite likely $\mathcal{N} \geq 2$ in most or all organisms having at least a limbic system, so free-will defined this way can be attributed to most or all vertebrates [69]. The computational requirements for exercising purposeful choices are not very challenging for artificial computers. Among human artefacts, including what most people would define as robots, the organisational closure condition is the major hurdle not yet leapt. This point has been recognised in the artificial intelligence literature, especially concerning "cognitive robotics" which emphasises embodiment [22,26]. For the time-being, it seems free-will, as defined here, is a unique property of living things, but the possibility of extending it to synthetic robots remains.

**Acknowledgments:** This work was unfunded, but was inspired by attendance at the workshop "Information, Causality and the Origin of Life" in Arizona State University at Tempe, AZ 30 September–2 October 2014, funded by the Templeton World Charity Foundation. The cost of open access publication was met by the Queen's University Belfast.

**Conflicts of Interest:** The author declares no conflict of interest.

## Abbreviations

The following abbreviations are used in this manuscript:

DFA     Determinate Finite State Automaton
DM     Determinate Machine
FSA     Finite State Automaton
NDM     Non-Determinate Machine
TM     Turing Machine
UTM     Universal Turing Machine

## References

1. Sharov, A.A. Functional Information: Towards Synthesis of Biosemiotics and Cybernetics. *Entropy* **2010**, *12*, 1050–1070.
2. Hazen, R.M. The emergence of patterning in life's origin and evolution. *Int. J. Dev. Biol.* **2009**, *53*, 683–692.
3. Farnsworth, K.; Nelson, J.; Gershenson, C. Living is Information Processing: From Molecules to Global Systems. *Acta Biotheor.* **2013**, *61*, 203–222.
4. Von Weizsäcker, C.F. *Die Einheit der Natur*; Deutscher Taschenbuch Verlag: Munich, Germany, 1974.
5. Stonier, T. Information as a basic property of the universe. *Biosystems* **1996**, *38*, 135–140.
6. Devlin, K.J. *Logic and Information*; Cambridge University Press: Cambridge, UK, 1992.
7. Floridi, L. Information. In *The Blackwell Guide to the Philosophy of Computing and Information*; Floridi, L., Ed.; Blackwell Publishing Ltd.: Hoboken, NJ, USA, 2003; pp. 40–61.
8. Szostak, J.W. Functional information: Molecular messages. *Nature* **2003**, *423*, 689.
9. Hazen, R.M.; Griffin, P.L.; Carothers, J.M.; Szostak, J.W. Functional information and the emergence of biocomplexity. *Proc. Natl. Acad. Sci.* **2007**, *104*, 8574–8581.
10. Hagiya, M.; Aubert-Kato, N.; Wang, S.; Kobayashi, S. Molecular computers for molecular robots as hybrid systems. *Theor. Comp. Sci.* **2016**, *632*, 4–20.

11. Pfeifer, R.; Bongard, J. *How the Body Shapes the Way We Think: A New View of Intelligence*; MIT Press: Boston, MA, USA, 2007.

12. Frankfurt, H. Freedom of the will and the concept of a person. *J. Philos.* **1971**, *68*, 5–20.

13. Kane, R. *A Contemporary Introduction to Free Will*; Oxford University Press: Oxford, UK, 2005.

14. McKenna, M.; Pereboom, D. *Free Will: A Contemporary Introduction*; Routledge: Abingdon, UK, 2016.

15. Coeckelbergh, M. *The Metaphysics of Autonomy*; Palgrave Macmillan: London, UK, 2004.

16. Westen, P. Getting the Fly out of the Bottle: The False Problem of Free Will and Determinism. *Buffalo Crim. Law Rev.* **2005**, *8*, 599–652.

17. Van Inwagen, P. *An Essay on Free Will*; Oxford University Press: Oxford, UK, 1983.

18. List, C. Free will, determinism, and the possibility of doing otherwise. *Noûs* **2014**, *48*, 156–178.

19. Van Inwagen, P. Some Thoughts on An Essay on Free Will. *Harvard Rev. Phil.* **2015**, *22*, 16–30.

20. Strawson, G. *Freedom and Belief*; Oxford University Press: Oxford, UK, 1986.

21. Kauffman, S.A. Autocatalytic sets of proteins. *J. Theor. Biol.* **1986**, *119*, 1–24.

22. Froese, T.; Virgo, N.; Izquierdo, E. Autonomy: A review and a reappraisal. In Proceedings of the European Conference on Artificial Life, Lisbon, Portugal, 10–14 September 2007; pp. 455–464.

23. Varela, F. *Principles of Biological Autonomy*; Elsevier: Amsterdam, The Netherlands, 1979.

24. Zeleny, M. What is autopoiesis? In *Autopoiesis: A Theory of Living Organization*; Elsevier North Holland: New York, NY, USA, 1981; pp. 4–17.

25. Rosen, R. *Life Itself*; Columbia University Press: New York, NY, USA, 1991.

26. Vernon, D.; Lowe, R.; Thill, S.; Ziemke, T. Embodied cognition and circular causality: On the role of constitutive autonomy in the reciprocal coupling of perception and action. *Front. Psychol.* **2015**, *6*, doi:10.3389/fpsyg.2015.01660.

27. Bich, L. Systems and organizations: Theoretical tools, conceptual distinctions and epistemological implications. In *Towards a Post-Bertalanffy Systemics*; Springer International Publishing: Cham, Switzerland, 2016; pp. 203–209.

28. Varzi, A. Mereology. Available online: http://philsci-archive.pitt.edu/12040/ (accessed on 19 May 2017).

29. Effingham, N. *An Introduction to Ontology*; Polity Press: Cambridge, UK, 2013.

30. Van Inwagen, P. When Are Objects Parts? *Phil. Perspect.* **1987**, *1*, 21–47.

31. Heylighen, F. Relational Closure: A mathematical concept for distinction-making and complexity analysis. *Cybern. Syst.* **1990**, *90*, 335–342.

32. Kauffman, S.; Clayton, P. On emergence, agency, and organization. *Biol. Philos.* **2006**, *21*, 501–521.

33. Ginsborg, H. Kant's biological teleology and its philosophical significance. In *A Companion to Kant*; Bird, G., Ed.; Blackwell: Oxford, UK, 2006; pp. 455–469.

34. Hordijk, W.; Steel, M. Detecting autocatalytic, self-sustaining sets in chemical reaction systems. *J. Theor. Biol.* **2004**, *227*, 451–461.

35. Hordijk, W.; Hein, J.; Steel, M. Autocatalytic Sets and the Origin of Life. *Entropy* **2010**, *12*, 1733–1742.

36. Segré, D.; Ben-Eli, D.; Lancet, D. Compositional genomes: Prebiotic information transfer in mutually catalytic noncovalent assemblies. *Proc. Natl. Acad. Sci.* **2000**, *97*, 4112–4117.

37. Luisi, P. Autopoiesis: A review and a reappraisal. *Naturwissenschaften* **2003**, *90*, 49–59.

38. Davies, J.A. *Life Unfolding: How the Human Body Creates Itself*; Oxford University Press: Oxford, UK, 2014.

39. Von Neumann, J.; Burks, A. *Theory of Self-Reproducing automata*; Illinois University Press: Chicago, IL, USA, 1966.

40. Varela, F.; Maturana, H.; Uribe, R. Autopoiesis: The organization of living systems, its characterization and a model. *Curr. Mod. Biol.* **1974**, *5*, 187–196.

41. Auletta, G.; Ellis, G.F.R.; Jaeger, L. Top-down causation by information control: From a philosophical problem to a scientific research programme. *J. R. Soc. Interface* **2008**, *5*, 1159–1172.

42. Jaeger, L.; Calkins, E.R. Downward causation by information control in micro-organisms. *Interface Focus* **2012**, *2*, 26–41.

43. Cummins, R. Functional Analysis. *J. Philos.* **1975**, *72*, 741–765.

44. Farnsworth, K.D.; Albantakis, L.; Caruso, T. Unifying concepts of biological function from molecules to ecosystems. *Oikos* **2017**, doi: 10.1111/oik.04171.

45. Lorenz, D.; Jeng, A.; Deem, M. The emergence of modularity in biological systems. *Phys. Life Rev.* **2011**, *8*, 129–160.

46. Butterfield, J. Laws, causation and dynamics at different levels. *Interface Focus* **2012**, *2*, 101–114.
47. List, C. Levels: Descriptive, Explanatory, and Ontological. Available online: http://philsci-archive.pitt.edu/12040/ (accessed on 19 May 2017).
48. Ellis, G.F.R. Top-down causation and emergence: Some comments on mechanisms. *Interface Focus* **2012**, *2*, 126–140.
49. Ellis, G.F.R. On the nature of causation in complex systems. *Trans. R. Soc. S. Afr.* **2008**, *63*, 1–16, doi:10.1080/00359190809519211.
50. Farnsworth, K.D.; Ellis, G.F.R.; Jaeger, L. Living through Downward Causation. In *From Matter to Life: Information and Causality*; Walker, S.I., Davies, P.C.W., Ellis, G.F.R., Eds.; Cambridge University Press: Cambridge, UK, 2017; pp. 303–333.
51. Gilbert, S. *Developmental Biology*; Sinauer Associates: Sunderland, MA, USA, 2013.
52. Seeley, T. *Honeybee Democracy*; Princeton University Press: Princeton, NJ, USA, 2010.
53. Lineweaver, C.H.; Egan, C. Life, gravity and the second law of thermodynamics. *Phys. Life Rev.* **2008**, *5*, 225–242.
54. Adami, C.; Ofria, C.; Collier, T. Evolution of biological complexity. *Proc. Natl. Acad. Sci.* **2000**, *97*, 4463–4468.
55. Hoel, E.; Albantakis, L.; Marshall, W. Can the macro beat the micro? Integrated information across spatiotemporal scales. *Neurosci. Conscious.* **2016**, *1*, niw012.
56. Marshall, W.; Albantakis, L.; Tononi, G. Black-boxing and cause-effect power. *arXivs* **2016**, arXiv:1608.03461.
57. Mayr, E. Teleological and Teleonomic: A New Analysis. In *Methodological and Historical Essays in the Natural and Social Sciences*; Springer: Dordrecht, The Netherlands, 1974; pp. 91–117.
58. Walker, S.I. Top-down causation and the rise of information in the emergence of life. *Information* **2014**, *5*, 424–439.
59. Wilson, D.; Wilson, E.O. Rethinking the Theoretical Foundation of Sociobiology. *Q. Rev. Biol.* **2007**, *82*, 327–348.
60. Wilson, E.O. *Sociobiology: The New Synthesis*; Harvard University Press: Cambridge, MA, USA, 1975.
61. Danchin, A. Bacteria as computers making computers. *FEMS Microbiol. Rev.* **2009**, *33*, 3–26.
62. Walker, S.; Davies, P. The algorithmic origins of life. *J. R. Soc. Interface* **2013**, *10*, 20120869.
63. Kauffman, S.A. *Investigations*; Oxford University Press: Oxford, UK, 2000.
64. Ptashne, M. Principles of a switch. *Nat. Chem. Biol.* **2011**, *7*, 484–487.
65. Moore, E.F. Gedanken-experiments on sequential machines. *Auto. Stud.* **1956**, *34*, 129–153.
66. Minsky, M.L. *Computation–Finite and Infinite Machines*; Prentice Hall: Englewood Cliffs, NJ, USA, 1967.
67. Zhang, J. Adaptive learning via selectionism and Bayesianism, Part 1: Connection between the two. *Neural Netw.* **2009**, *22*, 220228.
68. Krakauer, D. The inferential evolution of biological complexity: Forgetting nature by learning nurture. In *Complexity and the Arrow of Time*; Lineweaver, C.H., Davies, P.C.W., Ruse, M., Eds.; Cambridge University Press: Cambridge, UK, 2013; pp. 224–245.
69. Bruce, L.L.; Neary, T.J. The Limbic System of Tetrapods: A Comparative Analysis of Cortical and Amygdalar Populations. *Brain Behav. Evol.* **1995**, *46*, 224–234.

MDPI AG

St. Alban-Anlage 66

4052 Basel, Switzerland

Tel. +41 61 683 77 34

Fax +41 61 302 89 18

http://www.mdpi.com

*Entropy* Editorial Office

E-mail: entropy@mdpi.com

http://www.mdpi.com/journal/entropy

MDPI AG
St. Alban-Anlage 66
4052 Basel
Switzerland

Tel: +41 61 683 77 34
Fax: +41 61 302 89 18

www.mdpi.com

MDPI

ISBN 978-3-03842-515-1